Current Topics in Neurotoxicity

Series Editors
Richard M. Kostrzewa
Trevor Archer

For further volumes:
http://www.springer.com/series/8791

Giovanni Laviola · Simone Macrì
Editors

Adaptive and Maladaptive Aspects of Developmental Stress

Editors
Giovanni Laviola
Istituto Superiore di Sanità
299 Viale Regina Elena
Rome, Italy

Simone Macrì
Istituto Superiore di Sanità
299 Viale Regina Elena
Rome, Italy

ISBN 978-1-4614-5604-9 ISBN 978-1-4614-5605-6 (eBook)
DOI 10.1007/978-1-4614-5605-6
Springer New York Heidelberg Dordrecht London

Library of Congress Control Number: 2012954050

Printed on acid-free paper

Springer is part of Springer Science+Business Media (www.springer.com)

Preface

A few years ago (a time at which the words face and book were still independent from one another), we were ruminating on a set of experiments demonstrating that being a rat and being reared to a careful mother—a good rat mum spends hours feeding and licking her pups—increased the odds that a subsequent challenge or stressor would be handled without major hassles (i.e., with a small activation of all those biological systems mediating the increased heartbeat, vigilance, motor agitation, and sense of fear, common to all "stressed" mammals). "Lucky rat," did one of us inadvertently whisper. Why lucky? What is stress good for? What is stress bad for? Is being less sensitive to external stressors necessarily good? By the same token, is being particularly sensitive to external stressors a sign of poor welfare? What is the link between stress and pathology? Where does stress sensitivity root? In other words, what are the genetic and environmental determinants of individual reactivity to external stressors? If developmental contexts played a predominant role, would perinatal and teenage adversities inevitably relate to disease states?

All these questions became a refrain pervading our meals, coffee breaks, and structured meetings. Besides, under the persuasion that broad questions like these require a multidisciplinary approach, we've come across the work of eminent scientists that had analogous questions in mind, that proposed innovative approaches, and that provided original answers.

After a few years, exceptional scientific progress has been made and an intermediate digest, providing partial answers to the previous questions, can be proposed. We envisioned such a digest to combine the unique views of the aforementioned eminent scientists about the link between development and stress. We thus asked them whether they were willing to contribute such perspective into a book—likely to be of interest to scholars ranging from university students to established scientists—tackling the multifaceted nature of "stress" from different angles. The positive response we got from them resulted into *Adaptive and Maladaptive Aspects of Developmental Stress*. The book, featuring contributions from scientists working in different countries, is introduced by a general overview offered by Trevor Archer and Richard Kostrzewa and then subdivided into three sections aimed at providing introductory concepts about the stress response system and its adaptive role in evolutionary terms (Part I); experimental and clinical data about the maturation of the stress response system in reaction to contextual features and in association with functional adjustments or pathological derailments in our species (Part II); and experimental comparative data, obtained in animal species such as birds, rodents, and nonhuman primates, linking development (prenatal, early postnatal, adolescent), environmental conditions, stress, and adaptive/maladaptive outcomes (Part III).

In Part I, Chap. 2, Del Giudice, Ellis, and Shirtcliff set the general framework of the book introducing the stress response system and its development within a broad evolutionary-adaptive perspective; Maestripieri and Klimczuk (Chap. 3) then tighten the link between humans and other animal species by highlighting the similarities between the fundamental mechanisms behind the developmental regulation of the stress response system in different mammals. In Part II, Chap. 4, DiCorcia, Sravish, and Tronick describe a heuristic approach explaining the development of stress resilience (the ability to cope with repeated minor stressors) through the everyday confrontation with stressful events experienced by children; using an anthropological approach, Flinn, Ponzi, Nepomnaschy, and Noone then describe the evolutionary-adaptive underpinnings of the developmental plasticity of the stress response system, both from a theoretical and from an empirical (data-based) perspective (Chap. 5); subsequently, Nater and Skoluda elaborate on the consequences of developmental stressors (prenatal phase in Chap. 6 and childhood and adolescence in Chap. 7) in humans within the context of adaptive plasticity and pathology; their analyses rest on an applied approach, bridging individual maturation and environmental variables in the formation of the adult phenotype. In Part III, Chap. 8, Morley-Fletcher, Mairesse, and Maccari set the transition from clinical to preclinical data and describe the long-term influences of prenatal stress (in the form of psychophysiological stressors applied to rat mothers during gestation) on behavioral and neuroendocrine regulations in rats; building on this, Hauser describes the influences that an altered prenatal maturation of the stress response system (achieved through pharmacological interventions) may have on individual long-term cognitive and emotional adjustments in rats and primates (Chap. 9); the focus is shifted from prenatal to postnatal development by Cirulli and Berry who describe the fundamental mechanisms mediating the long-term consequences of early life stressors in rodents and their significance with respect to human health and disease (Chap. 10); Coutellier then elucidates the adaptive regulations shown by laboratory rodents in response to different forms of challenges encountered shortly after birth (Chap. 11); Macrì and Laviola subsequently describe the unique effects of stress during adolescence, summarizing literature showing that, compared to younger and older conspecifics, adolescent rodents show a differential short- and long-term stress reactivity (Chap. 12); finally, leveraging on data from avian research, Costantini offers a wide evolutionary-adaptive analysis of the effects of different stressors on embryonic maturation (Chap. 13).

During the production of this book, we were partly funded by the grant ECS-Emotion from the Department of Antidrug Policies c/o Presidency of the Council of Ministers, Italy.

Ultimately, we gratefully acknowledge a number of external reviewers for their expert help with the fine-tuning of the original contributions and, of course, *Springer* for welcoming our proposal and for assisting in publishing this work.

Rome, Italy Giovanni Laviola and Simone Macrì

Contents

Contributors

Trevor Archer Department of Psychology, University of Gothenburg, Gothenburg, Sweden

A. Berry Behavioural Neuroscience Section, Department of Cell Biology and Neurosciences, Istituto Superiore di Sanità, Rome, Italy

F. Cirulli Behavioural Neuroscience Section, Department of Cell Biology and Neurosciences, Istituto Superiore di Sanità, Rome, Italy

David Costantini Animal Health and Comparative Medicine, Veterinary and Life Sciences, Institute for Biodiversity, College of Medical, University of Glasgow, Glasgow, UK

Laurence Coutellier Behavioral and Functional Neuroscience Laboratory, School of Medicine, Stanford University, Palo Alto, CA, USA

Jennifer A. DiCorcia Department of Psychology, Child Development Unit, University of Massachusetts, Boston, MA, USA

Department of Newborn Medicine, Brigham and Women's Hospital, Harvard Medical School, Boston, MA, USA

Bruce J. Ellis John and Doris Norton School of Family and Consumer Sciences, University of Arizona, Tucson, AZ, USA

Mark V. Flinn Department of Anthropology, University of Missouri, Columbia, MO, USA

Marco Del Giudice Center for Cognitive Science, Department of Psychology, University of Turin, Torino, Italy

Jonas Hauser Institute of Behavioural Neuroscience, Department of Cognitive, Perceptual and Brain Sciences, University College London, Bedford, London, UK

Amanda C.E. Klimczuk Institute for Mind and Biology, The University of Chicago, Chicago, IL, USA

Richard M. Kostrzewa Department of Biomedical Sciences, Quillen College of Medicine, East Tennessee State University, Johnson City, TN, USA

Giovanni Laviola Section of Behavioral Neuroscience, Department of Cell Biology and Neurosciences, Istituto Superiore di Sanità, Rome, Italy

Stefania Maccari Neural Plasticity Team-UMR CNRS/USTL n° 8576 Structural and Functional Glycobiology Unit, North University of Lille, Lille, France

Simone Macrì Section of Behavioral Neuroscience, Department of Cell Biology and Neurosciences, Istituto Superiore di Sanità, Rome, Italy

Dario Maestripieri Institute for Mind and Biology, The University of Chicago, Chicago, IL, USA

Jérôme Mairesse Neural Plasticity Team-UMR CNRS/USTL n° 8576 Structural and Functional Glycobiology Unit, North University of Lille, Lille, France

Sara Morley-Fletcher Neural Plasticity Team-UMR CNRS/USTL n° 8576 Structural and Functional Glycobiology Unit, North University of Lille, Lille, France

Urs M. Nater Department of Psychology, Philipps University of Marburg, Marburg, Germany

Pablo Nepomnaschy Simon Fraser University, Burnaby, BC, Canada

Robert Noone Center for Family Consultation, Evanston, IL, USA

Davide Ponzi Division of Biological Sciences, University of Missouri, Columbia, MO, USA

Elizabeth A. Shirtcliff Department of Psychology, University of New Orleans, New Orleans, LA, USA

Nadine Skoluda Department of Psychology, Philipps University of Marburg, Marburg, Germany

Akhila V. Sravish Department of Psychology, Child Development Unit, University of Massachusetts, Boston, MA, USA

Ed Tronick Department of Psychology, Child Development Unit, University of Massachusetts, Boston, MA, USA

Chapter 1
The Inductive Agency of Stress: From Perinatal to Adolescent Induction

Trevor Archer and Richard M. Kostrzewa

Abstract The influence of stress agents, whether social, restraint, malnutrition or mild unpredictable, during the fetal–prenatal, infant–postnatal, adolescent or young adult phases of the lifespan generally, but not always, implies disruption of the normal process of development. Several notions of stress including the adaptive calibration model, adaptive emotional processes and arousability, GABAergic integrity and nutrient deficiency, and resilience influence the physiological and behavioural expressions of maternal stress, affecting nursing behaviour and offspring outcome. The adaptive/maladaptive effects of stress in humans are affected by developmental programming of the hypothalamic–pituitary–adrenal (HPA) axis and other neuroendocrine systems related to stress that may facilitate expressions of resilience. The adaptive/maladaptive effects of stress in animal models outline dysfunctional HPA axis and brain regional alterations of phenotypic expressions that interact with epigenetic mechanisms and developmental plasticity. Maladaptive stress regulation in adolescence is influenced by several factors, not the least being serotonergic, glucocorticoid and regional integrity pertaining to trauma in adolescence. The occurrence of oxidative stress may imply damage but the propensity for hormesis, a notion not unrelated to resilience, provides opportunities for long-lasting health benefits.

Individuals' particular reception, appraisal, and eventual functional and physiological responses to stressors are determined initially by those developmental pathways, of prenatal origin, that unfold postnatally, during infancy and in childhood. The concomitant expression of this neurodevelopment against a unique genetic inheritance will

T. Archer (✉)
Department of Psychology, University of Gothenburg, Box 500, 40530 Gothenburg, Sweden
e-mail: Trevor.archer@psy.gu.se

R.M. Kostrzewa
Department of Biomedical Sciences, Quillen College of Medicine, East Tennessee State University, Johnson City, TN 37614, USA

G. Laviola and S. Macrì (eds.), *Adaptive and Maladaptive Aspects of Developmental Stress*, Current Topics in Neurotoxicity 3, DOI 10.1007/978-1-4614-5605-6_1,

provide an eventual predisposition towards a capacity for life circumstance that is associated with varying proportions of health or ill health. Early life experiences, including those occurring in utero, have been associated consistently with increased risk for several adolescent/adult onset chronic disease states. The notion of "fetal programming", whereby fetal environmental variations leading to "predictive adaptive responses" predispose the individual to its life after birth, bears upon a lifespan (Gluckman et al. 2005, 2007) that, for example, may encompass behaviour governed by generally successful decision-making or those governed by hasty, risky decisions, that is, health or disorder (Barber et al. 1996; Glover 2011). Various epidemiological studies have shown strong associations between birth weight and risk for coronary heart disease, hypertension, type II diabetes and other diseases related to stress during adulthood. According to the Barker notion (Barker 2004) of programming during embryonic and fetal life, "set points" of physiologic and metabolic responses in adult life are determined. The "Barker Hypothesis", or Thrifty phenotype, states that conditions during pregnancy will have long-term effects on adult health (Barker 1998). Associated risk of lifelong diseases includes cardiovascular disease, type 2 diabetes, obesity and hypertension. There exists an underlying assumption that critical periods of developmental plasticity in utero provide the circumstance of selection of the fetal phenotype best adapted to the intrauterine environment. Lau and Rogers (2004) discuss the adversity of in utero conditions, for example, toxicants, inflammation, that predispose the developing individual to aberrant physiological functioning as an adult. Stress, chronic or traumatic, presents conditions that readily induce an aberrant development. Following birth, the influence of stressors, whether stimulating, minor, major or traumatic, positive or negative, contributes to the continued development of the individual, among other factors through epigenetic influences exerting greater or lesser determinants of the molecular basis of structure and function (Archer et al. 2011). To greater or lesser extent, the neurotoxic agency of stress exerts its detrimental influences upon neurodevelopmental processes during the period of the brain growth spurt and during other phases until adulthood thereby maintaining the notion of critical phases for the outcomes of treatment whether prenatal, postnatal or adolescent (Archer 2010).

1.1 Notions of Stress

Del Giudice et al. (Chap. 2) have presented an evolutionary–developmental framework for individual differences in stress responsiveness, termed the "Adaptive Calibration Model" (ACM), to provide an integrative biological analysis of the stress response system (Del Giudice et al. 2011) that incorporates mechanism, ontogeny, phylogeny and adaptation. It postulates three main biological functions: (1) coordination of organisms' allostatic response to physical and psychosocial challenges, (2) encoding and filtration of information about the social and physical environment that mediates individuals' receptiveness to environmental inputs, and (3) regulation of physiology and behaviour over a wide range of fitness-relevant dimensions involving defensive behaviours, competitive risk-taking, cognition, attachment, affiliation and reproductive functioning. The notion of ACM was applied to generate novel predictions on profiles of responsiveness among individuals and how these profiles unfold over the individual's lifespan. The influence of prenatal and maternal psychosocial stress presents a related issue. Maestripieri and Klimczuk (Chap. 3) describe pre- and postnatal psychosocial stress induced by the mother's behaviour in humans and other primates, through which the offspring's behavioural attributes are modulated by the mother. Adaptive emotional processes, for example, maternal attraction arousability and maternal anxiety arousability, serve to enhance and sustain female interaction motivation with their infants, to invest in them and to protect them following birth (Maestripieri 2011). Alterations to these emotional processes with concomitant changes in maternal motivation mediate the reduction and eventual termination of the maternal investment associated with infant weaning. It is indicated that prenatal and maternal stress may be mediated by similar physiological mechanisms with a central role of the hypothalamic–pituitary–adrenal (HPA) axis with the biomedical/clinical view, the stress inoculation model and the ACM providing different assumptions and predictions pertaining to adaptive/maladaptive developmental consequences. The notion of resilience in this regard implies that moderate levels of rejection may induce physiological and behavioural changes that enhance resilience in later life, whereas maternal abuse or high levels of rejection ought to preclude adaptive advantages.

Rodent studies have focused upon ethologically relevant stressors, such as chronic social stress (CSS) a type of early life stress allowing eventual investigation of transgenerational effects, to gain deeper insights into affective disorders (Brunton and Russell 2010; Herzog et al. 2009; Nephew and Bridges 2011). CSS during lactation, through presentation of a novel male intruder, affects the maternal behaviour and the growth of the dam and offspring, with accompanying decreases in maternal care and increases in maternal aggression, with enduring effects for the offspring. A wide range of studies have produced numerous alterations of the neuroendocrine–biobehavioural profiles of CSS-exposed rodents compared with those normally raised; and similar human studies point to similar differences between the offspring of mothers with securely attached children and insecure mothers (Strathearn et al. 2009). The experience of stress in childhood correlates negatively with plasma oxytocin concentrations in adult men (Opacka-Juffry and Mohiyeddini 2012). Other studies on the exposure of male rodents to maternal separation during the early life period have indicated that the depression-like behaviour and associated neuroendocrine alterations are mediated through epigenetic mechanisms (Der-Avakian and Markou 2010; Murgatroyd et al. 2009, 2010). Rat dams that had been exposed to CSS as infants displayed substantial increases in pup retrieval and nursing behaviour linked to attenuated oxytocin, prolactin and vasopressin gene expression in those brain nuclei involved in the control of maternal behaviour (Neumann et al. 2005). Thus, the early life CSS that dams were exposed to induced permanent deficits in nursing behaviour that predate the stress-induced deficits in affective status observed in both laboratory and clinical studies (Murgatroyd and Nephew 2012).

Early life malnutrition affects predisposition towards the effects of stress. The magnitude of the HPA axis response to stress is limited by the inhibitory neurotransmitter, gamma-aminobutyric acid (GABA), inhibitory circuit and the glucocorticoid (GC) negative feedback system (Darnaudery and Maccari 2008). In this respect, the lack of GABA is linked to stress and affective disorders (Skilbeck et al. 2010). Docosahexaenoic acid (DHA), richly distributed in brain tissue and necessary for normal development (Su 2010), accumulates rapidly in the fetus from the third trimester to the sixteenth postnatal day in the rat pup (Green et al. 1999) and to the second year after birth in the human infant forebrain (Lauritzen et al. 2001). Exposure to maternal malnutrition stress is associated with several chronic health conditions and neuropsychiatric disorders in humans and laboratory animals (Kajantie 2006). Chen and Su (2012) examined whether or not brain development was the critical period during which DHA deficiency induces HPA axis dysregulation in response to stress during later stages of the lifespan. They exposed rats to a, *n*-3 fatty acid-deficient diet or the same diet supplemented with fish oil as an *n*-3 fatty acid-adequate diet either through the preweaning period, from embryo to weaning at 3 weeks age, or during the postweaning period, from 3 weeks age to 10 weeks. DHA deficiency during the preweaning period caused, at weaning, reduced hypothalamic levels of DHA and body weight. Concurrently, it increased significantly and prolonged restraint stress-induced changes in colonic temperature and serum corticosterone concentrations, a significant $GABA_A$ antagonist-induced increased heart rate change, depression-like behaviour in the forced swim test and anxiety-like behaviour in the elevated plus maze; all effects not observed in the rats subjected to postweaning *n*-3 fatty acid deficiency. The authors concluded that the preweaning brain development period when DHA was deficient induced excessive HPA responses to stress and elevated levels of depression–anxiety behaviours in adult years and implicated GABAergic mechanism involvement.

1.2 Adaptive/Maladaptive Aspects of Stress in Humans

Resilience affects how individuals deal with daily hassles, major stressors, trauma and catastrophes. Functionally, it refers to the notion of an individual's tendency to cope with stress and adversity, "to be knocked down by life and come back stronger than ever". When confronted by traumatic, disastrous or catastrophic events, resilient individuals expect to find a way to overcome the stressor, since their self-reliance will produce a learning/coping reaction rather than a victim/blaming reaction; they are attuned to being "survivors", despite the circumstance. For example, it has been shown that highly resilient intensive care nurses or prisoners of war utilize their hardiness and positive coping skills and psychological attributes, such as character and optimism, to maintain high levels of work performance or repatriation in the major stress/trauma environment of intensive care units or POW camps (Mealer et al. 2012; Segovia et al. 2012). DiCorcia et al. (Chap. 4) have presented the "Everyday Stress Resilience Hypothesis" which notion offers resilience as a process of regulating everyday life stressors and analyses the issue from a systems perspective. Accordingly, successful regulation accumulates into regulatory resilience which emerges during early development from successful coping with the inherent stress in typical interactions. The authors maintain that quotidian stressful events result in the activation of behavioural and physiological systems. Infant–adult interactions provide the communicative and regulatory processes that must develop the regulatory resilience linked to infant–caregiver relationships. The psychosocial context of resilience development and regulation is intimately coupled to stress reactivity in social environmental contexts. The "fight-or-flight" response is a prototypic human stress response in both behavioural and physiological manifestations that express individual levels of resilience to major stressors through the type of responses initiated. Von Dawans et al. (2012) have observed that participants experiencing acute social stress engaged in substantially more prosocial behaviour, such as trust, trustworthiness and sharing, than the participants in the control condition who were not confronted by socioevaluative stress. These highly specific effects were not affected by readiness to exhibit antisocial behaviour or involvement in nonsocial risk behaviour but suggest that social stress triggers approach behaviour. Flinn et al. (Chap. 5) have focussed upon the issue of why social relationships may affect health and disease propensity with particular regard to the role of stress hormones. In this regard, the high level of responsiveness of the HPA axis to traumatic experiences involving social challenge and confrontation was assessed in field studies of childhood stress and family environment in a rural community in Dominica, West Indies, through the registration of the children's hormonal responses to everyday interactions with their parents and other care providers. They observed that the long-term effects of traumatic early experiences on cortisol profiles were complex with domain-specific effects. There was normal recovery from physical stressors but also heightened responding to negative-affect social challenges. By and large, their findings were consistent with the notion that developmental programming of the HPA axis and other neuroendocrine systems related to stress may facilitate cognitive–emotional coping with the social challenges or modulate expressions of resilience if considered in this manner.

Prenatal stress exerts a variety of pathophysiological alterations that generally imply prognostic disadvantage in a number of health considerations including neuroimmune functioning, cognitive–emotional domains, bone tissue integrity and developmental markers, not least stress reactivity (Abdeslam 2012; Dancause et al. 2012; Dyuzhikova et al. 2012; Werner et al. 2012). Placental 11-β-hydroxysteroid dehydrogenase type 2 (HSD11β2) buffers the impact of maternal GC exposure by converting cortisol/corticosterone to inactive metabolites. Jensen Pena et al. (2012) studied epigenetic mechanisms through induction of prenatal stress (chronic restraint stress during gestation days 14–20) and observed a significant decrease in HSD11β2mRNA, increased mRNA levels of methyltransferase DNMT3a and increased DNA methylation at specific CpG sites within the HSD11β2 gene promoter. Their findings impact both upon the tissue specificity of epigenetic effects and the epigenetic status of placenta to predict corresponding changes in the developing brain. The issue of prenatal stress with its concomitant influences on programming for the developing individual was addressed by Skoluda and Nater (Chap. 6) who have reviewed the impact of various sources of prenatal stress on various outcomes in infants. Their observations have shown that prenatal stress is associated predominantly with maladaptive consequences, including negative birth outcomes, altered physiological stress responses, behavioural and conduct problems and impaired cognitive and motor development. This theme is maintained in the chapter by Nater and Skoluda (Chap. 7) who have addressed the focus upon negative outcomes resulting from early life stress by reviewing the adaptive and maladaptive consequences of early stress. Their observations further reinforce notions concerning the links between early life stress and brain and CNS alterations, impairments in cognitive functioning, altered physiological responses to stress, increased risk for developing behavioural problems and outcomes leading to somatic and psychiatric illnesses, not least the overshadowing spectre of posttraumatic-determined direction of inadequate growth. In both of these chapters the authors have proposed a psychobiological stress model to integrate their observations.

1.3 Adaptive/Maladaptive Aspects of Stress in Animal Models

Prenatal and early life adversity was found to be mediated through the alterations of HPA axis reactivity that affected circulating levels of the markers for stress, the GCs. This exposure to early adverse environment is linked to enhanced cardio-metabolic disease risk through the programming phenomenon. Morley-Fletcher et al. (Chap. 8) have presented the prenatal restraint stress (PRS) rat model that induces long-lasting neurobiological and behavioural changes such as impaired feedback mechanisms due to a faulty HPA axis, disruption of circadian rhythms and disruptions in neuroplasticity with the concomitant deficits in glutamatergic system integrity. They show that the chronic treatment of PRS rats during adulthood has provided a high level of predictive validity. PRS has been found to induce anxiety-related behaviour, biomarkers and a range of other alterations, not least dopamine (DA) metabolism and biology (Baier et al. 2012; Diz-Chaves et al. 2012; Mairesse et al. 2012): these are, among other changes, expressed by lack of plasma peak in leptin, reduced expression of the γ_2 subunit of $GABA_A$ receptors in the amygdala, increased expression of mGlu5 receptors in the amygdala but reduced expression of mGlu5 receptors in the hippocampus and reduced expression of mGlu2/3 receptors in the hippocampus (Laloux et al. 2012). Prenatal exposure to GCs induces affective dysfunctional status to HPA axis alterations. Treatment of pregnant rats with dexamethasone (DEX) in the final trimester caused reduced exploratory behaviour and reduced sucrose consumption (Liu et al. 2012b), suggesting that GC exposure prenatally increased HPA axis activity and induced depression-like symptoms. Four weeks of swimming exercise reduced serum corticosterone levels and increased exploratory behaviour and sucrose intake. Furthermore, late gestational DEX administration (days 18–22) decreased significantly the density of calretinin immunoreactive cells in the lateral amygdala of adult female offspring (Zuloaga et al. 2012). Calretinin (29 kDa calbindin) is a vitamin D-dependent calcium-binding protein involved in Ca^{++} signalling that functions as a modulator of neuronal excitability; it offers a diagnostic marker for certain disorders, particularly those associated with affect and cognitive–emotional instability (Giachino et al. 2007; Medalla and Barbas 2010; Steullet and Cabungcal 2010). Drake et al. (2011) have demonstrated that multigenerational effects on birth weight and disease risk is associated with different processes in the exposed first generation (F1) and may be transmissible to a second generation (F2) with marked implications for the related etiopathogenesis of disorder (Bennett et al. 2008).

In the chapter by Hauser (Chap. 9), animal studies demonstrating two phenotypes are described as follows: (1) transient reduction in body weight and (2) alteration of HPA axis activity. Several studies reporting overall increased activity of serotoninergic and dopaminergic systems, the anxiety-enhancing and cognition-disrupting effects of prenatal GC administration were held as instances of epigenetic programming; by this account the alterations of gene expression that are induced by mechanisms not involving alteration of DNA sequence. Similar mechanisms were proposed to mediate the long-term effects of changes in maternal behaviour (Stuebe et al. 2011). Taken together, these converging lines of evidence imply that an ongoing continuity of epigenetic processes controlled by environmental contingencies is a major determinant of the course of each individual's development (Archer et al. 2010a; Palomo et al. 2007). The enduring quality, and related consequence, of early life experiences points to an influence upon multiple functional domains, cognitive, social, emotional and everyday activity, and biomarkers, tissue and neurotransmitter integrity and metabolism. In animal models, for example, variations in early maternal care are associated with differences in the HPA axis stress response in the offspring that are mediated via changes in the epigenetic regulation of the GC receptor (GR) gene. Oberlander et al. (2008) studied relationships between prenatal exposure to maternal mood and methylation status of a CpG-rich region in the promoter and exon 1F of the human GR gene (NR3C1) in newborn infants and HPA stress reactivity at 3 months of age. They observed that exposure to maternal depression/anxious mood during the third trimester was associated with increased methylation of NR3C1 at a predicted NGFI-A binding site as well as increased salivary cortisol stress responses at 3 months with controls for prenatal selective serotonin reuptake inhibitors (SSRIs), postnatal age and pre- and postnatal maternal mood. Thus, it was shown that the methylation status of the human NR3C1 gene in newborn infants was sensitive to prenatal maternal mood. Epigenetic process provided a link between antenatal maternal mood and HPA stress reactivity in the infant. Cirulli and Berry (Chap. 10) have discussed the crucial role of the prenatal and early postnatal period that impacts upon the developing systems with particular regard for the adverse conditions created by psychological or toxic stress. They have indicated that changes in the effectors of stress responses during critical developmental stages may favour varying degrees of vulnerability for obesity and/or type II diabetes, neurologic and psychiatric health and neurodegeneration. Mulligan et al. (2012) have observed a significant correlation between culturally relevant measures of maternal prenatal stress, newborn infant birth weight and newborn methylation in the NR3C1. They suggest that the increased methylation may constrain plasticity in subsequent gene expression and thereby restrict the range of stress adaptation responses available to so-afflicted individuals with concomitant risk for multidomain disorder. Comorbidity expressions between several of these disorders implicate the effects of stress in different pathophysiologies, thereby affecting individual longevity and quality of life during ageing. Candidate gene association studies suggest the influence of a range of genes in different disorders of CNS structure and function whereby indices of comorbidity both complicate the array of gene involvement by offering a substrate of hazardous interactivity. For instance, the putative role of the serotonin transporter gene in affective–dissociative spectrum disorders is consistent with genetic variation and complication of comorbidity (Palomo et al. 2007). In concurrence with other accounts, they point to the central contributions of genetic susceptibility, the permanent changes in gene expression incurred through epigenetic forces and the important search for a broad spectrum of interventions for damage and destruction afflicted by early adversity and stress. Nevertheless, the relative contributions of endophenotypes and epistasis that mediate epigenetic phenomena and eventual outcomes provide a multitude of interactive combinations that bedevil disorder diagnosis and prognosis (Archer et al. 2010b).

The epigenetic influences of prolonged stress experiences attest to permanent changes in both the behaviour and biomarkers associated with the afflicted individuals. Sterrenburg et al. (2011) exposed male and female rats to chronic variable mild stress (CVMS) following which immediate early gene products: corticotrophin-releasing factor (CRF) mRNA and peptide, various epigenetic-associated enzymes and DNA methylation of the *crf* gene were determined in several brain regions including the hypothalamic paraventricular nucleus, oval and fusiform parts of the bed nucleus of the stria terminalis and central amygdala. It was found that CVMS-induced site-specific *crf* gene methylation in all brain centres in male and female stria terminalis and central amygdala. The histone acetyltransferase, CREB-binding protein was increased in female stria terminalis while histone deacetylase 5 was decreased in the male central amygdala. All these changes were accompanied by increased levels of c-fos in the paraventricular nucleus, fusiform stria terminalis and central amygdala in males, FosB in the paraventricular nucleus of both males and females and in male oval and fusiform stria terminalis. CVMS increased CRF mRNA in males in the paraventricular nucleus whereas in females CRF peptide was decreased. It was concluded that CVMS-induced brain region-specific and gender-specific changes in epigenetic activity and neuronal activation were involved in the gender specificity to stress reactivity and susceptibility to depressiveness. In this respect, several factors linked to oxidative stress, for example, malnutrition, separation, heat shock, tobacco smoke, air pollutants, metals organic chemicals and various other sources (Qiu et al. 2004; Schury and Kolassa 2012; Su et al. 2012) contribute to the changes in epigenetic state thereby mediating the pathogenesis of disorders (Cortessis et al. 2012).

Developmental plasticity, a general term that refers to changes in neural connections due to environmental interactions and cognition, like neuroplasticity/brain plasticity is specific to the alterations affected in neurons, circuits and synaptic connections through the "forces/processes" of development. On the other hand, phenotypic plasticity reflects the capacity of individuals to alter their phenotypes in response to changes in the environment (Price et al. 2003), that is, the processes through which patterns of neurohumoral–biobehavioural expressions during adolescence and early and later adulthood are "programmed" by early life events/environments. Against this background, Coutellier (Chap. 11) has provided descriptions of developmental phenotypic plasticity and the manner through which it results in adaptive phenotypes in later life. In rodent studies, early experience influenced long-term development of behavioural, neuroendocrine and cognitive–emotional domains. Factors mediating the effects of perinatal conditions on the development through infancy include (a) maternal behaviour which modulates the consequences of early challenges on adaptive regulation, (b) non-maternal variables, such as gender, levels of GC and inflammatory effects due to neuroimmune mobilization. The complexity of developmental and phenotypic plasticity accrues from the multiplicity of interacting factors, both behavioural and biomarker. Gong et al. (2012), by studying 7-week-old female mice subjected to chronic mild stress (CMS) in establishing an animal model of depression, have observed that early adult depressive females may affect their offspring. It was shown that the female mice themselves presented normal fertility but their offspring exhibited lowered neonatal survival rate and body weight from birth to adulthood. These offspring expressed also deficits in neonatal reflex attainment and memory performance concurrent with a higher level of emotionality as adult mice. Astrocyte numbers, hippocampal volume and neurogenesis were reduced in the offspring of depressive dams whereas GR expression in the hippocampus was increased.

1.4 Adaptive/Maladaptive Stress Regulation in Adolescence

Adolescence introduces a variety of age-specific expressions remarkable for the discontinuity with earlier expressions, as the developing individual emerges to face challenges inherent to adulthood (Archer et al. 2010a). Crucial to cognitive–emotional domains in adolescence, when daily problems appear to be of a similar order of magnitude as major life events are viewed by normal healthy adults (McCullough et al. 2000), an increase in the frequency of positive affect accompanied by a decrease in negative affect may offer improved prognoses for management of emotional upheaval (Garcia and Siddiqui 2009a, b). Macrì and Laviola (Chap. 12) have discussed the situation that stress sensitivity during adolescence is remarkably different from earlier and later periods of the individual's life cycle, with an exaggerated prolongation (and probably intensity) of the neuroendocrine responses. Response profiles are to be affected by lack of synchronicity with the HPA axis as regulated by experiences attuned to an adult social environment. The authors relate that precocious experiences with drugs of abuse or exposure to adverse environments may favour the onset of conduct disorders, such as addictive behaviours, sensation-/novelty-seeking and emotional disturbances, and outline links between adolescent plasticity and long-term individual regulation that affect adaptive plasticity to the adult environment and/or eventual predisposition to pathological behavioural states. Studying the central role of affect and character–temperament attributes for propensity to emotional stability, Garcia et al. (2012) observed that "self-fulfilling" adolescents (high positive affect, low negative affect) expressed higher levels of sleep quality, less stress and more energy, higher levels of persistence and more mature character, as opposed to temperament, attributes (i.e. higher scores in self-directness and cooperativeness) than is expressed by "self-destructive" adolescents (expressing low positive affect, high negative affect).

As Macri and Laviola have outlined, the propensity for age-specific developmental plasticity entails both risk and opportunity. Inappropriate stimulation may lead to disorder pathology under conditions wherein the heightened sensitivity to extraneous variables favours the integration of external cues into mature function. Adolescent risk-taking may result from this heightened susceptibility to environmental cues with particular regard to emotional mobilization and potential rewards. Johnson et al. (2012) studied the impact of social stress on adolescent risking, accounting for individual differences in risk-taking under non-stressed conditions. They found that adolescents in the stress condition took more risks than those in the no-stress condition. Interestingly, differences in risk-taking under stress were linked to the individuals' baseline risk-taking tendencies. Three types of risk-takers among the adolescents were observed: conservative, calculated and impulsive. The impulsive risk-takers were less accurate and showed less planning ability under stress whereas calculated risk-takers took fewer risks and conservative risk-takers engaged in low risk-taking regardless of stress conditions. In general terms, adolescents appear to take risks in "hot cognitive" than in "cold cognitive". Nevertheless, the wide variability in adolescents' behavioural expression, discussed above with regard to temperament and character, to stress-modulated trait-level risk-taking tendencies points to further attention upon individual phenotypic plasticity. The study of life-course approaches to life experience neuromaturational relationships indicates the bidirectionality (positive–negative) enduring quality of interrelationships from the fetal phase through adolescence, and eventually over the life span (Johnson and Blum 2012).

Serotonin (5-HT), a "phylogenetically old" neurotransmitter distributed throughout the brain, exerts two major influences during early development: (1) it functions as a growth factor that regulates neural development (Whitaker-Azmitia et al. 1996) through cell division, differentiation, migration, myelination, synaptogenesis and dendritic pruning and arborisation (Gaspar et al. 2003); (2) it acts in modulating important domains including cognition, emotion, attention, sleep, reproduction, stress reactivity and arousal; and (3) it is central to HPA and locus coeruleus–noradrenaline system development (Chaouloff et al. 1999; Laplante et al. 2002). Oberlander (2012) has examined the developmental effects of 5-HT fetal signalling with regard to SSRI in utero exposure and genetic variations of the 5-HT transporter gene (SLC6A4). It was observed that prenatal 5-HT signalling exerted consequences dependent upon biological, that is, genetic/epigenetic variables, experiential, that is, prenatal drug or maternal mood exposure, and contextual, that is, postnatal social environment factors. Both the exposure to SSRIs and genetic variations affecting 5-HT signalling may increase sensitivity to negative social contexts in some individuals but confer sensitivity to positive life circumstances in other individuals. In this regard, the former bear resemblance in attribution and expression to the "self-destructive" individuals, described above, whereas the latter resemble "self-fulfilling" individuals (Archer et al. 2008). Genetic inactivation of brain-specific tryptophan hydroxylase 2 (Tph2) function was accompanied by phenotypic changes in laboratory studies involving enhanced fear reactivity and aggression, depression-like behaviour and growth retardation (Alenina et al. 2009; Gutknecht et al. 2008; Jacobsen et al. 2011). All these effects, amplified by epigenetic interactions (Abumaria et al. 2007, 2008; Arborelius and Eklund 2007; Gartside et al. 2003), implicate alterations in Tph2 functional domains governing personal attributes expressed by cognitive–emotional deficits and stress systems regulation in human subjects (Lesch et al. 2012; Waider et al. 2011).

Adolescents that experience traumatic stress and develop posttraumatic symptoms secrete higher levels of cortisol (Schechter et al. 2012) and comprised brain regional integrity, prefrontal cortex and total cerebral grey matter volume (Carrion et al. 2007, 2010a, b) than healthy control adolescents. It has been observed that those children adolescents exposed to a history of violence and trauma-related distress presented worse levels of academic performance, such as reading ability and IQ (Delaney-Black et al. 2002; Saigh et al. 1997, 2006). Among 23 disadvantaged and chronically stressed adolescents who also participated in functional magnetic resonance imaging during processing of emotional faces and structural magnetic resonance imaging, cortisol change on the Trier Social Stress Test-Child version showed a significant inverse relationship with left hippocampus response to fearful faces (Liu et al. 2012a). They concluded that the increased cortisol response to the Trier social stressor was associated with diminished response of the left hippocampus to faces depicting fear, implying that HPA–corticolimbic system mechanisms may underlie vulnerability to maladaptive responses to stress in adolescents that may contribute to development of stress-related disorders. Nevertheless, Bicanic et al. (2012) obtained dysregulated HPA axis in female adolescent victims of single sexual trauma with posttraumatic stress disorder (PTSD) that was expressed as reduced cortisol and dehydroepiandrosterone sulphate levels. Excessive corticosterone secretion, such as occurs under conditions of PTSD (Arbel et al. 1994; Weems and Carrion 2007), results in neurotoxicity in brain regions rich in GRs, most particularly the hippocampus and prefrontal cortex (Bremner et al. 1997; Woon and Hedges 2008). Carrion and Wong (2012) observed that adolescents who had been exposed to PTSD expressed higher levels of cortisol. Pre-bedtime levels of cortisol predicted reductions in hippocampal volume longitudinally. They found that cortisol levels were correlated negatively with prefrontal cortex volume with functional imaging studies indicating reduced hippocampal and prefrontal cortex activities on tasks of memory and executive functioning in the PTSD adolescents, compared with healthy control adolescents. Alternatively, elevated risk for narcissism, characterized by grandiosity, low empathy and entitlement, may be a developmental hazard for male adolescents. It appears that the HPA axis is chronically activated in males with unhealthy narcissism, stress-related condition of constant activation of the HPA axis with major health implications (Reinhard et al. 2012).

1.5 Early Life Oxidative Stress and Hormesis

Costantini (Chap. 13) has provided an account of the physiological processes underlying the long-term effects of environmental conditions during the phase of early life upon notions of Darwinian fitness. The extent to which early stress exposure was detrimental to fitness seems to be dependent upon levels of severity: mild stress levels appear to produce stimulatory, possibly beneficial effects, through hormetic responses to stressful stimuli. Costantini et al. (2012) studied zebra finches (*Taeniopygia guttata*) to test the hypothesis that individuals exposed to mild heat stress during early life will be less afflicted by oxidative stress when faced with higher levels of heat stress during adulthood than those individuals that were either not pre-exposed to mild heat stress or exposed to high heat stress during early life. They observed that the early exposure to mild heat stress primed the biological systems involved more effectively to withstand the oxidative stress induced by higher heat stress as an adult. Hormetic effects may be mediated through autophagy: for example low dose and low dose rate ionizing radiation in model systems can stimulate cell proliferation by altering the equilibrium between the phosphorylated and dephosphorylated forms of growth factor receptors (Szumiel 2012). *Hormesis* (from Greek *hórmēsis,* rapid motion, eagerness; also referred as Mithridatism after the Pontine King, Mithridates IV, who introduced the technique, around 100 bc, to avoid assassination through poisoning) is the term for generally favourable biological responses to low exposures to toxins and other stressors. A pollutant or toxin showing hormesis thus induces the opposite effect in small doses than it does in large doses. For example, radiation hormesis, referred to as radiation homeostasis also, is the hypothesis that low doses of ionizing radiation (within the region and just above natural background levels) are beneficial, stimulating the activation of repair mechanisms that protect against disease and that are not activated in absence of ionizing radiation. The "reserve repair" mechanisms are hypothesized to be sufficiently effective when stimulated as to not only cancel the detrimental effects of ionizing radiation but also inhibit disease not related to radiation exposure (Fig. 1.1).

Autophagy is the process of self-digestion by a cell through the action of enzymes originating within the same cell, an essential homeostatic process through which cells breakdown their own components. It is emerging as an important biological paradigm that potentially may exert a broad impact upon fundamental biological and translational processes within several domains including chronic inflammatory diseases, ageing and neurodegenerative disorders and adaptive immunity (Deretic and Tooze 2012). Radioadaptation is induced by various weak stress stimuli and is dependent upon signalling events that ultimately reduce the molecular damage expression at the cellular level upon subsequent exposure to a moderate radiation dose. Autophagy may be induced also by reactive oxygen species (ROS): the balance between signalling functions and damaging effects of ROS offers the crucial factor deciding the fate of biological cells under oxidative stress conditions, following exposure to ionizing radiation. Autophagy has been proposed to be a protective mechanism to overcome neurodegenerative processes that are modulated

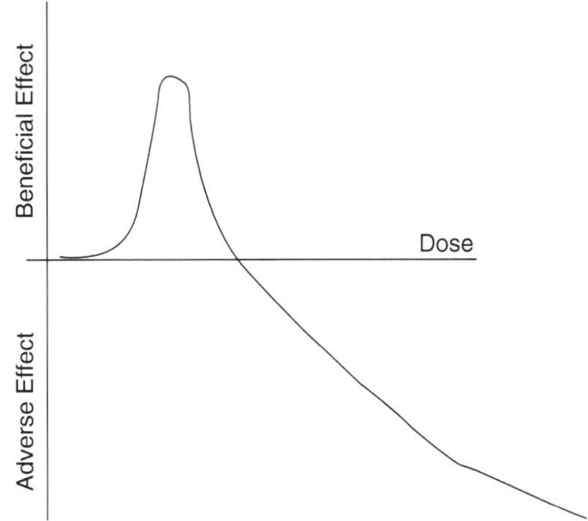

Fig. 1.1 An hormesis dose–response graph whereby a very low dose of a chemical agent (or of ionizing radiation) may trigger from an organism the opposite response to a very high dose (or of radiation). It is suggested that low doses of toxins or other stressors may activate the repair mechanisms of the body through autophagy (see accompanying text)

through endoplasmic reticulum stress (Matus et al. 2012); homeostatic "crosstalk" between endoplasmic reticulum stress and the autophagy pathway has been proposed to mediate the therapeutic effects whereby an "hormesis mechanism" moderates the degenerative propensity through preconditioning effects involving the dynamic balance between endoplasmic reticulum stress and autophagy. Lopez-Martinez and Hahn (2012) have demonstrated hormesis in the California fruit fly by prior exposure to a conditioning bout of anoxia stress that confers resistance to oxidative stress and enhances male sexual performance later. Anoxic conditioning of adults before emergence led to an increase in antioxidant capacity driven by mitochondrial superoxide dismutase and glutathione peroxidase. Males that had received anoxic conditioning and exposed to gamma irradiation as a strong oxidative stressor had lower lipid and protein damage at sexual maturity, thereby demonstrating the hormetic treatment effect of the short-term anoxic conditioning procedure.

The CMS animal model of depression (Willner 1997), which as a consequence of sequential, unpredictable exposure to series of mild stressors, induces reduced responsiveness to several normally rewarding stimuli (Henningsen et al. 2009), thereby producing the anhedonia aspect of depression spectrum disorder. The CMS paradigm induces anhedonic behaviour, a major symptom of disorder though not expressed by all the animals, by exposing rats to a series of mild stressors for 7 weeks, with antidepressant treatment during the last 4 weeks. The procedure of segregating CMS-exposed rats into two subgroups, confirmed by tests of place-preference conditioning and gene and protein analysis (Bisgaard et al. 2007), as well as HPA activation profiles results in (1) a "CMS-sensitive" group that develops anhedonia-like symptoms and (2) a "CMS-resilient" group that expresses resilience to the influences of CMS upon hedonic status, as assessed by sucrose consumption (Bergström et al. 2008). The neurobiological changes induced by the CMS animal model of depressive disorders have been examined critically in comparison with clinical expressions of this disorder spectrum with regard to neurochemistry, neurochemical receptor expression and functionality, neurotrophin expression and cellular plasticity (Hill et al. 2012) They have concluded that (a) the CMS paradigm evokes an array of neurobiological alterations that reflect those symptom profiles observed in clinical depression, and (b) the CMS model offers an investigatory tool to discover novel systems that may be rendered dysfunctional/dysregulated in this spectrum of disorders. The utility of the CMS model has been enhanced through studies aimed at the elucidating the pharmacogenomics of mood disorders and the treatment of stress-associated conditions. Christensen et al. (2011) performed a search for biomarkers involved in treatment resistance and stress resilience in order to investigate mechanisms underlying antidepressant drug refraction and stress-coping strategies. They identified four genes associated with recovery, two genes implicated in treatment resistance and three genes involved in stress resilience. The identified genes associated with mechanisms of cellular plasticity, including signal transduction, cell proliferation, cell differentiation and synaptic release. Hierarchical clustering analysis confirmed the subgroup segregation pattern in the CMS model. Wang et al. (2012) analysed differences in oxidative parameters among CMS-resilient, CMS-sensitive and normal control rats in the cortex, hippocampus and cerebellum pertaining to oxidative stress profiles. They obtained significant increases in protein peroxidation in the cortex and hippocampus and catalase activity in the all three regions, and a significant decrease in superoxide dismutase activity (all three regions) in the CMS-sensitive group compared to the CMS-resilient and control rats. There was an increased lipid peroxidation in the cerebellum of both the CMS-sensitive and the CMS-resilient groups compared to controls. They concluded that CMS-induced oxidative damage and alterations in the activity of antioxidants possibly leading to increased oxidative stress, independent of the rats' anhedonia status. Nevertheless, it would appear that anhedonia resilience was to some extent accompanied by a noteworthy degree of oxidative stress resilience. Finally, Henningsen et al. (2012) applied the CMS model to identify stress-susceptible and stress-resilient rats. Stress susceptibility was linked to increased expression of a sodium-channel protein (SCN9A), a potential antidepressant target. Differential protein profiling showed stress susceptibility to be associated with deficits in synaptic vesicle release involving SNCA (α-synuclein protein gene), SYN-1 (synapsin-1 gene) and AP-3 (adaptor protein complex gene). Their results suggested that increased oxidative phosphorylation (COX5A (cytochrome c oxidase gene), NDUFB7 (NADH dehydrogenase (ubiqu none) 1 beta subcomplex, 7 gene), NDUFS8 (NADH dehydrogenase (ubiquinone) Fe-S protein 8 gene), COX5B (cytochrome c oxidase subunit Vb gene) and UQCRB (ubiquinol-cytochrome c reductase binding protein gene)) within the hippocampal CA regions provides part of a stress-protection mechanism.

1.6 Conclusion

The agency of stress, dependent upon type, severity, duration and frequency, remains mainly a factor that disturbs the normal progress of neurodevelopment with chronic neurologic and neuropsychiatric consequence under more extreme extents.

The nature of the stressors together with their occurrence during the different phases of development will be critical for the adaptive/maladaptive expressions of outcome in both human studies and animal models. The notion of resilience to stressors and its eventual development and mobilization has been explored in the laboratory and in human studies. Nevertheless, there is accumulating a plethora of intriguing studies from a variety of model systems that indicate the presence of hormetic effects and the accompanying autophagic mechanisms involved.

References

Abdeslam M (2012) Prenatal immune stress in rats dampens fever during adulthood. Dev Neurosci. 34:318–326. DOI: 10.1159/000339852

Abumaria N, Rygula R, Hiemka C, Fuchs E, Havemann-Reinecke U, Ruther E, Flügge G (2007) Effect of chronic citalopram on serotonin-related and stress-related genes in the dorsal raphe nucleus of the rat. Eur Neuropsychopharmacol 17:417–429

Abumaria N, Ribic A, Anacker C, Fuchs E, Flügge G (2008) Stress upregulates TPH1 but not TPH2 mRNA in the rat dorsal nucleus: identification of two TPH2 mRNA splice variants. Cell Mol Neurobiol 28:331–342

Alenina N, Kikic D, Todiras M, Mosienko V, Qadri F, Plehm R, Boyé P, Vilianovitch L, Sohr R, Tenner K, Hörtnagl H, Bader M (2009) Growth retardation and altered autonomic control in mice lacking brain serotonin. Proc Natl Acad Sci U S A 106:10332–10337

Arbel I, Kadar T, Silbermann M, Levy A (1994) The effects of long-term corticosterone administration on hippocampal morphology and cognitive performance of middle-aged rats. Brain Res 657:227–235

Arborelius L, Eklund MB (2007) Both long and brief maternal separation produces persistent changes in tissue levels of brain monoamines in middle-aged female rats. Neuroscience 145:738–750

Archer T (2010) Effects of endogenous agents on brain development: stress, abuse and therapeutic compounds. CNS Neurosci Ther 17:470–489

Archer T, Adolfsson B, Karlsson E (2008) Affective personality as cognitive-emotional presymptom profiles regulatory for self-reported health predispositions. Neurotox Res 14:21–44

Archer T, Beninger RJ, Palomo T, Kostrzewa RM (2010a) Epigenetics and biomarkers in the staging of neuropsychiatric disorders. Neurotox Res 18:347–366

Archer T, Kostrzewa RM, Beninger RJ, Palomo T (2010b) Staging perspectives in neurodevelopmental aspects of neuropsychiatry: agents, phases and ages at expression. Neurotox Res 18:287–305

Archer T, Fredriksson A, Schütz E, Kostrzewa RM (2011) Influence of physical exercise on neuroimmunological functioning and health: aging and stress. Neurotox Res 20:69–83

Baier CJ, Katunar MR, Adrover E, Pallares ME, Antonelli MC (2012) Gestational restraint stress and the developing dopaminergic system: an overview. Neurotox Res 22:16–32

Barber K, Mussin E, Taylor DK (1996) Fetal exposure to involuntary maternal smoking and childhood respiratory disease. Ann Allergy Asthma Immunol 76:427–430

Barker DJP (1998) Intrauterine programming of chronic disease. Clin Sci 95:115–128

Barker DJP (2004) The developmental origins of adult disease. J Am Coll Nutr 23(6):588S–595S

Bennett A, Sain SR, Vargas E, Moore LG (2008) Evidence that parent-of-origin affects birthweight reductions at high altitude. Am J Hum Biol 20:592–597

Bergström A, Jayatissa MN, Mörk A, Wiborg O (2008) Stress sensitivity and resilience in the mild stress rat model of depression: an in situ hybridization study. Brain Res 1196:41–52

Bicanic IA, Postma RM, Sinnema G, De Roos C, Olff M, Van Wesel F, Van de Putte EM (2012) Salivary cortisol and dehydroepiandrosterone sulfate in adolescent rape victims with post traumatic stress disorder. Psychoneuroendocrinology pii:S0306-4530(12):00230–00232, PMID: 22867760

Bisgaard CF, Jayatissa MN, Enghild JJ, Sanchez C, Artemychyn R, Wiborg O (2007) Proteomic investigation of the ventral rat hippocampus links DRP-2 to escitalopram treatment resistance and SNAP to stress resilience in the mild stress model of depression. J Mol Neurosci 32:132–144

Bremner JD, Randall P, Vermetten E, Staib L, Bronen RA, Mazure C, Capelli S, McCarthy G, Innis RB, Charney DS (1997) Magnetic resonance imaging-based measurement of hippocampal volume in posttraumatic stress disorder related to childhood physical and sexual abuse—a preliminary report. Biol Psychiatry 41:23–32

Brunton PJ, Russell JA (2010) Prenatal social stress in the rat programmes neuroendocrine and behavioural responses to stress in the adult offspring. Sex-specific effects. J Neuroendocrinol 22:258–271

Carrion VG, Wong SS (2012) Can traumatic stress alter the brain? Understanding the implications of early trauma on brain development and learning. J Adolesc Health 51:S23–S28

Carrion VG, Weems CF, Reiss AL (2007) Stress predicts brain changes in children: a pilot longitudinal study on youth stress, posttraumatic stress disorder, and the hippocampus. Pediatrics 119:509–516

Carrion VG, Haas BW, Garrett A, Song S, Reiss AL (2010a) Reduced hippocampal activity in youth with posttraumatic stress symptoms: an fMRI study. J Pediatr Psychol 35:559–569

Carrion VG, Weems CF, Richert K, Hoffman BC, Reiss AL (2010b) Decreased prefrontal cortical volume associated with increased bedtime cortisol in traumatized youth. Biol Psychiatry 68:491–493

Chaouloff F, Berton O, Mormede P (1999) Serotonin and stress. Neuropsychopharmacology 21:28S–32S

Chen HF, Su HM (2012) Exposure to a maternal n-3 fatty acid-deficient diet during brain development provokes excessive hypothalamic-pituitary-adrenal axis responses to stress and behavioural indices of depression and anxiety in male rat offspring later in life. J Nutr Biochem. doi:10.1016/j.jnutbio.2012.02.006

Christensen T, Bisgaard CF, Wiborg O (2011) Biomarkers of anhedonic-like behavior, antidepressant drug refraction, and stress resilience in a rat model of depression. Neuroscience 196:66–79

Cortessis VK, Thomas DC, Levine AJ, Breton CV, Mack TM, Siegmund KD, Haile RW, Laird PW (2012) Environmental epigenetics: prospects for studying epigenetic mediation of exposure-response relationships. Hum Genet 131(10):1565–1589, PMID: 22740325

Costantini D, Monaghan P, Metcalfe NB (2012) Early life experience primes resistance to oxidative stress. J Exp Biol 215:2820–2826

Dancause KN, Cao XJ, Veru F, Xu S, Long H, Yu C, Laplante DP, Walker CD, King S (2012) Brief communication: prenatal and early postnatal stress influences long bone length in adult rat offspring. Am J Phys Anthropol 149(2):307–311. doi:10.1002/ajpa.22117

Darnaudery M, Maccari S (2008) Epigenetic programming of the stress response in male and female rats by prenatal restraint stress. Brain Res Rev 57:571–585

Del Giudice M, Ellis BJ, Shirtcliff EA (2011) The adaptive calibration model of stress responsivity. Neurosci Biobehav Rev 35:1562–1592

Delaney-Black V, Covington C, Ondersma SJ, Nordstrom-Klee B, Templin T, Ager J, Janisse J, Sokol RJ (2002) Violence exposure, trauma, and IQ and/or reading deficits among urban children. Arch Pediatr Adolesc Med 156(3):280–285

Der-Avakian A, Markou A (2010) Neonatal maternal separation exacerbates the reward-enhancing effect of acute amphetamine administration and the anhedonic effect of repeated social defeat in adult rats. Neuroscience 170:1189–1198

Deretic V, Tooze SA (2012) Autophagy in stress, development and disease. Gordon Research Seminar, Ventura, CA

Diz-Chaves Y, Pernia O, Carrero P, Garcia-Segura LM (2012) Prenatal stress causes alterations in the morphology of microglia and the inflammatory response of the hippocampus of adult female mice. J Neuroinflammation 9:71

Djordjevic J, Djordjevic A, Adzic M, Radojcic MB (2012) Effects of chronic social isolation on Wistar rat behaviour and brain plasticity markers. Neuropsychobiology 66:112–119

Drake AJ, Liu L, Kerrigan D, Meehan RR, Seckl JR (2011) Multigenerational programming in the glucocorticoid rat is associated with generation-specific and parent of origin effects. Epigenetics 6:1334–1343

Dyuzhikova NA, Shiryaeva NV, Pavlova MB, Vaido AL (2012) Long-term effects of prenatal stress on the characteristics of hippocampal neurons in rats with different excitability of the nervous system. Bull Exp Biol Med 152:568–570

Garcia D, Siddiqui A (2009a) Adolescents' affective temperaments: life satisfaction, interpretation, and memory of events. J Posit Psychol 4:155–167

Garcia D, Siddiqui A (2009b) Adolescents' psychological well-being and memory for life events: influences on life satisfaction with respect to temperamental dispositions. J Happiness Stud 10:407–419

Garcia D, Kerekes N, Andersson-Arntén AC, Archer T (2012) Temperament, character, and adolescents' depressive symptoms: focusing on affect. Depress Res Treat. doi:10.1155/2012/925372

Gartside SE, Johnson DA, Leitch MM, Troakes C, Ingram CD (2003) Early life adversity programs changes in central 5-HT neuronal function in adulthood. Eur J Neurosci 17:2401–2408

Gaspar P, Cases O, Maroteaux L (2003) The developmental role of serotonin: news from mouse molecular genetics. Nat Rev Neurosci 4:1002–1012

Giachino C, Canalia N, Capone F, Fasolo A, Alleva E, Riva MA, Cirulli F, Peretto P (2007) Maternal deprivation and early handling affect density of calcium binding protein-containing neurons in selected brain regions and emotional behaviour in periadolescent rats. Neuroscience 145:568–578

Glover V (2011) Annual research review: prenatal stress and the origins of psychopathology. An evolutionary perspective. J Child Psychol Psychiatry 52:356–367

Gluckman PD, Hanson MA, Spencer HG (2005) Predictive adaptive responses and human evolution. Trends Ecol Evol 20:527–533

Gluckman PD, Hanson MA, Beedle AS (2007) Early life events and their consequences for later disease: a life history and evolutionary perspective. Am J Hum Biol 19:1–19

Gong Y, Sun XL, Wu FF, Su CJ, Ding JH, Hu G (2012) Female early adult depression results in detrimental impacts on behavioural performance and brain development in offspring. CNS Neurosci Ther 18:461–470

Green P, Glozman S, Kamensky B, Yavin E (1999) Developmental changes in rat brain membrane lipids and fatty acids. The preferential prenatal accumulation of docosahexaenoic acid. J Lipid Res 40:960–966

Gutknecht L, Waider J, Kraft S, Kriegebaum C, Holtmann B, Reif A, Schmitt A, Lesch KP (2008) Deficiency of brain 5-HT synthesis but serotonergic neuron formation in Tph2 knockout mice. J Neural Transm 115:1127–1132

Henningsen K, Andreasen JT, Bouzinova EV, Jayatissa MN, Jensen MS, Redrobe JP, Wiborg O (2009) Cognitive deficits in the rat chronic mild stress model for depression: relation to anhedonia-like responses. Behav Brain Res 198:136–141

Henningsen K, Palmfeldt J, Christiansen S, Baiges I, Bak S, Jensen ON, Gregersen N, Wiborg O (2012) Candidate hippocampal biomarkers of susceptibility and resilience to stress in a rat model of depression. Mol Cell Proteomics 11(7):M111.016428

Herzog CJ, Czeh B, Corbach S, Wuttke W, Schulte-Herbrüggen O, Hellweg R, Flügge G, Fuchs E (2009) Chronic social instability stress in female rats: a potential animal model for female depression. Neuroscience 159:982–992

Hill MN, Hellemans KGC, Verma P, Gorzalka BB, Weinberg J (2012) Neurobiology of chronic mild stress: parallels to major depression. Neurosci Biobehav Rev 36(9):2085–2117. doi:org/10.1016/j.neubiorev.2012-07.001

Jacobsen JP, Siesser WB, Sachs BD, Peterson S, Cools MJ, Setole V, Folgering JH, Flik G, Caron MG (2011) Deficit serotonin neurotransmission and depression-like serotonin biomarker alterations in tryptophan hydroxylase-2 (Tph2) loss-of-function mice. Mol Psychiatry 17(7):694–704. doi:doi:10.1038/mp.2011.50

Jensen Pena C, Monk C, Champagne FA (2012) Epigenetic effects of prenatal stress on 11β-hydroxysteroid dehydrogenase-2 in the placenta and fetal brain. PLoS One 7(6):e39791

Johnson SB, Blum RW (2012) Stress and the brain: how experiences and exposures across the life span shape health, development, and learning in adolescents. J Adolesc Health 51:S1–S2

Johnson SB, Dariotis JK, Wang C (2012) Adolescent risk taking under stressed and non-stressed conditions: conservative, calculating and impulsive types. J Adolesc Health 51(Suppl 2):S34–S40

Kajantie E (2006) Fetal origins of stress-related adult disease. Ann N Y Acad Sci 1083:11–27

Laloux C, Mairesse J, Van Camp G, Giovine A, Branchi I, Bouret S, Morley-Fletcher S, Bergonzelli G, Malagodi M, Gradini R, Nicoletti F, Darnaudéry M, Maccari S (2012) Anxiety-like behaviour and associated neurochemical and endocrinological alterations in male pups exposed to prenatal stress. Psychoneuroendocrinology 37(10):1646–1658, PMID: 22444623

Laplante P, Diorio J, Meaney MJ (2002) Serotonin regulates hippocampal glucocorticoid receptor expression via a 5-HT receptor. Brain Res Dev Brain Res 139:199–203

Lau C, Rogers JM (2004) Embryonic and fetal programming of physiologic disorders in adulthood. Birth Defects Res C Embryo Today 72:300–312

Lauritzen L, Hansen HS, Jorgensen MH, Michaelsen KF (2001) The essentiality of long chain n-3 fatty acids in relation to development and function of the brain and retina. Prog Lipid Res 40:1–94

Lesch KP, Araragi N, Waider J, van den Hov D, Gutkhecht L (2012) Targeting brain serotonin synthesis: insights into neurodevelopmental disorders with long-term outcomes related to negative emotionality, aggression and antisocial behaviour. Philos Trans R Soc Lond B Biol Sci 367:2426–2443

Liu J, Chaplin TM, Wang F, Sinha R, Mayes LC, Blumberg HP (2012a) Stress reactivity and corticolimbic response to emotional faces in adolescents. J Am Acad Child Adolesc Psychiatry 51:304–312

Liu W, Xu Y, Lu J, Zhang Y, Sheng H, Ni X (2012b) Swimming exercise ameliorates depression-like behaviours induced by prenatal exposure to glucocorticoids in rats. Neurosci Lett 524(2):119–123, PMID: 22444623

Lopez-Martinez G, Hahn DA (2012) Short-term anoxic conditioning hormesis boosts antioxidant defenses, lowers oxidative damage following irradiation and enhances male sexual performance in the Caribbean fruit fly, Anastrepha suspensa. J Exp Biol 215:2150–2161

Maestripieri D (2011) Emotions, stress, and maternal motivation in primates. Am J Primatol 73:516–529

Mairesse J, Silletti V, Laloux C, Zuena AR, Giovine A, Consolazione M, van Camp G, Malagodi M, Gaetani S, Cianci S, Catalani A, Mennuni G, Mazzetta A, van Reeth O, Gabriel C, Mocaër E, Nicoletti F, Morley-Fletcher S, Maccari S (2012) Chronic agomelatine treatment corrects the abnormalities in the circadian rhythm of motor activity and sleep/wake cycle induced by prenatal restraint stress in adult rats. Int J Neuropsychopharmacol 6:1–16

Matus S, Castillo K, Herz C (2012) Hormesis: protecting neurons against cellular stress in Parkinson's disease. Autophagy 8(6):997–1001, PMID: 22858553

McCullough G, Huebner ES, Laughlin JE (2000) Life events, self-concept, and adolescents' positive subjective well-being. Psychol Sch 37:281–290

Mealer M, Jones J, Moss M (2012) A qualitative study of resilience and posttraumatic stress disorder in United States ICU nurses. J Intensive Care Med 38(9):1445–1451

Medalla M, Barbas H (2010) Anterior cingulated synapses in prefrontal areas 10 and 46 suggest differential influence in cognitive control. J Neurosci 30:16068–16081

Mulligan C, D'Errico N, Stees J, Hughes D (2012) Maternal changes at NR3C1 in newborns associate with maternal prenatal stress exposure and newborn birth weight. Epigenetics 7(8):853–857, PMID: 22810058

Murgatroyd CA, Nephew BC (2012) Effects of early life social stress on maternal behaviour and neuroendocrinology. Psychoneuroendocrinology. doi:org/10.1016/j.psyneuen.2012.05.020

Murgatroyd CA, PatchevAV WY, Micale V, Bockmuhl Y, Fischer D, Holsboer F, Wotjak CT, Almeida OFX, Spengler D (2009) Dynamic DNA methylation programs persistent adverse effects of early-life stress. Nat Neurosci 12:1559–1566

Murgatroyd CA, Wu Y, Bockmuhl Y (2010) Genes learn from stress: how infantile trauma programs us for depression. Epigenetics 5:194–199

Nephew BC, Bridges RS (2011) Effects of chronic social stress during lactation on maternal behaviour and growth in rats. Stress 14:677–684

Neumann ID, Kromer SA, Bosch OJ (2005) Effects of psychosocial stress during pregnancy on neuroendocrine and behavioral parameters in lactation depend on the genetically-determined stress vulnerability. Psychoneuroendocrinology 30:791–806

Oberlander TF (2012) Fetal serotonin signalling: setting pathways for early childhood development and behaviour. J Adolesc Health 51:S9–S16

Oberlander TF, Weinberg J, Papsdorf M, Grunau R, Misri S, Devlin AM (2008) Prenatal exposure to maternal depression, neonatal methylation of human glucocorticoid receptor gene (NR3C1) and infant cortisol stress responses. Epigenetics 3:97–106

Opacka-Juffry J, Mohiyeddini C (2012) Experience of stress in childhood negatively correlates with plasma oxytocin concentration in adult men. Stress 15:1–10

Palomo T, Kostrzewa RM, Beninger RJ, Archer T (2007) Genetic variation and shared biological susceptibility underlying comorbidity in neuropsychiatry. Neurotox Res 12:29–42

Price TD, Qvarnström A, Irwin DE (2003) The role of phenotypic plasticity in driving genetic evolution. Proc Biol Sci 270:1433–1440

Qiu GH, Tan LK, Loh KS, Lim CY, Srivasteva G, Tsai ST, Tao Q (2004) The candidate tumor suppressor gene BLU, located at the commonly deleted region 3p21.3 is an E2F-regulated, stress-responsive gene and inactivated by both epigenetic and genetic mechanisms in nasopharyngeal carcinoma. Oncogene 23:4793–4806

Reinhard DA, Konrath SH, Lopez WD, Cameron HG (2012) Expensive egos: narcissistic males have higher cortisol. PLoS One 7(1):e30858

Saigh PA, Mroueh M, Bremner JD (1997) Scholastic impairments among traumatized adolescents. Behav Res Ther 35:429–436

Saigh PA, Yasik AE, Oberfield RA, Halamandaris PV, Bremner JD (2006) The intellectual performance of traumatized children and adolescents with or without posttraumatic stress disorder. J Abnorm Psychol 115:332–340

Schechter JC, Brennan PA, Cunningham PB, Foster SL, Whitmore E (2012) Stress, cortisol, and externalizing behavior in adolescent males: an examination in the context of multisystemic therapy. J Abnorm Child Psychol 40:913–922

Schury K, Kolassa IT (2012) Biological memory of childhood maltreatment: current knowledge and recommendation for future research. Ann N Y Acad Sci 1262:93–100

Segovia F, Moore JL, Linnville SE, Hoyt RE, Hain RE (2012) Optimism predicts resilience in repatriated prisoners of war: a 37-year longitudinal study. J Trauma Stress 25:330–336

Skilbeck KJ, Johnston GA, Hinton T (2010) Stress and GABA receptors. J Neurochem 112:1115–1130

Sterrenburg L, Gaszner B, Boerrigter J, Santbergen L, Bramini M, Elliott E, Chen A, Peeters BW, Roubos EW, Kozicz T (2011) Chronic stress induces sex-specific alterations in methylation and expression of corticotrophin-releasing factor gene in the rat. PLoS One 6(11):e28128, PMID: 22132228

Steullet P, Cabungcal JH (2010) Redox dysregulation affects the ventral but not dorsal hippocampus: impairment of parvalbumin neurons, gamma oscillations, and related behaviours. J Neurosci 30:2547–2558

Strathearn L, Fonagy P, Amico J, Montague PR (2009) Adult attachment predicts maternal brain and oxytocin response to infant cues. Neuropsychopharmacology 34:2655–2666

Stuebe AM, Grewen K, Pedersen CA, Propper C, Meltzer-Brody S (2011) Failed lactation and perinatal depression: common problems with shared neuroendocrine mechanisms? J Womens Health 21:264–272

Su HM (2010) Mechanisms of n-3 fatty acid-mediated development and maintenance of learning memory performance. J Nutr Biochem 21(5):364–373

Su ZY, Shu L, Khor TO, Lee JH, Fuentes F, Kong AN (2012) A perspective on dietary phytochemicals and cancer chemoprevention: oxidative stress, Nrf2, and epigenomics. Top Curr Chem. PMID: 22836898

Szumiel I (2012) Radiation hormesis: autophagy and other cellular mechanisms. Int J Radiat Biol 88(9):619–628, PMID: 22702489

Von Dawans B, Fischbacher U, Kirschbaum C, Fehr E, Heinrichs M (2012) The social dimension of stress reactivity: acute stress increases prosocial behaviour in humans. Psychol Sci 23:651–660

Waider J, Araragi N, Gutknecht L, Lesch KP (2011) Tryptophan hydroxylase-2 (TPH2) in disorders of cognitive control and emotion regulation: a perspective. Psychoneuroendocrinology 36:393–405

Wang C, Wu HM, Jing WR, Meng Q, Zhang H, Gao GD (2012) Oxidative parameters in the rat brain of chronic mild stress model for depression: relation to anhedonia-like responses. J Membr Biol. doi:10.1007/s00232-012-9436-4

Weems CF, Carrion VG (2007) The association between PTSD symptoms and salivary cortisol in youth: the role of time since the trauma. J Trauma Stress 20:903–907

Werner E, Zhao Y, Evans L, Kinsella M, Kurzius L, Altincatal A, McDonough L, Monk C (2012) Higher maternal prenatal cortisol and younger age predict greater infant reactivity to novelty at 4 months: an observation-based study. Dev Psychobiol. DOI: 10.1002/dev.21066

Whitaker-Azmitia PM, Druse M, Walker P, Lauder JM (1996) Serotonin as a developmental signal. Brain Res 73:19–29

Willner P (1997) Validity, reliability and utility of the chronic mild stress model of depression: a 10-year review and evaluation. Psychopharmacology 134:319–329

Woon FL, Hedges DW (2008) Hippocampal and amygdala volumes in children and adults with childhood maltreatment-related posttraumatic stress disorder: a meta-analysis. Hippocampus 18:729–736

Zuloaga DG, Carbone DL, Handa RJ (2012) Prenatal dexamethaxone selectively decreases calretinin expression in the adult female lateral amygdale. Neurosci Lett 521:109–114

Part I
Introductory Concepts

Chapter 2
Making Sense of Stress: An Evolutionary—Developmental Framework

Marco Del Giudice, Bruce J. Ellis, and Elizabeth A. Shirtcliff

Abstract In this chapter we present an evolutionary–developmental framework for individual differences in stress responsivity, the Adaptive Calibration Model (ACM). We argue that the core propositions of the ACM provide a context for the integrative biological analysis of the stress response system, exemplified by Tinbergen's "four questions" of mechanism, ontogeny, phylogeny, and adaptation. We then show how the ACM can be used to generate novel predictions on responsivity profiles in humans and their development across the life span.

2.1 Tinbergen's Four Questions in Stress Research

If anything qualifies as a complex biological mechanism, the stress response system (SRS) certainly does. The stress response involves the hierarchical, coordinated action of the autonomic system and the hypothalamic–pituitary–adrenal axis (HPA), as well as multilevel feedback loops with cortical brain structures. Be it by direct innervation or endocrine signaling, the SRS regulates an astonishing range of physiological and behavioral processes, including bodily growth, metabolism,

M. Del Giudice (✉)
Center for Cognitive Science, Department of Psychology, University of Turin,
Via Po 14, Torino 10123, Italy
e-mail: marco.delgiudice@unito.it

B.J. Ellis
John and Doris Norton School of Family and Consumer Sciences, University of Arizona,
McClelland Park, 650 North Park Avenue, Tucson, AZ, USA
e-mail: bjellis@email.arizona.edu

E.A. Shirtcliff
Department of Psychology, University of New Orleans, 2000 Lakeshore Drive,
New Orleans, LA 70148, USA
e-mail: birdie.shirtcliff@uno.edu

G. Laviola and S. Macrì (eds.), *Adaptive and Maladaptive Aspects of Developmental Stress*, Current Topics in Neurotoxicity 3, DOI 10.1007/978-1-4614-5605-6_2,

reproductive functioning, attention and memory, learning, aggression, risk-taking, caregiving, and so forth.

Whereas the basic design of the SRS has been worked out reasonably well (see, e.g., Chap. 9), the intricate details of its functioning in ecological contexts still pose formidable puzzles to researchers. First, the ubiquity of SRS involvement adds to the difficulty of making sense of the system's function(s). Second, stress physiology exhibits remarkable individual variation, which is still not well understood and often interpreted from a pathologizing standpoint, especially in the case of humans (see Ellis et al. 2012b). Third, the logic by which individual patterns of stress responsivity develop over time and the role played by genes and environments in the process remain elusive despite the hundreds of empirical studies carried out every year.

Complex biological mechanisms like the SRS can be fully understood only by approaching them from multiple interlocking perspectives. Tinbergen (1963) famously described the four main types of explanation required for a complete understanding of a biological system. Tinbergen's "four problems" or "four questions" have since become a standard heuristic device in evolutionary biology. With updated terminology, the four problems of biology can be summarized as *mechanism* (What is the structure like? How does it work?); *ontogeny* or *development* (How does the structure come to be over developmental time, and how does it change across the lifespan?); *phylogeny* (What is the evolutionary history of the structure? How did it change across generations and species?); and *adaptation* (Why is the structure the way it is? What selective advantages does it confer, or did it confer, to the organism?). Ontogenetic and mechanistic explanations concern the way an organism works in the present, without reference to evolution and adaptation; collectively, they are called *proximate* explanations. In contrast, *ultimate* explanations (phylogenetic and adaptationist) consider the organism in relation to its past and to the evolutionary forces that shaped its body and behavior (Mayr 1963). It should be obvious that, as already stressed by Tinbergen, the four types of explanation are pragmatically distinct but not logically independent from one another. Even more important, they are not mutually exclusive but complementary and synergistic: adaptive function crucially informs the study of mechanism and development, while development and mechanism constrain the range of adaptive explanations (McNamara and Houston 2009; Scott-Phillips et al. 2011). A similar interplay occurs between adaptationist and phylogenetic questions. Starting a virtuous cycle between different levels of explanation is the best way to build a satisfactory model of a complex biological system.

2.1.1 The Need for an Integrative Framework

Precisely because Tinbergen's four questions produce the best answers when they are asked synergistically rather than in isolation, it is extremely useful to possess an integrative theoretical framework. An adequate framework for stress research should involve all four types of biological explanation, tying them together in a coherent narrative. Moreover, it should address both the species-specific functioning of the SRS and the origin of individual and sex differences. Ideally, it should be possible to apply it to different species (with the necessary changes and refinements), a reasonable requirement given the ancient and highly conserved structure of the SRS (Nesse 2007; Porges 2001, 2007). In recent years, considerable progress has been made toward this goal, and a number of evolutionary models of stress responsivity have appeared in the literature (e.g., Carere et al. 2010). Among the most notable are the hawk–dove model by Korte et al. (2005); the social plasticity model by Flinn (2006); the polyvagal theory by Porges (2001, 2007); the tend-and-befriend hypothesis by Taylor et al. (2000); and the theory of biological sensitivity to context (BSC) by Boyce and Ellis (2005). While each of these models provides crucial insights in the function, development, and phylogeny of the SRS, none of them has the scope of a truly integrative theory.

2.1.2 The Adaptive Calibration Model

In the remainder of this chapter, we will introduce the Adaptive Calibration Model (ACM), our recent attempt to provide the field with a comprehensive evolutionary–developmental framework (Del Giudice et al. 2011). The ACM extends and refines the BSC theory (Boyce and Ellis 2005; Ellis et al. 2005), while incorporating several key elements of other evolutionary models (e.g., Flinn 2006; Korte et al. 2005; Porges 2007; Taylor et al. 2000).

From the standpoint of Tinbergen's four questions, the main focus of the ACM is on adaptation and development, but the model also makes several novel predictions about the mechanism of the stress response. The main elements of the ACM are: (a) an evolutionary analysis of the functions of the SRS, defined as an integrated, hierarchically organized system comprising the autonomic nervous system and the HPA axis; (b) a theory of the adaptive match between environmental conditions and stress responsivity; and (c) a taxonomy of four prototypical responsivity patterns to be found in humans, their behavioral and neurobiological correlates, and their hypothesized developmental trajectories. Whereas (b) and (c) are tailored to human ecology and physiology, our evolutionary analysis (a) is based on general biological principles and has the potential to be applied (with minor adjustments) to many different species. Thus, the ACM in its present form is best conceived as consisting of a general theoretical "core," with a detailed theory of human development built on top of it.

In keeping with the broad scope of this volume, here we focus primarily on the theoretical core of our model. After providing a succinct overview of some key concepts in evolutionary biology (Sect. 2.2), we present an evolutionary analysis of the functions of the SRS and argue that it operates as a mechanism of conditional adaptation and a central mediator of the development of life history strategies (Sect. 2.3). Then, we briefly outline our main predictions about the development of individual and sex differences in humans (Sect. 2.4). Interested readers can refer to Del Giudice et al. (2011) for an extended treatment of the ACM, including many additional details and empirical predictions.

2.2 Biological Foundations

2.2.1 Life History Theory

Life history theory is a branch of evolutionary biology dealing with the way organisms allocate time and energy to the various activities that comprise their life cycle (see Ellis et al. 2009; Hill 1993; Roff 2002). All organisms live in a world of limited resources; the energy that can be extracted from the environment in a given amount of time, for example, is intrinsically limited. Time itself is also a limited good; the time spent by an organism looking for mates cannot be used to search for food or care for extant offspring. Since all these activities contribute to an organism's evolutionary fitness, devoting time and energy to one will typically involve both benefits and costs, thus engendering trade-offs between different fitness components. For example, there is a trade-off between bodily growth and reproduction because both require substantial energetic investment, and thus producing offspring reduces somatic growth. Life history theory concerns optimal allocation of time and energy toward competing life functions—bodily maintenance, growth, and reproduction—over the life cycle.

Life history *strategies* are adaptive solutions to a number of simultaneous fitness trade-offs. The most basic trade-offs are between *somatic effort* (i.e., growth, body maintenance, and learning) and *reproductive effort*; and, within reproductive effort, between *mating* (i.e., finding and attracting mates, conceiving offspring) and *parenting* (i.e., investing resources in already conceived offspring). From another perspective, the critical decisions involved in a life history strategy can be summarized by the trade-offs between *current* and *future reproduction* and between *quality* and *quantity of offspring* (see Ellis et al. 2009).

In sexual species, the two sexes predictably differ on life history-related dimensions; they thus can be expected to employ somewhat different strategies in response to the same cues in the environment (James et al. 2012). In most species, males tend to engage in higher mating effort and lower parental effort than females (Geary 2002; Kokko and Jennions 2008; Trivers 1972). In addition, males usually undergo stronger sexual selection (i.e., their reproductive success is more variable) and tend to mature more slowly in order to gain the competitive abilities and qualities needed for successful competition for mates. Sexual asymmetries in life history strategies can be attenuated in species with monogamous mating systems and when both parents contribute to offspring care.

One of the most important implications of life history theory is that no strategy can be optimal in every situation; more specifically, the optimal (i.e., fitness-maximizing) strategy for a given organism depends on its ecology and on a series of factors such as resource availability, mortality, and environmental uncertainty. Indeed, organisms usually embody mechanisms that allow them to fine-tune their life histories according to the environmental cues they encounter during development. For this reason, life history traits and strategies tend not to be genetically fixed but rather evolve to show developmental plasticity (Ellis et al. 2009). Developing organisms assess their local environments and adjust their strategic allocation choices, following evolved rules that maximize expected fitness in different ecological conditions. To the extent they result from evolved mechanisms of plasticity, individual differences in life history are examples of *conditional adaptation* (see below).

2.2.1.1 Factors in the Development of Life History Strategies

The key dimensions of the environment that affect the development of life history strategies are *resource availability, extrinsic morbidity–mortality*, and *unpredictability,* as signaled by observable cues. As explained in detail by Ellis et al. (2009), energetic stress (i.e., malnutrition, low energy intake, negative energy balance, and associated internal stressors such as disease) tends to cause the developing organism to shift toward *slow* strategies, characterized by slower growth and maturation and delayed reproduction. In contrast, both extrinsic (i.e., uncontrollable) morbidity–mortality and unpredictable fluctuations in environmental parameters tend to entrain the development of *fast* strategies, accelerating sexual maturation, promoting early reproduction, and reducing the amount of parental investment provided to the young.

Of course, genetic factors also contribute to determine individual life history strategies. Theoretical models suggest that one should often expect a balance between genetic and environmental determination of phenotypic individual differences. At the individual level, regulatory mechanisms should often evolve so as to integrate both genetic and environmental information in phenotypic determination (Leimar et al. 2006). At the population level, the opportunity for habitat choice plus heterogeneous environmental conditions can maintain a diverse population composed of both "specialists" (fixed phenotypes) and "generalists" (plastic phenotypes), as shown by Wilson and Yoshimura (1994). In a similar vein, differential susceptibility theory (Belsky 1997, 2005) maintains that, because the cues driving the development of conditional phenotypes are not completely reliable, children vary in their susceptibility to rearing influences. Such differential susceptibility underlies pervasive person-by-environment interactions, whereby individuals with given genotypes or phenotypes show higher sensitivity to environmentally induced effects on development (see Belsky 1997, 2005; Belsky and Pluess 2009; Boyce and Ellis 2005; Ellis et al. 2011).

2.2.1.2 Life History Strategies and the Organization of Behavior

When interpreted in a narrow sense, life history strategies refer mainly to growth- and reproduction-related traits such as maturation timing, age at first reproduction, fertility, and number of sexual partners. However, it is easy to see that the choice of a specific strategy will affect a much broader range of traits and behaviors (e.g., Belsky et al. 1991; Figueredo et al. 2004, 2005, 2006; Meaney 2007; Wolf et al. 2007). Imagine an organism that, following cues of extrinsic morbidity-mortality and unpredictability, adopts a strategy characterized by early reproduction and high mating effort. To succeed, the organism needs to outcompete same-sex conspecifics and be chosen by members of the other sex. Especially for males, this is likely to involve dominance-seeking behavior, plus considerable investment in traits and displays that the other sex finds attractive. The cues of environmental risk that drive the choice of the strategy will also prompt higher risk-taking in other domains (e.g., exploration, fighting, dangerous sexual displays), preference for immediate over delayed rewards, and impulsivity (Wolf et al. 2007).

Thus, life history strategies play a powerful role in the organization of behavior. Traits and behaviors that covary along life history dimensions form a broad cluster which includes exploration/learning styles, mating and sexual strategies, pair-bonding, parenting, status- and dominance-seeking, risk-taking, impulsivity, aggression, cooperation, and altruism. Correlations within this cluster have been documented in both humans (e.g., Del Giudice 2009; Figueredo et al. 2004, 2006; Kruger et al. 2008) and other animals (e.g., Dingemanse and Réale 2005; Korte et al. 2005).

2.2.2 Conditional Adaptation and Developmental Switch Points

Conditional adaptation is the evolved ability of an organism to modify its developmental trajectory (and the resulting phenotype) to match the local conditions of the social and physical environment. Conditional adaptation is a manifestation of adaptive developmental plasticity (Pigliucci 2001; West-Eberhard 2003) and is closely related to the concept of a *predictive adaptive response* (e.g., Gluckman et al. 2007). Mechanisms of developmental adaptation can be guided both by external environmental factors (e.g., predation pressures, quality of parental investment, seasonal change, diet) and by indicators of the individual's status or relative competitive abilities in the population (e.g., age, body size, health, history of wins and losses in agonistic encounters).

How do genetic and environmental factors drive conditional adaptation? West-Eberhard (2003) proposed that developmental change is coordinated by regulatory switch mechanisms, which serve as transducers (mediators) of genetic, environmental, and structural influences on phenotypic variation. These switch mechanisms control *developmental switch points:* "…[points] in time when some element of phenotype changes from a default state, action, or pathway to an alternative one—it is activated, deactivated, altered, or moved" (West-Eberhard 2003, p. 67). This can involve a discrete structural change or a change in the rates of a process. Genetic and environmental inputs interact with the extant phenotype to determine the functioning of regulatory switch mechanisms and influence their thresholds. Once a threshold is passed (i.e., the switch occurs), the regulatory mechanism coordinates the expression and use of gene products and environmental elements that mediate the species-typical transition to the new phenotypic stage as well as individually differentiated pathways within that stage.

Most critically, regulatory switch mechanisms provide a common locus of operations for genetic and environmental influences on phenotypic development; that is, these mechanisms are the vehicle through which gene–gene, environment–environment, and gene–environment interactions occur. These inputs structure the operation of regulatory switch mechanisms and may affect the threshold necessary for a developmental switch to occur and/or the organism's ability to cross that threshold (West-Eberhard 2003).

2.3 The Stress Response System as a Mechanism of Conditional Adaptation

2.3.1 Functions of the Stress Response System

The SRS has three main biological functions (see Fig. 2.1). We will now examine these functions in light of the biological concepts presented in the previous section.

2.3.1.1 Allostasis

A key function of the SRS is to coordinate the organism's physiological and behavioral response to environmental threats and opportunities. This includes any event that may have important (i.e., fitness-relevant) consequences for the organism and requires the organism to modify its current state in order to be dealt with effectively. In addition to threats and dangers, environmental opportunities may be represented by unexpected or novel events, and even highly pleasurable situations (e.g., signs of sexual availability in a potential mate). The whole-organism adjustment to environmental challenge is often termed *allostasis* (McEwen 1998; McEwen and Wingfield 2003; Sterling and Eyer 1988). The SRS mediates allostasis by coordinating brain/body changes in response to environmental challenges, both in the short and in the long term. Because allostasis is a broader concept than "stress response" and because many of the challenges that activate the sympathetic nervous system (SNS) and HPA are not "stressors" in the classical sense, the label "stress response system" is not entirely adequate to describe the function of the SRS. Here we employ it for lack of a widely accepted alternative; however, we want to make it clear that the SRS is a general interface with the environment, mediating the organism's adjustment to both positive and negative events (Boyce and Ellis 2005; Koolhaas et al. 2011).

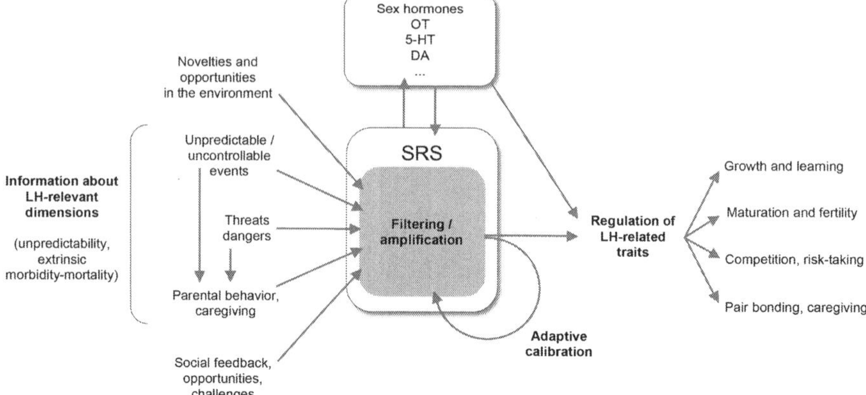

Fig. 2.1 The core theoretical structure of the Adaptive Calibration Model (ACM). *SRS* stress response system; *LH* life history; *OT* oxytocin; *5-HT* serotonin, *DA* dopamine

2.3.1.2 Information Encoding and Filtering

The second function of the SRS, closely connected to the first, is that of encoding and filtering *information* coming from the social and physical environment. The SRS receives complex information about the external environment through limbic structures, and complex information about the organism from interaction with other neuroendocrine systems (e.g., the HPG axis and the immune system; see Herman et al. 2003). Activation of the SRS components thus carries information about the likelihood of threats and opportunities in the environment, their type, and their severity. This information can be encoded by the SRS and, in the long run, provides the organism with a statistical "summary" of key dimensions of the environment, including the crucial life history-relevant dimensions of extrinsic morbidity–mortality and unpredictability. Indeed, unpredictable and uncontrollable events elicit the strongest SRS responses across species, especially at the level of the HPA axis (Dickerson and Kemeny 2004; Koolhaas et al. 2011).

The amount of information encoded by each component of the SRS depends on the specificity of its response. Parasympathetic withdrawal occurs frequently and is a relatively nonspecific response, so it comparatively conveys relatively little information about the local environment. Sympathetic activation, in contrast, is more specifically tied to challenges requiring fight-or-flight responses; patterns of SNS activation may thus provide reliable information about the dangerousness (or safety) of one's environment. The most information-rich response (and the one with the longest lasting effects) is that of the HPA axis, which is strongly activated in unpredictable and/or uncontrollable situations.

An important corollary of this informational view of SRS functioning is that the system's level of responsivity acts as an amplifier (when highly responsive) or filter (when unresponsive) of various types of environmental information. A highly responsive system makes an individual more informationally open and enhances his/her sensitivity to contextual influences, both "positive" and "negative" (Boyce and Ellis 2005; Ellis et al. 2006). An unresponsive system has a higher threshold for letting environmental signals in: many potential challenges will not be encoded as such, and many potentially relevant events will fail to affect the organism's physiology to a significant degree. This will result in a number of potential costs (e.g., reduced alertness, reduced sensitivity to social feedback) as well as potential benefits (e.g., resource economization, avoidance of immune suppression). In fact, many of the possible consequences of low responsivity can be read as either costs or benefits depending on context. Reduced sensitivity to feedback, for example, can be optimal in highly competitive contexts, or when taking deliberate risks. More generally, sometimes organisms do well to partially or totally shield themselves from the effects of environmental information.

A highly responsive SRS, by contrast, amplifies the signal coming from the environment and maximizes the chances that the organism will be modified by current experience. This, too, can have both costs and benefits. Potential costs of a highly responsive system include high physiological costs, hypersensitivity to social feedback, and exposure to psychological manipulation; in addition, the organism's action plans can get easily interrupted by minor challenging events, and the ability to deal with future events may be reduced if physiological resources are already overwhelmed. On the other hand, a highly responsive system facilitates some forms of learning, enhances mental activities in loc.lized domains, focuses attention, and primes memory storage, thus improving cognitive processes for dealing with environmental opportunities and threats (e.g., Barsegyan et al. 2010; Flinn 2006; Roozendaal 2000; van Marle et al. 2009).

2.3.1.3 Regulation of Life History-Relevant Traits

The role of the SRS extends way beyond mounting responses to immediate challenges. Profiles of SRS baseline activity and responsivity are associated with individual differences in a range of life history-relevant domains including competitive risk-taking, learning, self-regulation, attachment, affiliation, reproductive functioning, and caregiving. In the next paragraphs we will discuss some examples (for an extended treatment, see Del Giudice et al. 2011).

To begin with, the HPA is crucially involved in the regulation of metabolism, and chronic stress has been linked to individual differences in growth patterns (e.g., Hofer 1984; Schanberg et al. 1984). Physical growth is an important component of somatic effort, but, from the biological point of view, *learning* can also be conceptualized as a form of investment in "embodied capital." A learning organism spends time and energy accumulating knowledge and developing skills that may become useful in the future (e.g., Kaplan et al. 2000). The SRS modulates learning in a number of different ways: in humans, HPA and autonomic profiles have been associated with individual differences in cognitive functioning (e.g., Staton et al. 2009), memory (e.g., Stark et al. 2006), and self-regulation/executive function (e.g., Blair et al. 2005; Shoal et al. 2003; Williams et al. 2009).

The autonomic systems, HPA, and hypothalamic–pituitary–gonadal (HPG) axes are connected by extensive functional cross-talk (e.g., Ellis 2004; Viau 2002), and cortisol is a major regulator of fertility and sexual development. Given adequate bioenergetic resources to support growth and reproduction, exposures to chronic psychosocial stressors generally provoke early or accelerated development of the HPG axis but suppressed ovarian functioning in mature individuals (reviewed in Ellis 2004). The effects of *acute* response to challenge are much more variable; males and females do not respond in the same way, and whether acute stress suppresses or enhances fertility depends on individual characteristics such as dominance status (e.g., Chichinadze and Chichinadze 2008; Tilbrook et al. 2000). Especially in females, reproductive suppression can be an evolved response to temporary shortages of social or energetic resources (e.g., Brunton et al. 2008; see Wasser and Barash 1983), and there is evidence linking HPA functioning to fertility and pregnancy outcomes in human females (e.g., Nepomnaschy et al. 2004, 2006; Wasser and Place 2001).

Competition among same-sex individuals is the inevitable outcome of sexual reproduction. Dominance-seeking, aggression, and risk-taking are all functionally connected to mating competition, and all are associated with SRS functioning. In humans, there is a huge literature linking HPA and autonomic functioning to aggression, antisociality, and externalizing behavior (e.g., Alink et al. 2008; Lorber 2004; Shirtcliff et al. 2009; van Goozen et al. 2007). Given the centrality of risk-taking and impulsivity in life history models of behavior, it is noteworthy that HPA functioning has also been linked to risk-taking behavior in standardized laboratory tasks (e.g., Lighthall et al. 2009; van den Bos et al. 2009). Moreover, executive function and self-regulation play a key role as (negative) mediators of risky and impulsive behavior (Figueredo and Jacobs 2009). Stress exposure can also regulate mating behavior more directly by, for example, altering mate preferences and affecting the perceived attractiveness of potential sexual partners (e.g., Lass-Hennemann et al. 2010).

In the modulation of risky competition, the SRS interacts with sex hormones, serotonin (5-HT), and dopamine (DA). Studies of aggression and antisocial behavior often report interactions between cortisol, testosterone (T), and adrenal androgens such as dehydroepiandrosterone (DHEA) and dehydroepiandrosterone sulfate (DHEAS) (e.g., Popma et al. 2007; van Goozen et al. 2007). The general function of 5-HT is to regulate avoidance of threat, withdrawal from dangerous or aversive cues, and behavioral inhibition/restraint. Serotonergic activity is thus crucially involved in risk aversion and self-regulation (Cools et al. 2008; Fairbanks 2009; Tops et al. 2009). Serotonin is an upstream modulator of SRS activity through its action on the amygdala and hypothalamus; serotonergic neurotransmission, in turn, is reciprocally affected by cortisol (Porter et al. 2004; van Goozen et al. 2007). Dopaminergic activity is also tightly linked to SRS functioning (Alexander et al. 2011; Gatzke-Kopp 2011) recently argued that reduced dopaminergic activity can be adaptive in highly dangerous and unstable environments (and especially so for males) by promoting sensation-seeking, risk-taking, and preference for immediate rewards.

Finally, the SRS is involved in the regulation of parental investment, both directly (e.g., caregiving) and indirectly by affecting the mechanisms of pair-bonding. In humans, individual differences in SRS functioning have been associated with differences in romantic attachment styles (e.g., Quirin et al. 2008; Laurent and Powers 2007; Oskis et al. 2011; Powers et al. 2006); in turn, romantic attachment predicts relationship stability, commitment, and investment (reviewed in Del Giudice 2009). The key molecules that can be expected to interact with the SRS in the regulation of pair-bonding and parental investment are sex hormones, vasopressin, oxytocin, serotonin, and endogenous opioids. Oxytocin secretion, in particular, has been related to individual differences in romantic attachment styles (e.g., Marazziti et al. 2006). Differences in SRS functioning (as well as in oxytocin- and serotonin-related genes) have been also linked to individual differences in maternal sensitivity and parenting behavior (e.g., Bakermans-Kranenburg and van IJzendoorn 2009; Martorell and Bugental 2006).

2.3.2 The Developmental Role of the Stress Response System

As discussed in the last section, the SRS has a pervasive role in the regulation—and, most importantly, the *integration*—of physiology and behavior across the whole spectrum of life history-relevant traits. In a life history framework, this is no coincidence: we argue that—together with sex hormones and relevant neurotransmitter systems—the SRS is a critical mediator of life history development, gathering information from the environment and translating it into broadband individual differences in behavior and physiology (Fig. 2.1; see also Korte et al. 2005; Worthman 2009). In other words, the SRS interacts with other neurobiological systems so as to enable conditional adaptation. Following the logic of West-Eberhard's theory (Sect. 2.2), it should be possible to identify a number of developmental switch points in an organism's life cycle when plasticity is preferentially expressed and environmental cues are integrated with genotypic information to adjust the organism's developmental trajectory. Of course, long-lived organisms can be expected to have more switch points than short-lived ones so as to permit sequential adjustment of life history decisions as environmental conditions change (Del Giudice and Belsky 2011).

Crucially, different strategies may require different calibrations of the SRS itself (curved arrow in Fig. 2.1); for example, a slow strategy in a safe environment could be optimally served by a responsive HPA axis and parasympathetic system, coupled with moderate sympathetic reactivity. SRS calibration can be expected to depend on the system's previous history of activity (Adam et al. 2007), in interaction with factors such as the individual's sex and developmental stage (Miller et al. 2007). The analysis presented in this section can be summarized in the first three points of the ACM (adapted from Del Giudice et al. 2011), which embody the model's theoretical core:

1. The SRS has three main biological functions: to coordinate the organism's allostatic response to physical and psychosocial challenges; to encode and filter information from the environment, thus mediating the organism's openness to environmental inputs; and to regulate a broad range of life history-relevant traits and behaviors.
2. The SRS works as a mechanism of conditional adaptation, regulating the development of alternative life history strategies. Different patterns of activation and responsivity in early development modulate differential susceptibility to environmental influence and shift susceptible individuals on alternative pathways, leading to individual differences in life history strategies and in the adaptive calibration of stress responsivity (Fig. 2.1).
3. Activation of the SRS during the initial life stages provides crucial information about life history-relevant dimensions of the environment. Frequent, intense SNS/HPA activation carries information about extrinsic morbidity–mortality and environmental unpredictability; consequently, it tends to shift life history strategies toward the fast end of the life history continuum. In contrast, a safe environment (and/or the buffer provided by investing parents and alloparents) results in infrequent and low-intensity activation of the SNS and HPA axis and shifts development toward slow strategies oriented to high somatic effort and parental investment.

2.4 The Development of Stress Responsivity in Humans

The theoretical core of the ACM (Fig. 2.1) can be employed as the foundation for a detailed model of the development of stress responsivity in humans. We tried to accomplish this in two steps: first, we advanced some general predictions on the relation between environmental conditions and responsivity; second, we derived a (provisional) taxonomy of four prototypical patterns of SRS responsivity labeled *sensitive* [I], *buffered* [II], *vigilant* [III], and *unemotional* [IV]. The four patterns are characterized by combinations of physiological parameters indexing the functioning of the parasympathetic and sympathetic branches of the autonomic system and of the HPA axis. Our predictions can be summarized in the remaining four points of the ACM, as follows:

4. At a very general level, a nonlinear relationship exists between environmental stress during ontogenetic development and the optimal level of stress responsivity (Fig. 2.2). Note that the environment-responsivity relationship need not be the same for all the components of the SRS (for details see Del Giudice et al. 2011). Furthermore, stress responsivity is expected to show domain-specific effects; for example, a generally unresponsive component of the SRS may respond strongly to some particular type of challenge.
5. Because of sex differences in life history trade-offs and optimal strategies, sex differences are expected in the distribution of responsivity patterns and in their

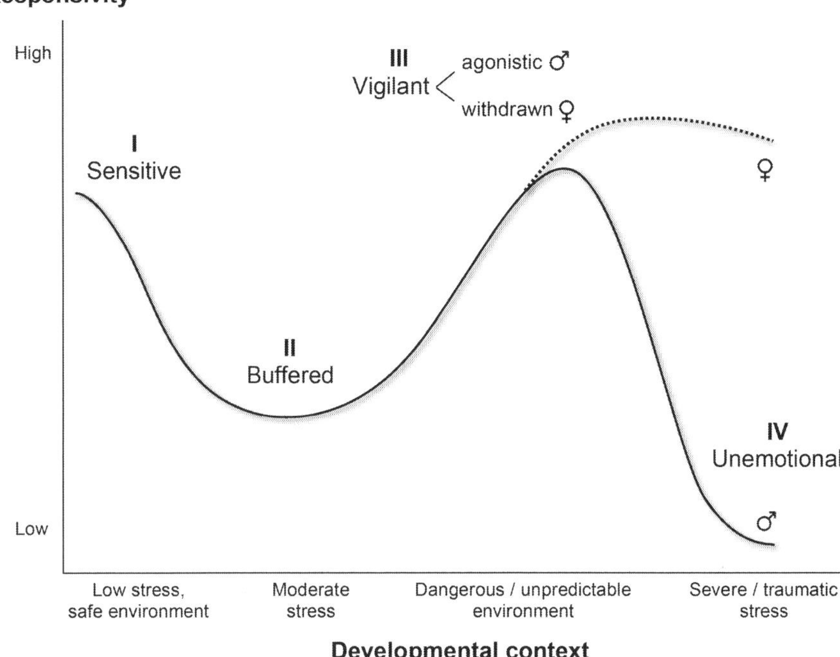

Fig. 2.2 Environmental effects on the development of stress responsivity in humans, according to the ACM. At a very general level, a nonlinear relation exists between exposures to environmental stress and support during development and optimal levels of stress responsivity. The figure does not imply that all components of the SRS will show identical responsivity profiles nor that they will activate at the same time or over the same time course. Male/female symbols indicate sex-typical patterns of responsivity, but substantial within-sex differences in responsivity are expected as well

specific behavioral correlates. Sex differences should become more pronounced toward the fast end of the life history continuum; in environments characterized by severe/traumatic stress, we predict the emergence of a male-biased pattern of low responsivity.

6. Pre- and early postnatal development, the juvenile transition (see below), and puberty are likely switch points for the calibration of stress responsivity. Individual and sex differences in SRS functioning are predicted to emerge according to the evolutionary function of each developmental stage.
7. Responsivity profiles develop under the joint effects of environmental and genetic factors. Genotypic variation may have directional effects on stress responsivity and associated strategies, thus predisposing some individuals to follow a certain developmental trajectory. Genotypic variation, in part through effects on the SRS, may also affect their sensitivity to environmental inputs, resulting in gene–environment interactions whereby some individuals display a broader range of possible developmental outcomes (i.e., broader reaction norms) than others.

2.4.1 Environmental Stress and Responsivity

In safe, protected, low-stress environments, a highly responsive SRS enhances social learning and engagement with the external world, allowing the child to benefit more fully from social resources and opportunities, thus favoring development of a *sensitive* phenotype (pattern I). A sensitive phenotype in this context may make children better at detecting positive opportunities and learning to capitalize on them (e.g., seeing a teacher as a prospective mentor, taking advice from a parent). Social learning and sensitivity to context are especially adaptive in the context of slow life history strategies, as a form of protracted somatic investment. It is important to note that in very safe and protected settings, sensitive individuals will *rarely* experience strong, sustained activation of the sympathetic and HPA systems; precisely because of the high quality of the environment, they will most likely experience a pattern of low-key, short-lived activations followed by quick recovery. Thus, the individual enjoys the benefits of responsivity without paying significant fitness costs (e.g., immune, energetic, and so on). At moderate levels of environmental stress, however, the cost/benefit balance begins to shift; the optimal level of HPA and sympathetic responsivity falls downward, leading to *buffered* phenotypes (pattern II).

The benefits of increased responsivity rise again when the environment is perceived as dangerous and/or unpredictable. A responsive SRS enhances the individual's ability to react appropriately to dangers and threats while maintaining a high level of engagement with the social and physical environment. Moreover, engaging in fast strategies should lead the individual to allocate resources in a manner that discounts the long-term physiological costs of the stress response in favor of more immediate advantages. In this context, the benefits of successful defensive strategies outweigh the costs of frequent, sustained HPA and sympathetic activation, leading to *vigilant* phenotypes (pattern III). High HPA and sympathetic responsivity, however, can be associated with rather different behavioral patterns, leaning toward the "fight" (*vigilant-agonistic*, III-A) or "flight" (*vigilant-withdrawn*, III-W) side of the sympathetic response. Furthermore, evolutionary theory provides reasons to expect males and females to differ in the distribution of agonistic vs. withdrawn patterns (see below). Increased SRS responsivity in dangerous environments can be expected to go together with increased responsivity in other neurobiological systems; for example, hyper-dopaminergic function may contribute to the vigilant phenotype by boosting attention to threat-related cues and fast associative learning (Gatzke-Kopp 2011).

What happens in extremely dangerous environments characterized by severe or traumatic stress? We argue that the balance shifts again toward low responsivity, especially for males who adopt a fast, mating-oriented strategy characterized by antagonistic competition and extreme risk-taking. Such a strategy requires outright *insensitivity* to threats, dangers, social feedback, and the social context. For an extreme risk-taker, informational insulation from environmental signals of threat is an asset, not a weakness. In particular, adopting an exploitative/antisocial interpersonal style requires one to be shielded from social rejection, disapproval, and feelings of shame (all amplified by heightened HPA responsivity). In summary, an *unemotional* pattern of generalized low responsivity (pattern IV) can be evolutionarily adaptive (i.e., fitness-maximizing) at the high-risk end of the environmental spectrum, despite its possible negative consequences for the social group and for the individual's subjective well-being. The same principle applies to other neurobiological systems involved in the regulation of risk-taking; for example, hypodopaminergic function is likely adaptive in severely stressful environments (Gatzke-Kopp 2011).

Figure 2.2 depicts the overall predicted relations between developmental context and stress responsivity, extending the original BSC curve to the right and showing the male-biased pattern of low responsivity in high-risk environments. This broadband analysis can be supplemented with a more fine-grained description of the profiles of basal activity and responsivity of the various SRS components (see Del Giudice et al. 2011).

2.4.2 Sex Differences

Because the costs and benefits associated with life history trade-offs are not the same for males and females, life history strategies show consistent differences between the sexes (Sect. 2.2). On average, men engage in faster strategies and invest more in mating effort (and less in parenting effort) than women. The extent of sex differences in life history-related behavior, however, is not fixed but depends in part on the local environment.

At the slow end of the life history continuum, both sexes engage in high parental investment, and male and female interests largely converge on long-term, committed pair bonds; sex differences in behavior are thus expected to be relatively small. As environmental danger and unpredictability increase, males benefit by shifting to low-investment, high-mating strategies; females, however, do not have the same flexibility since they benefit much less from mating with multiple partners and incur higher fixed costs through childbearing. Thus, male and female strategies should increasingly diverge at moderate to high levels of environmental danger/unpredictability. In addition, sexual competition takes different forms in males and females, with males engaging in more physical aggression and substantially higher levels of risk-taking behavior (e.g., Archer 2009; Byrnes et al. 1999; Kruger and Nesse 2006; Wilson et al. 2002). As life history strategies become faster, sexual competition becomes stronger, and sex differences in competitive strategies become more apparent. For these reasons, sex differences in responsivity patterns and in the associated behavioral phenotypes should be relatively small at low to moderate levels of environmental stress (patterns I and II) and increase in stressful environments (pattern III). Finally, males should be overrepresented as high-risk, low-investment strategists (pattern IV) because of the larger potential benefits of extreme mating-oriented behavior.

2.4.3 Developmental Stages and Switch Points in Human Development

The human life history can be described as a sequence of stages and transitions (Bogin 1999). Life history strategies unfold progressively according to the evolutionary function of each life stage. Del Giudice and Belsky (2011) proposed that the major switch points in the development of human life history strategies are (a) pre- and early postnatal development, (b) the juvenile transition, and (c) puberty. The juvenile transition (Del Giudice et al. 2009) is the transition from early to middle childhood, taking place at around 6–8 years of age in Western societies. This developmental transition is marked by the event of "adrenal puberty" or *adrenarche* (Auchus and Rainey 2004; Ibáñez et al. 2000), whereby the cortex of the adrenal glands begins to secrete increasing quantities of androgens, mainly DHEA and DHEAS. The onset of human juvenility (i.e., middle childhood) witnesses massive changes in children's social behavior, cognitive abilities, and the emergence or intensification of sex differences in aggression, attachment, play, language use, and so forth (reviewed in Del Giudice et al. 2009).

The juvenile transition can be expected to be a critical turning point in the development of stress responsivity. First, we predict that sex differences in the developmental trajectories of stress responsivity will become apparent starting from the beginning of middle childhood, with a further increase at puberty. Second, we expect that individual changes in responsivity will be especially frequent in the transition from early to middle childhood. Early childhood affords an "evaluation" period in which the child can sample the environment—both directly and through the mediation of parents. With juvenility, however, stress responsivity becomes an integral component of the child's emerging life history strategy. Indeed, the SRS is crucially involved in the biological functions of juvenility—including social learning and peer competition. For this reason, it may be adaptive for some children to adjust their levels of responsivity when transitioning from early to middle childhood, possibly under the effect of adrenal androgens.

With the onset of puberty, sexual behavior and romantic attachment come to the forefront, and social competition further intensifies (see Ellis et al. 2012a; Weisfeld 1999). Puberty affords another opportunity to "revise" one's strategy, depending for example on the success enjoyed—or the level of competition experienced—during juvenility. The activation of sex hormone pathways also provides a source of novel genetic effects on life history-related behavior. Thus, adolescence is expected to witness the further intensification of both individual and sex-related differences.

2.5 Conclusion

The ACM offers an integrative view of the evolved functions of the SRS and its role in development. We believe this perspective will prove useful both in organizing and systematizing existing knowledge and in suggesting novel questions for empirical research. In our opinion, what the field needs is more fundamental theory, rather than a multitude of alternative micro-models without a common frame of reference. Although the original model was developed to capture individual differences in humans, we are excited at the prospect of extending the core of the ACM to deal with different species and different ecologies. Adding a phylogenetic and comparative dimension to the ACM would be extremely valuable, in keeping with the spirit of Tinbergen's four questions. Many stimulating reflections on the ACM from the perspective of behavioral ecology can be found in Sih (2011).

To conclude, we wish to stress that the ACM is a work in progress and that many theoretical and empirical gaps still have to be filled in. For example, much more work is needed on domain-specificity in SRS functioning and on the mechanistic basis of genetic effects and G×E interactions in development. Furthermore, the initial focus of our model was skewed toward adaptive variation; while the ACM recasts many supposedly "pathological" processes in an adaptive framework, it still lacks an explicit treatment of actual dysfunction and pathology (Ellis et al. 2012b). Mathematical models of the developmental processes hypothesized in the ACM would also help refine the theory and test the robustness of its assumptions. We anticipate that, in the near future, substantial portions of the model will have to be updated, revised, and possibly rejected. If so, the model will have served his goal of moving the field forward, promoting theoretical advance, and increasing the vitality of an important and exciting field of research.

References

Adam EK, Klimes-Dougan B, Gunnar MR (2007) Social regulation of the adrenocortical response to stress in infants, children and adolescents: implications for psychopathology and education. In: Coch D, Dawson G, Fischer KW (eds) Human behavior, learning, and the developing brain: atypical development. Guilford, New York, pp 264–304

Alexander N, Osinsky R, Mueller E, Schmitz A, Guenthert S, Kuepper Y, Hennig J (2011) Genetic variants within the dopaminergic system interact to modulate endocrine stress reactivity and recovery. Behav Brain Res 216:53–58

Alink LRA, van IJzendoorn MH, Bakermans-Kranenburg MJ, Mesman J, Juffer F, Koot HM (2008) Cortisol and externalizing behavior in children and adolescents: mixed meta-analytic evidence for the inverse relation of basal cortisol and cortisol reactivity with externalizing behavior. Dev Psychobiol 50:427–450

Archer J (2009) Does sexual selection explain human sex differences in aggression? Behav Brain Sci 32:249–266

Auchus RJ, Rainey WE (2004) Adrenarche—physiology, biochemistry and human disease. Clin Endocrinol 60:288–296

Bakermans-Kranenburg MJ, van IJzendoorn MH (2009) Oxytocin receptor (*OXTR*) and serotonin transporter (*5-HTT*) genes associated with observed parenting. Soc Cogn Affect Neurosci 3:128–134

Barsegyan A, Mackenzie SM, Kurose BD, McGaugh JL, Roozendaal B (2010) Glucocorticoids in the prefrontal cortex enhance memory consolidation and impair working memory by a common neural mechanism. Proc Natl Acad Sci 107:16655–16660

Belsky J (1997) Variation in susceptibility to rearing influences: an evolutionary argument. Psychol Inq 8:182–186

Belsky J (2005) Differential susceptibility to rearing influences: an evolutionary hypothesis and some evidence. In: Bjorklund D, Ellis B (eds) Origins of the social mind: evolutionary psychology and child development. Guilford, New York, pp 139–163

Belsky J, Pluess M (2009) Beyond diathesis-stress: differential susceptibility to environmental influences. Psychol Bull 135:885–908

Belsky J, Steinberg L, Draper P (1991) Childhood experience, interpersonal development, and reproductive strategy: an evolutionary theory of socialization. Child Dev 62:647–670

Blair C, Granger D, Razza RP (2005) Cortisol reactivity is positively related to executive function in preschool children attending head start. Child Dev 76:554–567

Bogin B (1999) Evolutionary perspective on human growth. Annu Rev Anthropol 28:109–153

Boyce WT, Ellis BJ (2005) Biological sensitivity to context: I. An evolutionary-developmental theory of the origins and functions of stress reactivity. Dev Psychopathol 17:271–301

Brunton PJ, Russell JA, Douglas AJ (2008) Adaptive responses of the maternal hypothalamic-pituitary-adrenal axis during pregnancy and lactation. J Neuroendocrinol 20:764–776

Byrnes JP, Miller DC, Schafer WD (1999) Gender differences in risk taking: a meta-analysis. Psychol Bull 125:367–383

Carere C, Caramaschi D, Fawcett TW (2010) Covariation between personalities and individual differences in coping with stress: converging evidence and hypotheses. Curr Zool 56:728–740

Chichinadze K, Chichinadze N (2008) Stress-induced increase of testosterone: contributions of social status and sympathetic reactivity. Physiol Behav 94:595–603

Cools R, Roberts AC, Robbins TW (2008) Serotoninergic regulation of emotional and behavioural control processes. Trends Cogn Sci 12:31–40

Del Giudice M (2009) Sex, attachment, and the development of reproductive strategies. Behav Brain Sci 32:1–67

Del Giudice M, Belsky J (2011) The development of life history strategies: toward a multi-stage theory. In: Buss DM, Hawley PH (eds) The evolution of personality and individual differences. Oxford University Press, New York, pp 154–176

Del Giudice M, Angeleri R, Manera V (2009) The juvenile transition: a developmental switch point in human life history. Dev Rev 29:1–31

Del Giudice M, Ellis BJ, Shirtcliff EA (2011) The Adaptive Calibration Model of stress responsivity. Neurosci Biobehav Rev 35:1562–1592

Dickerson SS, Kemeny ME (2004) Acute stressors and cortisol responses: a theoretical integration and synthesis of laboratory research. Psychol Bull 130:355–391

Dingemanse NJ, Réale D (2005) Natural selection and animal personality. Behaviour 142:1165–1190

Ellis BJ (2004) Timing of pubertal maturation in girls: an integrated life history approach. Psychol Bull 130:920–958

Ellis BJ, Essex MJ, Boyce WT (2005) Biological sensitivity to context: II. Empirical explorations of an evolutionary-developmental theory. Dev Psychopathol 17:303–328

Ellis BJ, Jackson JJ, Boyce WT (2006) The stress response system: universality and adaptive individual differences. Dev Rev 26:175–212

Ellis BJ, Figueredo AJ, Brumbach BH, Schlomer GL (2009) The impact of harsh versus unpredictable environments on the evolution and development of life history strategies. Hum Nat 20:204–268

Ellis BJ, Boyce WT, Belsky J, Bakermans-Kranenburg MJ, van IJzendoorn MH (2011) Differential susceptibility to the environment: a neurodevelopmental theory. Dev Psychopathol 23:7–28

Ellis BJ, Del Giudice M, Dishion TJ, Figueredo AJ, Gray P, Griskevicius V et al (2012a) The evolutionary basis of risky adolescent behavior: implications for science, policy, and practice. Dev Psychol 48(3):624–627

Ellis BJ, Del Giudice M, Shirtcliff EA (2012b) Beyond allostatic load: the stress response system as a mechanism of conditional adaptation. In: Beauchaine TP, Hinshaw SP (eds) Child and adolescent psychopathology, 2nd edn. Wiley, New York

Fairbanks LA (2009) Hormonal and neurochemical influences on aggression in group-living monkeys. In: Ellison PT, Gray PB (eds) Endocrinology of social relationships. Harvard University Press, Cambridge, MA, pp 159–195

Figueredo AJ, Jacobs WJ (2009) Aggression, risk-taking, and alternative life history strategies: the behavioral ecology of social deviance. In: Frias-Armenta M, Corral-Verdugo V (eds) Biopsychosocial perspectives on aggression. Nova Science Publishers, Hauppauge, NY

Figueredo AJ, Vásquez G, Brumbach BH, Schneider SMR (2004) The heritability of life history strategy: the K-factor, covitality, and personality. Soc Biol 51:121–143

Figueredo AJ, Vásquez G, Brumbach B, Sefcek JA, Krisner BR, Jacobs WJ (2005) The K-factor: individual differences in life-history strategy. Pers Indiv Differ 39:1349–1360

Figueredo AJ, Vásquez G, Brumbach B, Schneider SMR, Sefcek JA, Tal IR et al (2006) Consilience and life history theory: from genes to brain to reproductive strategy. Dev Rev 26:243–275

Flinn MV (2006) Evolution and ontogeny of the stress response to social challenges in the human child. Dev Rev 26:138–174

Gatzke-Kopp LM (2011) The canary in the coalmine: the sensitivity of mesolimbic dopamine to environmental adversity during development. Neurosci Biobehav Rev 35:794–803

Geary DC (2002) Sexual selection and human life history. Adv Child Dev Behav 30:41–101

Gluckman PD, Hanson MA, Beedle AS (2007) Early life events and their consequences for later disease: a life history and evolutionary perspective. Am J Hum Biol 19:1–19

Herman JP, Figueiredo H, Mueller NK, Ulrich-Lai Y, Ostrander MM, Choi DC, Cullinan WE (2003) Central mechanisms of stress integration: hierarchical circuitry controlling hypothalamo–pituitary–adrenocortical responsiveness. Front Neuroendocrinol 24:151–180

Hill K (1993) Life history theory and evolutionary anthropology. Evol Anthropol 2:78–88

Hofer M (1984) Relationships as regulators. Psychosom Med 46:183

Ibáñez L, Dimartino-Nardi J, Potau N, Saenger P (2000) Premature adrenarche—normal variant or forerunner of adult disease? Endocr Rev 21:671–696

James J, Ellis BJ, Schlomer GL, Garber J (2012) Sex-specific pathways to early puberty, sexual debut and sexual risk-taking: tests of an integrated evolutionary-developmental model. Dev Psychol 48(3):687–702

Kaplan H, Hill K, Lancaster J, Hurtado AM (2000) A theory of human life history evolution: diet, intelligence, and longevity. Evol Anthropol 9:156–185

Kokko H, Jennions M (2008) Parental investment, sexual selection and sex ratios. J Evol Biol 21:919–948

Koolhaas JM, Bartolomucci A, Buwalda B, de Boer SF, Flügge G, Korte SM, Meerlo P et al (2011) Stress revisited: a critical evaluation of the stress concept. Neurosci Biobehav Rev 35:1291–1301

Korte SM, Koolhaas JM, Wingfield JC, McEwen BS (2005) The Darwinian concept of stress: benefits of allostasis and costs of allostatic load and the trade-offs in health and disease. Neurosci Biobehav Rev 29:3–38

Kruger DJ, Nesse RM (2006) An evolutionary life-history framework for understanding sex differences in human mortality rates. Hum Nat 17:74–97

Kruger DJ, Reischl T, Zimmerman MA (2008) Time perspective as a mechanism for functional developmental adaptation. J Soc Evol Cult Psychol 2:1–22

Lass-Hennemann J, Deuter CE, Kuehl LK, Schulz A, Blumenthal TD, Schachinger H (2010) Effects of stress on human mating preferences: stressed individuals prefer dissimilar mates. Proc R Soc B 277:2175–2183

Laurent H, Powers S (2007) Emotion regulation in emerging adult couples: temperament, attachment, and HPA response to conflict. Biol Psychol 76:61–71

Leimar O, Hammerstein P, Van Dooren TJM (2006) A new perspective on developmental plasticity and the principles of adaptive morph determination. Am Nat 167:367–376

Lighthall NR, Mather M, Gorlick MA (2009) Acute stress increases sex differences in risk seeking in the balloon analogue task. PLoS One 4:e6002

Lorber MF (2004) Psychophysiology of aggression, psychopathy, and conduct problems: a meta-analysis. Psychol Bull 130:531–552

Marazziti D, Dell'Osso B, Baroni S, Mungai F, Catena M, Rucci P et al (2006) A relationship between oxytocin and anxiety of romantic attachment. Clin Pract Epidemiol Ment Health 2:28

Martorell GA, Bugental DB (2006) Maternal variations in stress reactivity: implications for harsh parenting practices with very young children. J Fam Psychol 20:641–647

Mayr E (1963) Animal species and evolution. Harvard University Press, Cambridge, MA
McEwen BS (1998) Stress, adaptation, and disease: allostasis and allostatic load. Ann NY Acad Sci 840:33–44
McEwen BS, Wingfield JC (2003) The concept of allostasis in biology and biomedicine. Horm Behav 43:2–15
McNamara JM, Houston AI (2009) Integrating function and mechanism. Trends Ecol Evol 24:670–675
Meaney MJ (2007) Environmental programming of phenotypic diversity in female reproductive strategies. Adv Genet 59:173–215
Miller GE, Chen E, Zhou ES (2007) If it goes up, must it come down? Chronic stress and the hypothalamic-pituitary-adrenocortical axis in humans. Psychol Bull 133:25–45
Nepomnaschy PA, Welch K, McConnell D, Strassmann BI, England BG (2004) Stress and female reproductive function: a study of daily variations in cortisol, gonadotropins, and gonadal steroids in a rural Mayan population. Am J Hum Biol 16:523–532
Nepomnaschy PA, Welch KB, McConnell DS, Low BS, Strassmann BI, England BG (2006) Cortisol levels and very early pregnancy loss in humans. Proc Natl Acad Sci USA 103:3938–3942
Nesse RM, Bhatnagar S, Young EA (2007) Evolutionary origins and functions of the stress response. In: Fink J (ed) Encyclopedia of stress, 2nd edition. Academic Press, San Diego, CA, pp 965–970
Oskis A, Loveday C, Hucklebridge F, Thorn L, Clow A (2011) Anxious attachment style and salivary cortisol dysregulation in healthy female children and adolescents. J Child Psychol Psychiatry 52:111–118
Pigliucci M (2001) Phenotypic plasticity: beyond nature and nurture. Johns Hopkins University Press, Baltimore, MD
Popma A, Vermeiren R, Geluk CAML, Rinne T, van den Brink W, Knol DL, Jansen LMC, van Engeland H, Doreleijers TAH (2007) Cortisol moderates the relationship between testosterone and aggression in delinquent male adolescents. Biol Psychiatry 61:405–411
Porges SW (2001) The polyvagal theory: phylogenetic substrates of a social nervous system. Int J Psychophysiol 42:123–146
Porges SW (2007) The polyvagal perspective. Biol Psychol 74:116–143
Porter RJ, Gallagher P, Watson S, Young AH (2004) Corticosteroid-serotonin interactions in depression: a review of the human evidence. Psychopharmacology (Berl) 173:1–17
Powers SI, Pietromonaco PR, Gunlicks M, Sayer A (2006) Dating couples' attachment styles and patterns of cortisol reactivity and recovery in response to a relationship conflict. J Pers Soc Psychol 90:613–628
Quirin M, Pruessner JC, Kuhl J (2008) HPA system regulation and adult attachment anxiety: individual differences in reactive and awakening cortisol. Psychoneuroendocrinology 33:581–590
Roff DA (2002) Life history evolution. Sinauer, Sunderland, MA
Roozendaal B (2000) Glucocorticoids and the regulation of memory consolidation. Psychoneuroendocrinology 25:213–238
Schanberg S, Evoniuk G, Kuhn C (1984) Tactile and nutritional aspects of maternal care: specific regulators of neuroendocrine function and cellular development. Proc Soc Exp Biol Med 175:135
Scott-Phillips TC, Dickins TE, West SA (2011) Evolutionary theory and the ultimate-proximate distinction in the human behavioral sciences. Perspect Psychol Sci 6:38–47
Shirtcliff EA, Vitacco MJ, Graf AR, Gostisha AJ, Merz JL, Zahn-Waxler C (2009) Neurobiology of empathy and callousness: implications for the development of antisocial behavior. Behav Sci Law 27:137–171
Shoal GD, Giancola PR, Kirillovac GP (2003) Salivary cortisol, personality, and aggressive behavior in adolescent boys: a 5-year longitudinal study. J Am Acad Child Adolesc Psychiatry 42:1101–1107
Sih A (2011) Effects of early stress on behavioral syndromes: an integrated adaptive perspective. Neurosci Biobehav Rev 35:1452–1465

Stark R, Wolf OT, Tabbert K, Kagerer S, Zimmermann M, Kirsch P, Schienle A, Vaitl D (2006) Influence of the stress hormone cortisol on fear conditioning in humans: evidence for sex differences in the response of the prefrontal cortex. Neuroimage 32:1290–1298

Staton L, El-Sheikh M, Buckhalt JA (2009) Respiratory sinus arrhythmia and cognitive functioning in children. Dev Psychobiol 51:249–258

Sterling P, Eyer J (1988) Allostasis: a new paradigm to explain arousal pathology. In: Fisher S, Reason J (eds) Handbook of life stress, cognition and health. Wiley, New York, pp 629–649

Taylor SE, Klein LC, Lewis BP, Gruenenwald TL, Gurung RA, Updegraff JA (2000) Biobehavioral responses to stress in females: tend-and-befriend, not fight-or-flight. Psychol Rev 107:411–429

Tilbrook AJ, Turner AI, Clarke IJ (2000) Effects of stress on reproduction in non-rodent mammals: the role of glucocorticoids and sex differences. Rev Reprod 5:105–113

Tinbergen N (1963) On the aims and methods of ethology. Z Tierpsychol 20:410–433

Tops M, Russo S, Boksem MAS, Tucker DM (2009) Serotonin: modulator of a drive to withdraw. Brain Cogn 71:427–436

Trivers RL (1972) Parental investment and sexual selection. In: Campbell B (ed) Sexual selection and the descent of man 1871–1971. Aldine, Chicago, IL, pp 136–179

van den Bos R, Harteveld M, Stoop H (2009) Stress and decision-making in humans: performance is related to cortisol reactivity, albeit differently in men and women. Psychoneuroendocrinology 34:144–1458

van Goozen SHM, Faichild G, Snoek H, Harold GT (2007) The evidence for a neurobiological model of childhood antisocial behavior. Psychol Bull 133:149–182

van Marle HJF, Hermans EJ, Qin S, Fernández G (2009) From specificity to sensitivity: how acute stress affects amygdala processing of biologically salient stimuli. Biol Psychiatry 66:649–655

Viau V (2002) Functional cross-talk between the hypothalamic–pituitary–gonadal–adrenal axes. J Neuroendocrinol 14:506–513

Wasser SK, Barash DP (1983) Reproductive suppression among female mammals: Implications for biomedicine and sexual selection theory. Q Rev Biol 58:513–538

Wasser SK, Place NJ (2001) Reproductive filtering and the social environment. In Ellison P (ed) Reproductive ecology and human evolution. Aldine De Gruyter, New Brunswick, NJ, pp 137–157

Weisfeld GE (1999) Evolutionary principles of human adolescence. Basic Books, NY

West-Eberhard MJ (2003) Developmental plasticity and evolution. Oxford University Press, New York

Williams PG, Suchy Y, Rau HK (2009) Individual differences in executive functioning: implications for stress regulation. Ann Behav Med 37:126–140

Wilson DS, Yoshimura J (1994) On the coexistence of specialists and generalists. Am Nat 144:692–707

Wilson M, Daly M, Pound N (2002) An evolutionary psychological perspective on the modulation of competitive confrontation and risk-taking. In: Pfaff D et al (eds) Hormones, brain and behavior. Academic, San Diego, pp 381–408

Wolf M, van Doorn GS, Leimar O, Weissing FJ (2007) Life-history trade-offs favour the evolution of animal personalities. Nature 447:581–585

Worthman CM (2009) Habits of the heart: life history and the developmental neuroendocrinology of emotion. Am J Hum Biol 21:772–781

Chapter 3
Prenatal and Maternal Psychosocial Stress in Primates: Adaptive Plasticity or Vulnerability to Pathology?

Dario Maestripieri and Amanda C.E. Klimczuk

Abstract In many species of vertebrates, prenatal and early postnatal stress can have long-lasting consequences for neuroanatomical, neuroendocrine, or behavioral development. In primates including humans, prenatal psychosocial stress and postnatal psychosocial stress induced by the mother's behavior represent important sources of nongenetic maternal effects through which mothers can modify their offspring's phenotype. Prenatal and maternal psychosocial stress are probably mediated by similar physiological mechanisms and primarily including the HPA axis. The biomedical/clinical view, the stress-inoculation model, and the adaptive calibration model make different assumptions and predictions concerning the adaptive or maladaptive developmental consequences of prenatal and maternal psychosocial stress. Studies of experimentally induced prenatal psychosocial stress in primates indicate that fetal programming occurs with characteristics similar to those observed in laboratory rodents and in humans. Studies of naturally occurring maternal psychosocial stress in primates have focused on maternal abuse and rejection of offspring. Although the developmental consequences of exposure to maternal abuse or high rates of maternal rejection are unlikely to be adaptive, exposure to moderate levels of rejection appears to result in physiological and behavioral changes that enhance resilience later in life. It is possible that some aspects of normal parenting in nonhuman primates and humans are designed to be stress inducing to prepare offspring to deal with the psychosocial stress that is an inevitable part of life in complex and competitive social environments.

D. Maestripieri (✉) • A.C.E. Klimczuk
Institute for Mind and Biology, The University of Chicago, 5730 South Woodlawn Avenue, Chicago, IL 60637, USA
e-mail: dario@uchicago.edu

G. Laviola and S. Macrì (eds.), *Adaptive and Maladaptive Aspects of Developmental Stress*, Current Topics in Neurotoxicity 3, DOI 10.1007/978-1-4614-5605-6_3,

3.1 Introduction

Maternal effects are influences a mother's phenotype has on her offspring's phenotype that occur independent of the offspring's genotype (Mousseau and Fox 1998; Maestripieri and Mateo 2009). Maternal phenotypic traits that influence the offspring's phenotype are subject to natural selection so long as they are both variable and heritable. Such traits are *genetic* maternal effects, and the genes underlying them are called *maternal effect genes*. In contrast, *environmental* maternal effects are nonheritable, because variation in these traits results from extrasomatic rather than genetic differences. A maternal phenotype that is maladapted to the environment—as manifested, for example, in pathological alterations in nutritional state, key physiological parameters, or behavior—negatively impacts the offspring's ability to survive or reproduce. However, environmental maternal effects can sometimes be adaptive for a mother or her offspring. On the one hand, maternal effects can help to maximize the mother's fitness by allowing her to adjust her level of parental investment in accordance with prevailing conditions (e.g., by reducing offspring size or growth rate when food is scarce). On the other hand, maternal effects can also benefit offspring by providing preemptive information about the environment they will likely be born into, thereby enhancing their abilities to survive and reproduce in such an environment. Because the time and energy a mother invests in her current offspring is unavailable for future reproductive effort, the mother and the offspring have different investment optima. Environmental maternal effects are one arena in which this mother–offspring conflict can be staged (see Uller and Pen 2011). (As discussed below, whether maternal effects primarily benefit mothers or their offspring has resulted in different adaptive interpretations of prenatal stress.)

Maternal effects are classified as prenatal or postnatal depending on whether parental modification of the offspring phenotype occurs before or after birth. Both types of effects have been documented in many vertebrate species. Prenatal maternal effects can be especially strong in birds because mothers can affect offspring development by depositing varying amounts of nutrients, hormones, and other biological substances in their eggs. In placental mammals, prenatal maternal effects are collectively referred to as *fetal programming*. Fetal programming can be very powerful because the mother's body serves as the fetus's environment for an extended period of gestation, opening many opportunities for maternal influence of fetal development through nutritional and other physiological mechanisms.

Postnatal maternal effects in vertebrates can be quite heterogeneous; they include food provisioning and other forms of parental care that alter offspring body condition, metabolism, and, later, behavior. Maternal effects can also occur in the social domain. For example, in cercopithecine monkeys, a female's dominance rank can affect her offspring's growth rate, age at first reproduction, and adult behavior (see Maestripieri 2009 for a review). In both birds and mammals, maternal effects may facilitate learning and imprinting of social, habitat, and food preferences that match early experiences (Mateo 2009). Other effects are more indirect: for example, parental nest site choice determines the offspring's social environment, which in turn affects the production of hormones that may have long-term consequences for behavior (Price 1998).

The question of whether maternal effects are adaptive or maladaptive for the offspring is especially relevant for issues of prenatal or maternal psychosocial stress. Prenatal psychosocial stress refers to environmental psychosocial stress the mother experiences during pregnancy and *communicates* to the fetus via transfer of hormones and other physiological substances through the placenta. In utero, maternal hormones can directly affect the fetus's hormones, body, and brain. After birth, the mother's hormones are transferred only through breast milk; however, because she is still the most important aspect of the offspring's early postnatal environment, the mother herself can be a significant source of environmental stress. Therefore, we refer to the psychosocial postnatal stress induced in offspring by the mother's behavior as *maternal stress*. Maternal stress may be only a subset of all psychosocial stress experienced by a young individual, but it is clearly an important source of maternal effects (see the *maternal mediation hypothesis* of environmental stress; e.g., Macrì and Würbel 2006, 2007).

Prenatal and maternal psychosocial stress have been extensively studied in laboratory rodents and in humans. Less is known about these maternal effects in primates (but see Maestripieri 2009; Groothuis and Maestripieri 2013). Studies of nonhuman primates provide important links between the research literature in rodents and humans. If the findings of rodent studies are replicated in nonhuman primates, there is a greater probability that they also apply to humans. Conversely, if processes occurring in humans can also be demonstrated in nonhuman primates, it is likely that such processes have a biological basis and can be studied in other animal models as well.

The effects of prenatal and maternal psychosocial stress are likely mediated by common physiological mechanisms, the most important of which is the hypothalamic–pituitary–adrenal (HPA) axis. Many studies of prenatal stress in laboratory rodents and humans suggest that the most likely mediator of fetal programming is maternal cortisol (e.g., Welberg and Seckl 2001; Glover et al. 2010; Oitzl et al. 2010). Cortisol increases significantly and predictably in relation to a wide range of acute psychosocial stressors. In addition, the difference in concentration between maternal and fetal cortisol is so large that even small fluctuations in maternal cortisol can exert significant effects on fetal physiology (see Chap. 2; Flinn et al. 2011; Del Giudice 2012). Maternal cortisol levels during gestation have been shown to predict behavioral reactivity and HPA functioning in infants and children (Glover et al. 2010; Del Giudice 2012). A number of additional stress-related hormones and neurotransmitters have also been proposed as possible mediators of fetal programming, including maternal and placental corticotrophin-releasing hormone (CRH), maternal adrenocorticotropic hormone (ACTH), adrenal steroid hormone dehydroepiandrosterone (DHEA), serotonin (5-HT), and norepinephrine (NE) (Talge et al. 2007; Glover et al. 2010).

Long-term HPA axis alterations (involving basal secretion of ACTH or cortisol, or hormonal secretion in response to stress or CRH/ACTH challenges or to dexamethasone-induced glucocorticoid negative feedback) induced by prenatal or maternal psychosocial stress could underlie adaptive adjustments in reactivity to environment (e.g., emotional reactivity or metabolic responsiveness), or they could reflect chronic stress-related pathologies such as posttraumatic stress disorder (PTSD). Both phenomena have been well studied in humans and have their parallels in nonhuman primates. In this chapter, we will review and discuss the literature on the effects of prenatal and maternal psychosocial stress in nonhuman primates, addressing whenever possible both their potential adaptive significance and their underlying mechanisms. However, we will first address some conceptual issues regarding the interpretation of these forms of stress.

3.2 Prenatal Stress

3.2.1 Conceptual Interpretations of Prenatal Psychosocial Stress

A great deal of developmental research conducted by psychologists, psychiatrists, and biomedical scientists is based on a normative view of brain, neuroendocrine, behavioral, social, emotional, and cognitive development in which any significant deviations from the norm are construed as pathological and maladaptive for the developing organism. Such deviations may include alterations of basic physiological parameters outside their normal ranges for a particular age and gender, of the timing of particular events during development, or of developmental trajectories such as growth or maturation curves. From this biomedical/clinical perspective, all forms of stress have a negative connotation by definition, and prenatal psychosocial stress is considered a significant risk factor for abnormal fetal and childhood development and health. Studies informed by this perspective and conducted with humans and laboratory rodents have reported a host of adverse developmental consequences of prenatal psychosocial stress (e.g., Dodic et al. 1999; Kofman 2002; Maccari et al. 2003; Talge et al. 2007; Glover 2011; see Chap. 8).

In contrast to the biomedical/clinical view, an evolutionary perspective on development posits that since the environment in which organisms develop can be highly variable, developmental variation may represent adaptation to different environmental circumstances rather than pathology. Evolutionary scientists realize that psychosocial stress is an integral part of the lives of all social organisms and that these organisms possess a number of strategies to cope with such stress. Some of these strategies represent short-term responses to acute perturbations of the environment, while others represent long-term adjustments to chronic stressors and other stable features of the environment. Early life, especially in utero, is characterized by high plasticity in brain structure and physiological function and therefore is an ideal period in which to make long-term adjustments to stable characteristics of the environment.

Evolutionary interpretations of prenatal psychosocial stress recognize that the maternal body and the placenta acquire information about stressful features of the mother's social environment and shape the neural and neuroendocrine development of the fetus to be well suited for that environment. One version of these evolutionary interpretations, the *adaptive tuning hypothesis*, suggests that prenatal environmental stress mediated by the mother's body acts as a developmental cue to offspring, predictively programming their future phenotypes to better survive in suboptimal environmental conditions (Gluckman and Hanson 2004; Horton 2005). The adaptive tuning hypothesis assumes that, with the exception of the extremely high levels of stress that tend to result in maladaptive outcomes, the stress an organism experiences prenatally will optimize its postnatal phenotype in an environment featuring that level of stress. However, pathologies can arise if the later environment does not match the early prenatal environment that induced tuning. An extended and more sophisticated version of the adaptive tuning hypothesis of prenatal stress, which also explains postnatal stress, is represented by the adaptive calibration model of stress responsivity (Del Giudice et al. 2011; see below).

A different evolutionary hypothesis proposed by Hayward and Wingfield (2004) maintains that maternally mediated prenatal stress programs the offspring so as to reduce the need for parental investment. In other words, if a mother is stressed and her ability to invest resources in her offspring is diminished, prenatal programming will produce a *thrifty* (smaller, slower growing, or less demanding) offspring to match her offspring's needs to her current provisioning ability. In this case, the maternal/fetal matching benefits the mother at the expense of the fetus. This hypothesis was tested in a series of studies conducted with birds in which both maternal ability and prenatal exposure to stress hormones were experimentally manipulated: mothers had their wing feathers clipped to reduce their foraging ability, and chicks were exposed to higher than average doses of corticosterone through injections into the eggs (Love and Williams 2008; see also Breuner 2008). Increased prenatal corticosterone exposure resulted in higher brood mortality and in the production of lighter offspring, thus matching offspring demand to maternal condition. The results of these studies are consistent with Hayward and Wingfield's hypothesis and showed stress-related increases in egg corticosterone to be an important mechanism underlying *selfish* maternal effects.

Taken together, these different evolutionary hypotheses suggest that developmental programming can be adaptive for offspring as well as mothers. Because maternal and fetal interests overlap significantly, we expect some level of cooperation in the extent to which the fetus is susceptible to the effects of prenatal stress. However, reducing investment in the offspring through brood or litter reduction or by producing smaller and less demanding offspring can provide additional benefits to mothers at a cost to an individual offspring. Insofar as the interests of mothers and offspring diverge, there should be conflict and competition over the extent of fetal programming. The idea that fetal programming could represent an important arena for mother–offspring conflict has been elaborated by Del Giudice (2012).

Del Giudice (2012) challenges the main assumption of the adaptive tuning hypothesis, which views the process of fetal programming as a fully cooperative enterprise in which the mother supplies environmental information via her stress hormone levels and the fetus passively accepts it. He argues instead that the mother and the fetus should be in conflict over the extent of postnatal plasticity, the process through which the after-birth environment can shape or modify the offspring's phenotype (Ellis et al. 2011). By definition, high postnatal plasticity implies increased susceptibility to the effects of maternal behavior. Thus, the mother would benefit if she were able to increase the offspring's susceptibility to her own behavior beyond the offspring's optimum, as this would give her increased leverage in all subsequent instances of parent–offspring conflict. Conversely, the offspring should avoid becoming too plastic and too susceptible to maternal influence.

A growing body of research has shown that there is a great deal of individual variation in vulnerability to early environmental influences, and that this variation is in part genetic and in part environmental (Belsky and Pluess 2009; Ellis et al. 2011). Since prenatal stress increases HPA responsiveness and emotional reactivity, postnatal plasticity can be programmed by prenatal exposure to psychosocial stress, especially in those offspring who carry *plasticity* genes. Therefore, mothers may be selected to amplify the physiological effects of prenatal stress by releasing increasing amounts of stress hormones during pregnancy, while fetuses may be selected to reduce them by limiting the levels of maternal hormones that cross the placental barrier. One mechanism for such filtering is the conversion of 50–90% of maternal cortisol into its inactive form, cortisone, by the enzyme placental dehydrogenase 11β-HSD2, which normally serves to protect the fetus from excessive cortisol exposure. Additional mechanisms also exist for filtering other maternal hormones and neurotransmitters (Del Giudice 2012).

3.2.2 *Prenatal Psychosocial Stress in Primates*

There are only a handful of studies of the long-term effects of prenatal stress in nonhuman primates. Most of them have been conducted by the same group of researchers, utilizing one of two species (squirrel monkeys or rhesus macaques) and similar experimental procedures. Schneider and Coe (1993) investigated the effects of chronic prenatal stress in squirrel monkeys by removing pregnant females from their groups and rehousing them with other pregnant females. One group of subjects was rehoused only once; another group was relocated three times into groups with shifting social compositions. After birth, infants were subjected to a standardized battery of neuromotor tests. The offspring of chronically stressed mothers did not differ in body weight from non-stressed controls, but they showed a host of other abnormalities including delayed motor maturation, reduced activity, shortened attention spans, less visual orienting, and poorer balance.

A subsequent series of studies by Schneider, Coe, and collaborators investigated the effects of prenatal stress on offspring behavioral and neuroendocrine development in rhesus macaques (see Coe et al. 2010 for a review). In these experiments, pregnant females were removed from their home cages and exposed to loud, unpredictable noise bursts once per day, 5 days per week, for approximately 25% of pregnancy. The prenatally stressed infants were reared together with non-stressed controls in a nursery to eliminate possible confounds from postnatal maternal behavior. Stressed infants tended to have lower birth weight than controls despite normal gestational length. Additionally, they performed more poorly on neurobehavioral outcomes than controls. Some of these effects appeared to vary based on the timing of prenatal stress: infants stressed early in gestation (days 45–90) performed more poorly on measures of attention (visual orienting) and motor maturity (head posture) than those stressed later in gestation (days 120–134).

The researchers also observed behavioral effects. Prenatally stressed individuals showed reduced exploratory behavior (less climbing and play), increased reactivity to novelty (e.g., higher emotionality and anxiety, more disturbance behavior, stereotypies, clinging and self-clasping, and freezing), and lower sociability (e.g., less time playing and grooming, and less time in proximity to cagemates) in various testing situations. Some of these effects persisted until 4 years of age. These behavioral modifications were accompanied by enhanced basal activity and stress responsivity of the HPA axis (mainly ACTH, not cortisol), as well as by alterations in brain monoamine neurotransmitters: higher cerebrospinal fluid (CSF) levels of MHPG and DOPAC under basal conditions and higher MHPG and NE levels in response to stress. Fetally stressed infants also demonstrated a prolonged HPA axis response to a pharmacological challenge, suggesting impairment in the glucocorticoid negative feedback system. Finally, at 4 years of age, fetally stressed infants showed decreased neurogenesis in the dentate gyrus and significant decreases in hippocampal volume and in the size of the corpus callosum, indicating long-term effects of prenatal stress on brain structure. Some of the effects of prenatal stress were replicated by administering ACTH for 14 days mid-gestation, confirming hormonal etiologies for the observed changes. Like their stressed counterparts, fetuses experimentally exposed to higher ACTH levels had delayed motor development, shorter attention spans, and increased anxiety and irritability after birth (Coe et al. 2010).

Further evidence that prenatal alterations in HPA axis function can result in long-term neuroanatomical and physiological consequences comes from studies in which pregnant female monkeys were treated with the synthetic glucocorticoid hormone dexamethasone (dex). An early study by Uno et al. (1990) reported that administering dex to pregnant rhesus monkeys as late as 72 h before delivery significantly reduced the density of the newborns' pyramidal neurons, as well as the thickness and circumference of their Ammon's horn and dentate gyrus in the hippocampus. Concordantly, infants whose mothers were treated for 30 days with dexamethasone had smaller hippocampi than controls, with a dose-related loss of neurons. More recently, DeVries et al. (2007) treated pregnant female vervet monkeys with three doses of dex and found that it reduced maternal cortisol in a dose-dependent manner at 22 weeks of pregnancy without affecting gestation length or birth weight. However, the dex treatment did delay postnatal growth. Furthermore, high-dose infants had comparatively heightened cortisol responses to the mild stress of blood sampling when tested at 8 months of age, and all dex-treated infants showed cardiovascular signs of hypertension such as increased heart rate and blood pressure. At 12–14 months of age, dex-treated infants were subjected to a dexamethasone suppression test, which normally suppresses cortisol levels in the evening before they return to basal levels in the morning. The infants showed no difference in morning cortisol in relation to prenatal treatment relative to controls, suggesting no alterations in the glucocorticoid negative feedback mechanism.

Taken together, the findings of the limited research on the developmental effects of prenatal psychosocial stress in nonhuman primates demonstrate that fetal programming does indeed occur, and that it produces long-term alterations in emotional and behavioral reactivity, neuroendocrine function, and, in some cases, neuroanatomy, similar to those observed in studies of laboratory rodents and humans. Studies mimicking prenatal stress effects with administration of exogenous ACTH or synthetic cortisol have confirmed that maternal corticosteroids are functionally important for fetal programming. Unfortunately, all studies of prenatal psychosocial stress in primates to date have tested individuals housed in artificial experimental laboratory conditions (i.e., adults housed in single cages or small groups, or infants who were permanently separated from their mothers and reared with peers) and have utilized nonnaturalistic psychosocial stressors. Although neuroanatomical alterations induced by severe prenatal stress or the administration of large doses of hormones should probably be interpreted as pathological, no studies have been designed that explicitly test the possible adaptive value of behavioral and physiological changes induced by more moderate levels of stress. Lack of data concerning the social, mating, and reproductive success, or overall health and survivorship of prenatally stressed individuals makes even post hoc tests of these hypotheses impossible.

3.3 Maternal Stress

3.3.1 Conceptual Interpretations of Maternal Psychosocial Stress

The clinical/biomedical view of prenatal stress, which interprets all prenatal perturbations as potentially pathological, can be extended to encompass postnatal psychosocial stress as well. The idea is that it is best for the organism to develop in a safe and supportive environment in which all stressors are absent. This view assumes that stressors encountered in the early environment generate damage in a dose-dependent manner such that exposure to moderate stress results in a moderately negative developmental outcomes, while exposure to severe or intense stress results in serious negative developmental outcomes. That is to say, it proposes a linear relationship between the degree of early stress and the severity of its maladaptive consequences.

In contrast to the clinical/biomedical view, the *stress-inoculation* model suggests that there is a J-shaped relationship between early stress and unfavorable developmental outcomes (Parker et al. 2006; Parker and Maestripieri 2011; Seery 2011). Like the clinical/biomedical model, the inoculation hypothesis predicts that severe or intense early stress will result in serious maladaptation. However, it argues that too little stress exposure in early life leaves the organism unprepared for future stressful situations, while moderate stress exposure results in adaptive physiological and behavioral adjustments that better prepare the individual to cope with future challenges. In other words, exposure to moderate stress *inoculates* the individual against subsequent exposure, just as exposure to moderate numbers of specific bacteria or viruses allows the body to build an immune response to them in preparation for future encounters.

While the inoculation model makes relatively simple predictions concerning the effects of low, moderate, and high stress on the organism's subsequent resilience and vulnerability to unfavorable circumstances, the *adaptive calibration model* (Del Giudice et al. 2011) makes more complex predictions about the phenotypic consequences of early stress exposure using *conditional adaptation* and life-history theory as its guiding principles. This model predicts that in relatively non-stressful environments, organisms are at low risk of mortality and are thus free to exhibit *slow* lifestyles characterized by unhurried physical and sexual maturation, low anxiety, increased time devoted to learning, less risk taking, and more delayed gratification. Exposure to moderate stress should result in increased resilience, including low anxiety and reactivity to challenges but high sensitivity to social feedback. Finally, exposure to high stress, which indicates that the external environment is dangerous or unpredictable, should produce phenotypes adapted for *fast* lifestyles with high vigilance and anxiety, riskier behavior, and less delayed gratification. Even physiological and psychological alterations induced by severe early stress, such as hyporeactivity of the HPA axis, hyperaggressiveness, and reduced empathy, could be adaptations to life in a dangerous environment.

The three models discussed above—clinical, stress inoculation, and adaptive calibration—apply to all forms of postnatal psychosocial stress, including psychosocial stress induced by the mother's behavior. Human empirical research, which is usually conducted from the biomedical/clinical perspective, has tended to interpret all deviations in maternal responsiveness and parenting style from what is considered the norm or the optimum as damaging for children. For example, the insecure–ambivalent, insecure–avoidant, and especially the insecure–disorganized–disoriented patterns of attachment are all conceived as pathologies that result from suboptimal parental responsiveness or even from abusive or neglectful parenting behavior. However, the adaptive calibration model acknowledges that all insecure attachment patterns might have adaptive value (Del Giudice et al. 2011). For example, ambivalently attached children display patterns of HPA axis, sympathetic, and behavioral reactivity that would be useful in a socially unpredictable environment, while the insecure–avoidant attachment pattern is associated with patterns of HPA axis, sympathetic, and behavioral reactivity that suggest adaptation to a harsh and unsupportive environment (Loman and Gunnar 2010). Even the disorganized–disoriented attachment pattern, with its reactive neuroendocrine profile characterized by extremely elevated and sustained cortisol responses to psychosocial stress (Loman and Gunnar 2010), could have functional value if it successfully prepares children to deal with hostile and dangerous relationships (Del Giudice et al. 2011).

3.3.2 Maternal Psychosocial Stress in Primates

A great deal of research on early psychosocial stress and development in primates has focused on stress experimentally induced by separating infants from their mothers and rearing them under conditions of social deprivation (see Parker and Maestripieri 2011 for a review). Much less attention has been devoted to psychosocial stress naturally induced through maternal behavior. Such studies have mainly been conducted with cercopithecine monkeys and have focused on two aspects of stress-inducing maternal behavior: abusive behavior and rejection.

3.3.2.1 Maternal Abuse

Among rhesus macaques and other cercopithecine monkeys living in large captive groups, 5–10% of all infants born in a given year are physically abused by their mothers (Maestripieri et al. 1997; Maestripieri and Carroll 1998a, b). In rhesus macaques, abusive mothers may drag their infants by their tail or leg, or throw them in the air. Abuse bouts last only a few seconds, and the rest of the time abusive mothers show competent patterns of maternal behavior. Abuse is most frequent in the first month of infant life and rare or nonexistent after the third month, when infants are more independent from their mothers (Maestripieri 1998). Rhesus mothers can give birth once a year, and abusive mothers generally maltreat all of their infants with similar rates and patterns of behavior (Maestripieri et al. 1999). The contributions of infant behavior to the occurrence of abuse are negligible, whereas abusive behavior appears to be a stable maternal trait that is transmitted across generations, from mothers to daughters. As a result, it is concentrated in particular families and absent in others (Maestripieri and Carroll 1998a). Cross-fostering experiments demonstrated that early experience plays an important role in the intergenerational transmission of infant abuse (Maestripieri 2005). Approximately half of cross-fostered and non-cross-fostered females abused early in life exhibit abusive parenting with their first-born offspring (Maestripieri 2005), and those who do so have lower CSF concentrations of the serotonin metabolite 5-HIAA than those who do not (Maestripieri et al. 2006a, 2007).

Maternal abuse is both physically and psychologically stressful for a monkey infant. Moreover, even if abuse is limited to the first months of infant life, continuous coexistence with the abusive mother and observation of abuse being repeated with younger siblings could contribute to reinforce and perpetuate the traumatic effects of abuse into adulthood. Observations of social and behavioral development have suggested that abused infants may be delayed in the acquisition of independence from their mothers and in the development of peer relations in the first year of life (Maestripieri and Carroll 1998b). In addition, the stressful experience of being abused early in life results in both acute and long-term alterations in HPA axis function.

In a preliminary study, 10 abused and 10 control infants were studied during their first 6 months of life (McCormack et al. 2009). Basal morning levels of cortisol were measured at 1, 3, and 6 months of age, and ACTH and cortisol responses to stress were measured in month 6. In addition, infants were genotyped for the serotonin transporter (SERT) gene, and individuals carrying one or two copies of the short allele of this gene were compared to those carrying two copies of the long allele. During the first month, when physical abuse rates were the highest, abused infants had elevated basal morning cortisol levels compared to controls and showed greater distress responses to handling. In addition to a main effect of abuse on basal cortisol levels, there was also a significant interaction between early experience and SERT genotype: the effects of abuse on basal cortisol levels were especially strong in infants carrying the short SERT allele. After the first month, abused infants' basal HPA axis function recovered to levels similar to controls. Despite the normalization of basal activity, there were group by sex effects on the HPA axis stress response in month 6: abused males showed significantly higher ACTH stress responses than control males when exposed to novelty stress in the absence of the mother. The heightened ACTH stress responses were associated with higher levels of anxious behaviors at that age. Thus, abused infants—especially those with genetic vulnerabilities—exhibited both increased HPA axis activity and increased emotional reactivity not only during, but also a few months following, abuse.

In a larger follow-up study comparing 22 abused and 21 nonabused rhesus monkey infants over the first 3 years of life, plasma cortisol responses to psychosocial stress (a novel environment test) were assessed at 6-month intervals, and behavioral measures were assessed at 1-month intervals. Infants showed a significant increase in cortisol in response to stress test at all ages, and infants that were physically abused by their mothers showed a higher cortisol response than control infants at 1 year of age (Koch et al. 2013). Furthermore, abused infants showed significantly different responses to CRF challenges performed at 6-month intervals during their first 3 years of life when compared to nonabused infants (Sanchez et al. 2010). Specifically, the administration of exogenous CRF resulted in a greater increase in plasma cortisol concentrations in both male- and female-abused infants at 6, 12, 18, 24, 30, and 36 months of age. Abused infants also showed a blunted plasma ACTH response to CRF, but this difference was observed only at 6 months of age. This dampened ACTH response may be the result of negative feedback glucocorticoid inhibition, which normally inhibits the secretion of ACTH when there is a rapid increase in circulating cortisol, or of dysfunctional mechanisms that regulate the anterior pituitary's response to hypothalamic CRF. Altogether these results suggest that early maternal abuse results in greater adrenocortical (and possibly pituitary) responsiveness to challenges later in life.

All of the published research on the developmental effects of maternal abuse in macaques has been conducted with infants exposed to relatively low rates of abuse, whose life was not in jeopardy and who, in many cases, suffered only minor bruises and scratches that did not require external intervention. Since no data are available on the behavioral and neuroendocrine development of infants exposed to much more severe levels of abuse, the effects of variation in abuse intensity on the development of stress vulnerability vs. resilience are not well understood. The data reviewed above suggest that infants exposed to moderate levels of abuse exhibit increased vulnerability to stress later in life. This effect, however, may be driven by the high rates of maternal rejection that typically accompany maternal abuse rather than to abuse itself. Further research is needed to examine the effects of different levels of abuse on development and to disentangle the effects of high maternal rejection and abuse when they co-occur.

3.3.2.2 Maternal Rejection

Maternal rejection describes behaviors a mother performs that prevent the infant from making contact with her body or gaining access to her nipples (such as holding the infant at a distance or blocking her chest with an arm), as well as behaviors that forcefully remove the infant from the nipple and interrupt physical contact (such as pushing the infant away). Mothers also reject their infants by administering painful hits or bites. Being denied bodily contact and access to the nipples, even in the absence of physical aggression, causes significant distress to infants, who respond with loud and persistent vocalizations (screams and geckers) and temper tantrums (lying on the ground and acting as if they are having seizures) (Maestripieri 2002). Frequently rejected infants also show behavioral signs of depression. Maternal rejection is clearly a physically and psychologically stressful experience for primate infants.

Although maternal rejection rates change as a function of infant age and the mother's own age and experience, individual differences in rejection rates are generally consistent over time and across infants (Fairbanks 1996). In rhesus monkeys, infants are generally rejected in the third or fourth week of life at the rate of 1 episode every 2 h, or less (Maestripieri 1998). The rate of rejection gradually increases as infants grow older, peaking at 6 months of age when mothers resume their mating activities. However, some infants do not experience rejection at all, while others are rejected at the rate of 3–4 or more episodes per hour as early as in their first week of life (Maestripieri 1998).

Several studies of macaques and vervet monkeys have examined variation in infants' independence from their mothers and their tendency to explore the environment or respond to challenges, at various ages in relation to exposure to variable levels of maternal rejection in early infancy. An early study of rhesus monkeys reported that, following a 2-week separation from their mothers, infants whose mothers had been highly rejecting prior to the separation exhibited elevations in cortisol at reunion, while infants with non-rejecting mothers showed a marked decrease in cortisol levels (Gunnar et al. 1981). This finding may suggest that highly rejected infants are anxious about their relationships with their mothers and that they are not soothed by a reunion following separation because they may anticipate rejection. Consistent with the hypothesis that highly rejected infants are anxious, rhesus monkeys exposed to high levels of maternal rejection in the first few months of life tend to explore their environments less (Simpson 1985; see also Maestripieri et al. 2009). Other studies, however, found that infants reared by highly rejecting mothers generally develop independence (e.g., spend more time out of contact with their mothers, explore the environment more, and play more with their peers) at an earlier age than infants reared by mothers with low rejection levels (Simpson et al. 1989; Simpson and Datta 1990; Bardi et al. 2005; Bardi and Huffman 2006). These seemingly conflicting results can be reconciled by the notion that maternal rejection has opposite short- and long-term effects on infant dependence; highly rejected infants initially respond with increased clinginess and reluctance to leave their mothers, but eventually resign themselves to independence (Maestripieri et al. 2009).

Long-term effects of maternal rejection on reactivity to the environment can be observed in adolescence and also in adulthood. In vervet monkeys, adolescent males reared by highly rejecting mothers were more willing to approach and challenge a strange adult male (Fairbanks and McGuire 1988). Similarly, in Japanese macaques, Schino et al. (2001) found that individuals that were rejected more by their mothers early in life were less likely to respond with submissive signals or with avoidance to an approach from another individual and exhibited lower rates of scratching in the 5-min period following the receipt of aggression. Finally, Maestripieri et al. (2006b) showed that rhesus macaques that were rejected more by their mothers in the first 6 months of life engaged more in solitary play and showed greater avoidance of other individuals at age 2. In this study, the association between maternal behavior and offspring behaviors later in life was also reported in infants that were cross-fostered at birth and reared by unrelated adult females, which rules out potential confounds of inherited temperamental similarities between mothers and offspring.

Developmental differences in reactivity to novel stimuli or responsiveness to other individuals are likely to be accompanied by differences in the neurochemical and neuroendocrine substrates that regulate emotional and social processes. Maestripieri et al. (2006a, b) reported that offspring reared by mothers with higher levels of maternal rejection exhibited lower CSF levels of the serotonin metabolite 5-HIAA, the norepinephrine metabolite MHPG, and the dopamine metabolite HVA in the first 3 years of life than offspring reared by mothers with lower levels of rejection. These differences were observed in both non-fostered and cross-fostered infants. Furthermore, CSF MHPG levels in the second year of life were negatively correlated with solitary play and avoidance of other individuals, while CSF 5-HIAA levels were negatively correlated with scratching rates, suggesting that individuals with low CSF 5-HIAA had higher anxiety. A significant association between exposure to high maternal rejection and low CSF levels of 5-HIAA in the offspring was also reported in another population of free-ranging rhesus monkeys (Maestripieri et al. 2009). In this study, rhesus mothers who rejected their infants at high rates exhibited higher cortisol responses to stress, suggesting that these may be individuals under chronic stress. Altogether, these studies suggest that exposure to maternal rejection early in life may affect the development of different neural circuits underlying emotion regulation, ranging from fear to anxiety to impulse control.

When the behavioral and physiological effects of maternal rejection are considered together, they are generally consistent with the predictions of the stress-inoculation model. In fact, the data reviewed above suggest that infants that experience little or no rejection become fearful and behaviorally inhibited later in life, whereas those exposed to extremely high rates of rejection become highly anxious and impulsive. The behavior of infants exposed to moderate levels of rejection in the first few months of life suggests that they show adaptive responses to challenges and resilience to stress later on. An ongoing longitudinal study in our laboratory is investigating the development of the HPA axis and of the brain monoaminergic systems in three groups of rhesus monkey infants exposed to low, moderate, and high rates of maternal rejection early in life. Following the inoculation model, we predict that infants exposed to high and low levels of maternal rejection will exhibit stress-vulnerable neurobiological phenotypes (e.g., high basal cortisol levels, increased HPA axis responses to social and pharmacological challenges, dysregulation of peptide and monoamine systems involved in arousal and affective responses) compared to infants that received moderate levels of maternal rejection and exhibit resilient neurobiological phenotypes (e.g., comparatively lower basal cortisol levels; diminished HPA axis responses to social and pharmacological challenges; normative CSF levels of peptides and monoamine metabolites involved in arousal and affective responses). To assess whether these effects of early maternal rejection on offspring emotional and stress reactivity are adaptive and will help these individuals cope with psychosocial stressors later in life, we will analyze the different patterns of emotional and stress reactivity in relation to differences in dominance rank, aggression performed and received, mating success, health not heath ultimately longevity and reproductive success.

While the fitness consequences remain unproven, exposure to variable maternal rejection has definite phenotypic consequences for grown offspring for which adaptive hypotheses can be advanced. Two different studies so far have reported positive significant correlations between the rejection rates of rhesus mothers and those of their adult daughters (Berman 1990; Maestripieri et al. 2007), indicating that rejection rates are transmitted across generations. Maestripieri et al. (2007) found significant similarities in maternal rejection rates between mothers and daughters for both non-fostered and cross-fostered rhesus females, suggesting that the daughters' behavior was affected by exposure to their mothers' rejection in their first 6 months of life. Both non-fostered and cross-fostered rhesus females reared by mothers with high rates of maternal rejection had significantly lower CSF concentrations of the serotonin metabolite 5-HIAA in their first 3 years of life than females reared by mothers with lower (below the median) rates of maternal rejection, and low CSF 5-HIAA was associated with high rejection rates when the daughters produced and reared their first offspring (Maestripieri et al. 2006a, 2007). Therefore, the lower serotonergic function resulting from exposure to high maternal rejection rates in infancy contributes to the expression of high maternal rejection rates in adulthood in these females.

Different maternal rejection rates may represent adaptations to particular maternal characteristics (e.g., dominance rank, body condition, or age) or demographic and ecological circumstances (e.g., availability of food or social support from relatives) (Hauser and Fairbanks 1988; Fairbanks and McGuire 1995). For example, rejection rates are generally high in females of high dominance rank and good body condition who are under pressure to wean their infants quickly and produce an infant every year. Rejection rates, however, are also high in extremely old females in poor body condition, or in females under severe nutritional or social stress, because these females must reduce investment in their offspring to concentrate on their own survival and future reproduction. In cercopithecine monkeys, mothers and daughters have very similar dominance ranks and share their environment as well. Both dominance ranks and patterns of rank-related psychosocial stress are extremely stable for rhesus females, not only during their life spans but also across generations within their families. Therefore, insofar as a particular rate of maternal rejection represents an adaptation to a stressful social microenvironment, the transgenerational conservation of parenting style via stress effects and social learning represents a likely avenue for non-genomic transmission of behavioral adaptations.

3.4 Conclusions

Prenatal psychosocial stress, mediated by the mother's body and her hormones, and postnatal psychosocial stress, induced by the mother's behavior, can have similar programming effects on the infant's or child's social, emotional, and neuroendocrine development. These effects represent one particular type of evolutionary maternal effects, in which the mother's phenotype influences and shapes the offspring's phenotype, without direct transmission of genetic information or modification of the offspring's genotype. The physiological mechanisms underlying the prenatal and postnatal effects of psychosocial stress are probably similar, or the same, and involve the HPA axis and other neuroendocrine systems involved in emotion regulation and reactivity to the environment. Although stress-related maternal effects and their underlying mechanisms have been investigated to a lesser extent in nonhuman primates than in laboratory rodents or in humans, the available evidence suggests that these processes are largely similar across these different species of mammals.

Rodents and primates (including humans), however, differ in some key aspects of their life histories and social environments, and this has potentially important implications for the occurrence, characteristics, and possible adaptive significance of prenatal and postnatal stress-related maternal effects. First, many rodent species have a relatively short life span: they become reproductively active rapidly and reproduce quickly, and through the production of large litters, the processes of pregnancy and lactation are brief, and mothers have relatively few opportunities to shape the phenotype of the offspring, aside from the windows of time provided by pregnancy and the immediate postpartum period. Second, although it is relatively easy to experimentally induce psychosocial stress in pregnant or lactating laboratory rodent females, for example, by altering their housing conditions, it is not immediately clear what the naturalistic equivalents of these laboratory stressors may be and what are the most common forms of psychosocial stress encountered by pregnant or lactating females in the wild. Rodents do not live in complex societies similar to those of some primate species, in which psychosocial stress is an integral part of life; the extent to which prenatal or postnatal psychosocial stress is an important source of maternal effects in wild population of rodents is unclear. Moreover, even though laboratory rodents exhibit naturally occurring differences in maternal care styles that have important consequences for offspring development (see Champagne and Curley 2009 for a review), it is not clear that these effects are mediated by stress-related mechanisms. For example, to our knowledge, rodent mothers do not spontaneously exhibit stress-inducing behaviors toward their offspring that are structurally or functionally similar to maternal rejection/neglect or abusive mothering in monkeys and humans.

In primates that live in complex and highly competitive societies such as humans, chimpanzees, and many cercopithecine monkeys, the mother–infant relationship is deeply embedded in the social environment so that, for example, other social relationships the mother has with her partner, or previous children, or other family members, friends, coworkers, and competitors can have a direct and profound influence on her psychological well-being, her neuroendocrine function, her health, and therefore also on the quantity and quality of her interactions with her fetus, infant, or child. The long periods of gestation and lactation and the correspondingly long and slow process of growth and maturation of the offspring provide many opportunities for prenatal and postnatal maternal effects to operate.

The question of whether prenatal and postnatal maternal psychosocial stress have adaptive or maladaptive consequences in primates remains open to empirical investigation and to debate. It is possible to hypothesize that, during the evolutionary history of humans and other primates, psychosocial stress was such an inevitable and significant feature of the pregnancy and lactation periods that the maternal body at some point became selected to use the mechanisms and effects of such stress to influence and shape the phenotype of the developing offspring in a manner that was adaptive to herself or the offspring or both. In other words, since psychosocial stress during pregnancy and lactation is inevitable and may also be a good predictor of stressors encountered later in life, mothers began to *prepare* their fetuses, infants, and children and endow them with physiological, behavioral, and emotional/cognitive adaptations that would allow them to cope with stress in an optimal way throughout their life. Exposure to moderate stress early in life can promote the acquisition and fine-tuning of these physiological, behavioral, and emotional/cognitive mechanisms to cope with stress similar to the process through which exposure to moderate amounts of pathogens early in life strengthens the immune system and inoculates the body against future exposure to the same or similar pathogens. Thus, it is possible that mothers are not simply vehicles for passively transferring information from a surrounding stressful environment to their offspring, but in some cases they actively generate a moderate amount of psychosocial stress through their behavior so as to give their offspring the opportunity to develop the tools to deal with it.

Nonhuman primates are ideal animal models with which to test this hypothesis and other similar hypotheses concerning the adaptive significance of prenatal and postnatal maternal psychosocial stress. Primate mothers encourage the nutritional and social independence of their infants through behaviors such as rejection, which have the long-term effects of reducing the amount and frequency with which infants seek to be in contact and gain access to their mothers' nipples for suckling. We believe that the fact that maternal rejection generates significant psychosocial stress in the offspring is not an inevitable and inconsequential by-product of the weaning process. Rather, this stress is a phenomenon that needs an explanation. And this explanation may be that when mothers reject their infants, they simultaneously accomplish different goals: they encourage their infants to be nutritionally and socially independent, and they also give them the opportunity to develop the appropriate tools to deal with psychosocial stress and be inoculated against future exposures later in life.

The notion that parenting behavior can shape the physiological, behavioral, and emotional/cognitive mechanisms with which children react to stress is clearly applicable to humans. Authoritative parents often discipline their children in ways that generate a moderate amount of stress in the children. The traditional interpretation of parental discipline is that it is aimed at shaping the behavior of the child in way that conforms to the norms and expectations that society and parents have about child's behavior. The hypothesis that parentally induced psychosocial stress has the adaptive effect of enhancing the child's mechanisms for coping with stress is empirically testable and, if supported by data, would significantly increase our understanding of parent–child relationships, of the development of stress reactivity over the life span, and of the role of maternal effects in the process of adaptation to the environment.

References

Bardi M, Huffman MA (2006) Maternal behavior and maternal stress are associated with infant behavioral development. Dev Psychobiol 48:1–9

Bardi M, Bode AE, Ramirez SM (2005) Maternal care and development of stress responses in baboons. Am J Primatol 66:263–278

Belsky J, Pluess M (2009) Beyond diathesis-stress: differential susceptibility to environmental influences. Psychol Bull 135:885–908

Berman CM (1990) Intergenerational transmission of maternal rejection rates among free-ranging rhesus monkeys. Anim Behav 39:329–337

Breuner C (2008) Maternal stress, glucocorticoids, and the maternal/fetal match hypothesis. Horm Behav 54:485–487

Champagne FA, Curley JP (2009) The trans-generational influence of maternal care on offspring gene expression and behavior in rodents. In: Maestripieri D, Mateo JM (eds) Maternal effects in mammals. The University of Chicago Press, Chicago, pp 182–202

Coe CL, Lubach GR, Crispen HR, Shirtcliff EA, Schneider ML (2010) Challenges to maternal wellbeing during pregnancy impact temperament, attention, and neuromotor responses in the infant rhesus monkey. Dev Psychobiol 52:625–637

Del Giudice M (2012) Fetal programming by maternal stress: insights from a conflict perspective. Psychoneuroendocrinology 37:1614–1629

Del Giudice M, Ellis BJ, Shirtcliff EA (2011) The adaptive calibration model of stress responsivity. Neurosci Biobehav Rev 35:1562–1592

DeVries A, Holmes MC, Heijnis A, Seier JV, Heerden J, Louw J, Wolfe-Coote S, Meaney MJ, Levitt NS, Seckl JR (2007) Prenatal dexamethasone exposure induces changes in nonhuman primate offspring cardiometabolic and hypothalamic-pituitary-adrenal axis function. J Clin Invest 117:1058–1067

Dodic M, Peers A, Coghlan JP, Wintour M (1999) Can excess glucocorticoid, in utero, predispose to cardiovascular and metabolic disease in middle age? Trends Endocrinol Metab 10:86–91

Ellis BJ, Boyce WT, Belsky J, Bakermans-Kranenburg MJ, van Ijzendoorn MH (2011) Differential susceptibility to the environment: an evolutionary-neurodevelopmental theory. Dev Psychopathol 23:7–28

Fairbanks LA (1996) Individual differences in maternal styles: causes and consequences for mothers and offspring. Adv Study Behav 25:579–611

Fairbanks LA, McGuire MT (1988) Long-term effects of early mothering behavior on responsiveness to the environment in vervet monkeys. Dev Psychobiol 21:711–724

Fairbanks LA, McGuire MT (1995) Maternal condition and the quality of maternal care in vervet monkeys. Behaviour 132:733–754

Flinn MV, Nepomnaschy PA, Muehlenbein MP, Ponzi D (2011) Evolutionary functions of early social modulation of hypothalamic-pituitary-adrenal axis development in humans. Neurosci Biobehav Rev 35:1611–1629

Glover V (2011) Prenatal stress and the origins of psychopathology: an evolutionary perspective. J Child Psychol Psychiatry 52:356–367

Glover V, O'Connor TG, O'Donnell K (2010) Prenatal stress and the programming of the HPA axis. Neurosci Biobehav Rev 35:17–22

Gluckman PD, Hanson MA (2004) Living with the past: evolution, development, and patterns of disease. Science 305:1733–1736

Groothuis TTG, Maestripieri D (2013). Parental influences on offspring personality traits in oviparous and placental vertebrates. In: Carere C, Maestripieri D (eds) Animal personalities: behavior, physiology, and evolution. The University of Chicago Press, Chicago

Gunnar MR, Gonzalez C, Goodlin C, Levine S (1981) Behavioral and pituitary-adrenal responses during a prolonged separation period in infant rhesus macaques. Psychoneuroendocrinology 6:65–75

Hauser MD, Fairbanks LA (1988) Mother-offspring conflict in vervet monkeys: variation in response to ecological conditions. Anim Behav 36:802–813

Hayward LS, Wingfield JC (2004) Maternal corticosterone is transferred to avian yolk and may alter offspring growth and adult phenotype. Gen Comp Endocrinol 135:365–371

Horton TH (2005) Fetal origins of developmental plasticity: animal models of induced life history variation. Am J Hum Biol 7:34–43

Koch H, McCormack KM, Sanchez MM, Maestripieri D (2013) The development of the hypothalamic-pituitary-adrenal axis function in rhesus monkeys: effects of age, sex, and early experience. Dev Psychobiol, in press

Kofman O (2002) The role of prenatal stress in the etiology of developmental behavioural disorders. Neurosci Biobehav Rev 26:457–470

Loman MM, Gunnar MR (2010) Early experience and the development of stress reactivity and regulation in children. Neurosci Biobehav Rev 34:867–876

Love OP, Williams TD (2008) The adaptive value of stress-induced phenotypes: effects of maternally derived corticosterone on sex-biased investment, cost of reproduction, and maternal fitness. Am Nat 172:135–149

Maccari S, Darnaudery M, Morley-Fletcher S, Zuena AR, Cinque C, Van Reeth O (2003) Prenatal stress and long-term consequences: implications of glucocorticoid hormones. Neurosci Biobehav Rev 27:119–127

Macrì S, Würbel H (2006) Developmental plasticity of HPA and fear responses in rats: a critical review of the maternal mediation hypothesis. Horm Behav 50:667–680

Macrì S, Würbel H (2007) Effects of variation in postnatal maternal environment on maternal behaviour and fear and stress responses in rats. Anim Behav 73:171–184

Maestripieri D (1998) Parenting styles of abusive mothers in group-living rhesus macaques. Anim Behav 55:1–11

Maestripieri D (2002) Parent-offspring conflict in primates. Int J Primatol 23:923–951

Maestripieri D (2005) Early experience affects the intergenerational transmission of infant abuse in rhesus monkeys. Proc Natl Acad Sci U S A 102:9726–9729

Maestripieri D (2009) Maternal influences on offspring growth, reproduction, and behavior in primates. In: Maestripieri D, Mateo JM (eds) Maternal effects in mammals. The University of Chicago Press, Chicago, pp 256–291

Maestripieri D, Carroll KA (1998a) Risk factors for infant abuse and neglect in rhesus monkeys. Psychol Sci 9:143–145

Maestripieri D, Carroll KA (1998b) Behavioral and environmental correlates of infant abuse in group-living pigtail macaques. Infant Behav Dev 21:603–612

Maestripieri D, Mateo JM (eds) (2009) Maternal effects in mammals. The University of Chicago Press, Chicago

Maestripieri D, Wallen K, Carroll KA (1997) Infant abuse runs in families of group-living pigtail macaques. Child Abuse Negl 21:465–471

Maestripieri D, Tomaszycki M, Carroll KA (1999) Consistency and change in the behavior of rhesus macaque abusive mothers with successive infants. Dev Psychobiol 34:29–35

Maestripieri D, Higley JD, Lindell SG, Newman TK, McCormack KM, Sanchez MM (2006a) Early maternal rejection affects the development of monoaminergic systems and adult abusive parenting in rhesus macaques. Behav Neurosci 120:1017–1024

Maestripieri D, McCormack K, Lindell SG, Higley JD, Sanchez MM (2006b) Influence of parenting style on the offspring's behavior and CSF monoamine metabolites levels in crossfostered and noncrossfostered female rhesus macaques. Behav Brain Res 175:90–95

Maestripieri D, Lindell SG, Higley JD (2007) Intergenerational transmission of maternal behavior in rhesus monkeys and its underlying mechanisms. Dev Psychobiol 49:165–171

Maestripieri D, Hoffman CL, Anderson GM, Carter CS, Higley JD (2009) Mother-infant interactions in free-ranging rhesus macaques: relationships between physiological and behavioral variables. Physiol Behav 96:613–619

Mateo JM (2009) Maternal influences on development, social relationships, and survival behaviors. In: Maestripieri D, Mateo JM (eds) Maternal effects in mammals. The University of Chicago Press, Chicago, pp 133–158

McCormack K, Newman TK, Higley JD, Maestripieri D, Sanchez MM (2009) Serotonin transporter gene variation, infant abuse, and responsiveness to stress in rhesus macaque mothers and infants. Horm Behav 55:538–547

Mousseau TA, Fox CW (eds) (1998) Maternal effects as adaptations. Oxford University Press, Oxford

Oitzl MS, Champagne DL, Van der Veen R, de Kloet ER (2010) Brain development under stress: hypotheses of glucocorticoid actions revisited. Neurosci Biobehav Rev 34:853–866

Parker KJ, Maestripieri D (2011) Identifying key features of early stressful experiences that produce stress vulnerability and resilience in primates. Neurosci Biobehav Rev 35:1466–1483

Parker KJ, Buckmaster CL, Sundlass K, Schatzberg AF, Lyons DM (2006) Maternal mediation, stress inoculation, and the development of neuroendocrine stress resistance in primates. Proc Natl Acad Sci U S A 103:3000–3005

Price T (1998) Maternal and paternal effects in birds. In: Mousseau TA, Fox CW (eds) Maternal effects as adaptations. Oxford University Press, Oxford, pp 202–226

Sanchez MM, McCormack K, Grand AP, Fulks R, Graff A, Maestripieri D (2010) Effects of early maternal abuse and sex on ACTH and cortisol responses to the CRH challenge during the first 3 years of life in group-living rhesus monkeys. Dev Psychopathol 22:45–53

Schino G, Speranza L, Troisi A (2001) Early maternal rejection and later social anxiety in juvenile and adult Japanese macaques. Dev Psychobiol 38:186–190

Schneider ML, Coe CL (1993) Repeated social stress during pregnancy impairs neuromotor development of the primate infant. J Dev Behav Pediatr 14:81–87

Seery MD (2011) Resilience: a silver lining to experiencing adverse life events? Curr Dir Psychol Sci 20:390–394

Simpson MJA (1985) Effects of early experience on the behaviour of yearling rhesus monkeys (*Macaca mulatta*) in the presence of a strange object: classification and correlation approaches. Primates 26:57–72

Simpson MJA, Datta SB (1990) Predicting infant enterprise from early relationships in rhesus macaques. Behaviour 116:42–63

Simpson MJA, Gore MA, Janus M, Rayment FDG (1989) Prior experience of risk and individual differences in enterprise shown by rhesus monkey infants in the second half of their first year. Primates 30:493–509

Talge NM, Neal C, Glover V et al (2007) Antenatal maternal stress and long-term effects on child neurodevelopment: how and why? J Child Psychol Psychiatry 48:245–261

Uller T, Pen I (2011) A theoretical model of the evolution of maternal effects under parent-offspring conflict. Evolution 65:2075–2084

Uno H, Lohmiller L, Thieme C, Kemnitz JW, Engle MJ, Roecker EB, Farrell PM (1990) Brain damage induced by prenatal exposure to dexamethasone in fetal rhesus macaques. I. Hippocampus. Dev Brain Res 53:157–167

Welberg LA, Seckl JR (2001) Prenatal stress, glucocorticoids, and the programming of the brain. J Neuroendocrinol 132:113–128

Part II
Adaptive and Maladaptive Consequences Of Developmental Stress In Humans

Chapter 4
The Everyday Stress Resilience Hypothesis: Unfolding Resilience from a Perspective of Everyday Stress and Coping

Jennifer A. DiCorcia, Akhila V. Sravish, and Ed Tronick

Abstract Resilience is often associated with extreme trauma or overcoming extraordinary odds. This way of thinking about resilience leaves most of the ontogenetic picture a mystery. In this chapter, we put forth the Everyday Stress Resilience Hypothesis where resilience is seen as a process of regulating everyday life stressors and is analyzed from a systems perspective. The hypothesis argues that successful regulation accumulates into regulatory resilience which emerges during early development from successful coping with the inherent stress in typical interactions. These quotidian stressful events lead to the activation of behavioral and physiologic systems. Stress that is effectively resolved in the short run and with reiteration over the long term increases children's as well as adults' capacity to cope with more intense stressors. Infants, however, lack the regulatory capacities to take on this task by themselves. Therefore, through communicative and regulatory processes during infant–adult interactions, we demonstrate that the roots of regulatory resilience originate in infants' relationship with their caregivers and that infant reactivity, maternal sensitivity, and the nature of the stressor can help or hinder the growth of resilience.

J.A. DiCorcia, PhD (✉) • E. Tronick, PhD
Department of Psychology, Child Development Unit, University of Massachusetts, 100 Morrissey Blvd, Boston, MA 02125, USA

Department of Newborn Medicine, Brigham and Women's Hospital, Harvard Medical School, Boston, MA 02115, USA
e-mail: jennifer.dicorcia@umb.edu; ed.tronick@umb.edu

A.V. Sravish, MA
Department of Psychology, Child Development Unit, University of Massachusetts, 100 Morrissey Blvd, Boston, MA 02125, USA
e-mail: akhila.sravish@umb.edu

G. Laviola and S. Macrì (eds.), *Adaptive and Maladaptive Aspects of Developmental Stress*, Current Topics in Neurotoxicity 3, DOI 10.1007/978-1-4614-5605-6_4,

4.1 Introduction

Resilience is often referred to as a trait that develops from an individual's experience with extreme adversity. For this reason, much of the research includes high-risk and traumatized individuals (Cicchetti et al. 1993; Egeland et al. 1993; Haglund et al. 2007; Luthar et al. 2000; Nomura et al. 2006). We disagree and put forth the idea that resilience can also be a regulatory or coping capacity that develops from infants' experiences with everyday stress. Our perspective on the development of resilience is that all individuals, regardless of age, intermittently and frequently experience stressors in varying degrees and intensities by simply living in a world of complex social relationships and ever-changing, volatile situations. It is how individuals successfully or unsuccessfully regulate these everyday stresses that affects the development of resilience; that is how stress is reiteratively and chronically regulated at different psychobiologic levels which then molds individuals' regulatory capacity.

Stress may not be quickly associated with infancy. What could be stressful for infants? In home observations of healthy, typical infants and their mothers, we have found that infants at 3 and 6 months of age were in distressed states 11% of the time that lasted on average about 3 minutes. They were in heightened, highly aroused but affectively positive states 13% of the time with an average duration of 4 min. Even when playing with their mothers in face-to-face interactions, infants expressed sad or negative affect about 3% of the time, fussy vocalizations about 3% of the time, and distress indicators (e.g., spitting up) about 1% of the time (Weinberg et al. 1999). Research findings, however, are hardly needed to demonstrate the ubiquitousness of infants' experiencing stress. Supporting evidence is everywhere from infants crying for a bottle, fussing because they are wet or because they can't reach an object, crying in protest when their mother leaves them alone, or for no apparent reason (i.e., the mythical gas). Infants also get highly aroused while playing a game that is too exciting, such as peek-a-boo, where over-arousal transforms laughter into tears or spitting up. These common bouts of distress, however, are limited in duration by infants' self-regulation of the distress (e.g., thumb sucking, attending to an interesting object) or by a caretaker's intervention (e.g., picking the infant up). Although these observations may seem boringly quotidian, they are not because it is not only coping that is at stake. Development depends on infants' active engagement with the world of people, and stressed states preclude that engagement.

Although infants may be able to turn away from a bright light or an intrusive face, there are obvious physical and emotional limitations that infringe on infants' regulatory capacities such that infants' capacity to effectively deal with stressors is not entirely in their hands. It is dependent on their main caregiver, typically the mother, to intervene when necessary. The mother's ability to attend to her infant's signals and to respond appropriately is instrumental to the development of stress regulation and resilience.[1] But perfect contingency between mother and infant is not

[1]In infant development research, the terms "synchrony," "matching," and "mutual engagement" are typically used to describe the coordination of behaviors or contingency within the mother–infant dyad (Feldman et al 2007; Harrist and Waugh 2002). In this chapter, we discuss these concepts as being similar. We are aware, however, that the terms may be conceptually and analytically distinct.

implied for optimal growth. Rather, we propose that it is the very missteps in communication with their attendant stress and their reparation—the mismatches of intentions and affect and their re-coordination—between the mother and her infant that lead to resilience. In typical development perfect matching (perfect contingency; perfect coordination) is in-and-of-itself impossible, but experience with stress, while necessary, is not sufficient to build the infant's developing capacity for resilience. Specifically, missteps in communication within the dyad may be followed by a reparatory process, a dyadic coping mechanism that focuses on the process of transforming stressful mismatching states into non-stressful states (Gianino and Tronick 1988; Tronick 2006; Tronick and Beeghly 2011). When reparation is successful, the infant's stress level decreases and matching returns; the reiteration of successful reparation builds the infant's capacity for resilience. By contrast when reparation fails, dysregulation occurs and precludes the infant's engagement with the animate and inanimate world. Engagement is foundational for normal development. When engagement is chronically disrupted, negative cascading processes have the potential to disrupt development in a number of different domains, including biological, relational, and behavioral realms. Thus stress regulation is critical to typical development.

Recognizing the centrality of stress regulation leads to a number of questions. In this chapter, we present a reconceptualization of the concept of resilience by framing its development using a systems perspective. Specifically we address the important role of the mother–infant dyad and how the regulatory nature of the dyad can prevent infants from being chronically overwhelmed by stress in addition to fostering the growth of their regulatory capacities.

4.2 The Everyday Stress Resilience Hypothesis

Our approach to questions about the emergence of a resilient, biobehavioral phenotype during the first years of life is formulated in the Everyday Stress Resilience (ESR) Hypothesis. The hypothesis states that coping with everyday stressors influences infants' regulatory capacities for these typical stressors and prepares them to cope with later, more taxing stressors. In short, everyday coping experiences enable the development of regulatory capability and build a reservoir of capacity or a "regulatory resilience." Furthermore, based on human and animal research, we also argue that successful regulation of stress and the growth of regulatory resilience is not solely dependent on infants' internal self-organized regulatory capacities (Tronick 2006; Calkins and Hill 2007; Hofer 1987, 2006; Kopp 1989). Rather, stress regulation and its potential growth toward resilience are critically dependent on the quality of the infant–caretaker relationship. Although theories have emphasized the importance of the caregiver's regulatory role (Hofer 2006; Field 1994), the unique contribution of the ESR Hypothesis furthers this notion by contextualizing the early development of resilience in the typical, everyday process of dyadic regulation. In particular, of critical importance is the infant–caregiver dyad's capacity for continual, mutually coordinated regulation of infants' psychobiological states of stress—quotidian and intense—into non-stressful states. Thus, elaboration of the ESR Hypothesis requires placing it in a broader context of typical macro-development (developmental changes in capacities that can have a regulatory function in ontogenetic time such as babbling to speech or immobility to running) and real-time everyday activities (e.g., caretaking), the stress that travels with both, and the mutual regulatory interactive processes that regulate the stress. Critically, we will elucidate how mutual regulatory processes in-and-of-themselves generate micro-stressors. It is the regulation of these micro-stressors through a process of reparation that is critical for building resilience. The process of reparation within the dyad depends on many factors including context and individual differences in reactivity and regulation.

But first we present an analogy for the ESR Hypothesis: training for a marathon (Tronick 2006). Runners do not run marathon distances to train for a marathon. Instead they run a specific distance each day and increase that distance over the course of weeks. However, it is not until they actually run the marathon that they complete the full distance. Training within a marathoner's extant capacity does not lead to improvement, whereas progressive training develops the runner's stamina or coping capacities. Progressive training leads to a bit-by-bit accumulation of capacity culminating in the capacity to go the full distance. The increase in capacity is not related to a singular change but to changes in many different metabolic and muscular characteristics. Effective training is specifically aimed at processes that relate to running the marathon. This training does not prepare one for a triathlon or long-distance skiing. Of course, without the training, had runners tried to go the full daunting distance they would surely fail; the stress would exceed their capacity. Or had they overtrained their capacity would actually diminish because different systems would not have been able to recover from the inherent stress of training. Their capacity also diminishes when training has ended. Thus a progressive increase of training and reiterated chronic training is needed to maintain and grow one's capacity; with it one becomes resilient and without it resilience is lost. This analogy is similar to the inoculation analogy for stress (Parker et al. 2006) but differs in that it is not an all-or-none model but an ontogenetic model. It allows for a consideration of the loss of capacity and the need for recovery or reparation.

4.3 Stress at the Macro-Developmental Level as Regulated by Micro Real-Time Processes

We frame our understanding of the development of resilience using a dynamic systems perspective. Dynamic self-organizing biological systems have a hierarchical organization operating at multiple levels and temporal scales. They are information-rich with specific, intense, and continuous dynamic interactions with local contexts. Complex systems exhibit emergent properties at different levels. Self-organizing processes generate these emergent properties and lead to an increase in the complexity and coherence of the system. Prigogine and Stengers (1984) states that a primary principle governing the activities of open biological systems is that they must acquire energy and information from the environment to maintain and increase their coherence and complexity. The developing infant is one such a system. Its impressive features of very rapid development of emergent capacities, striking increases in complexity, and almost continuous informational exchanges with the external environment are reflections of continuously active, powerful self-organizing capacities.

Critically, it is necessary to recognize that ontogenetic change requires disorganization and reorganization. Stress travels with this process. Despite the smooth, step-by-step characterization of development seen in graphs charting developmental milestones, development does not proceed so smoothly. Development proceeds in an irregularly serrated pattern. Periods of stability (sometimes thought of as periods of practicing) in developmental domains are followed by periods of dismantling an already organized capacity and reorganizing it into a more complex and coherent form of organization. The transitions between periods of stability (attractor states) are inherently stressful not only because they are energetically demanding but because the transitions are unstable. During these transitions infants may actually lose complexity and coherence until the new organization emerges. Crawling, for example, needs to be dismantled to allow for the self-organized emergence of walking (van de Rijt-Plooij and Plooij 1992; Trevarthen 1982). Or how infants who learn how to crawl across a risky slope must then relearn how to cross the same slope while walking (Adolph 1997). Furthermore, because the disorganization of one system often disorganizes other systems, the stress may be exacerbated in intensity and duration. For example, infants beginning to transition from crawling to walking are not only disorganized motorically but are also emotionally and diurnally disorganized (van de Rijt-Plooij et al. 1993; Brazelton 1992).

A consequence of this developmental disorganization is that the moment-by-moment biobehavioral organization of the infant is threatened. Thus during periods of instability, infants are less able to maintain homeostasis and are more likely to become fatigued, overaroused, and distressed. A primary feature of the model, however, is that disorganization is an essential part of the developmental process. Disorganization is necessary for the emergence of something new and for an increase in complexity and coherence; it is the wellspring of change and the new. By contrast, fixed systems don't develop. Nonetheless for all of its benefits, the process of macro-development is costly and stressful.

Of course it is not only the process of development that is stressful for infants. There are everyday internal stressors, such as hunger, fatigue, metabolic processes, lack of diurnal regulation, and a myriad of others. There are common external forces that stress infants: a wet diaper, too bright a light, or a loud noise. Furthermore there are also quotidian stressful interchanges with the environment such as desiring an out of reach object, not getting a caretaker's attention, and playing with a frustrating toy. In essence there is a veritable ubiquitousness of stressors which can amplify each other and cumulate to create cascades of stress which in turn make infants more vulnerable. Thus one can only wonder how are infants able to regulate stress in the face of such demands?

4.4 Dyadic Regulatory Systems

Our view is that to overcome the ubiquitous stressor problem humans evolved an exceptional, though hardly unique, method for regulating this stress. Humans form a dyadic regulatory system in which the infant's regulatory capacity is supplemented and scaffolded by an external regulator—a caretaker—typically the mother. The dyadic regulatory process is referred to as the Mutual Regulation Model (Beebe et al. 2010; Brazelton et al. 1974; Hofer 1994; Tronick 1989; Fogel 1993). The Mutual Regulation Model stipulates that mothers and infants are linked subsystems that form a larger, more integrated dyadic regulatory system responsible for regulating infants' biobehavioral organization.

The regulatory functioning of the infant–caregiver dyadic system is guided by communicative processes (Tronick 1989; Fogel 1993; Trevarthen et al. 2001, 2006). Communicative signals convey infants' biobehavioral status to a receptive caregiver who responds to the signal. However, the communication within even typical mother–infant dyads is far from perfect. As seen in Fig. 4.1, in typical interactions the dyad oscillates from matching (synchronous) to mismatching (asynchronous) and back to matching states through the process of reparation (Tronick and Beeghly 2011; Tronick and Gianino 1986). When the regulatory function of the dyad operates successfully, the maternal regulatory input fulfills infants' signaled regulatory needs. For example, a maternal smile in response to her infant's attentional bid or an empathic frown when her infant is distressed.

As a consequence of successful regulatory inputs from the caregiver, the infant becomes more coherently organized and is better able to negotiate the complexity of the world. On the one hand, dysregulation can be overcome when the caretaker appropriately scaffolds the infant's regulatory capacity by responding to their cues. The infant's stress is reduced, and homeostatic balance is restored. Consequently the infant can continue to engage in the world and its challenges. For example, consider a caregiver who uses a crooning voice and gentle patting with a crying infant who is coping with the stress of dismantling of crawling in the service of eventually walking. The mother's compassionate intervention helps transform the infant's distressed state into a calmer, organized, alert state leading to reparatory success. On the other hand, continual mismatching of regulatory input from the caregiver and infant needs results in reparatory failure (see Fig. 4.1). For example, giving a hungry infant an object to play with will not repair their distressed state whereas if the dysregulated hunger state is instead repaired with maternal nursing, then the infant is likely to progress into a non-stressful state. The caregiver who ignores or misinterprets the infant's distress conveys a message to the infant that their regulatory demands cannot be met by the regulatory resources available. The stressful state continues, and engagement is precluded.

Paradoxically, mutual regulation in real time is in-and-of-itself stressful (Tronick 2006; Tronick and Cohn 1989). The stressors that occur during real-time mutual regulatory processes are micro-stressors, mismatches between external input and infant needs. They occur because regulation in the real time cannot be perfect.

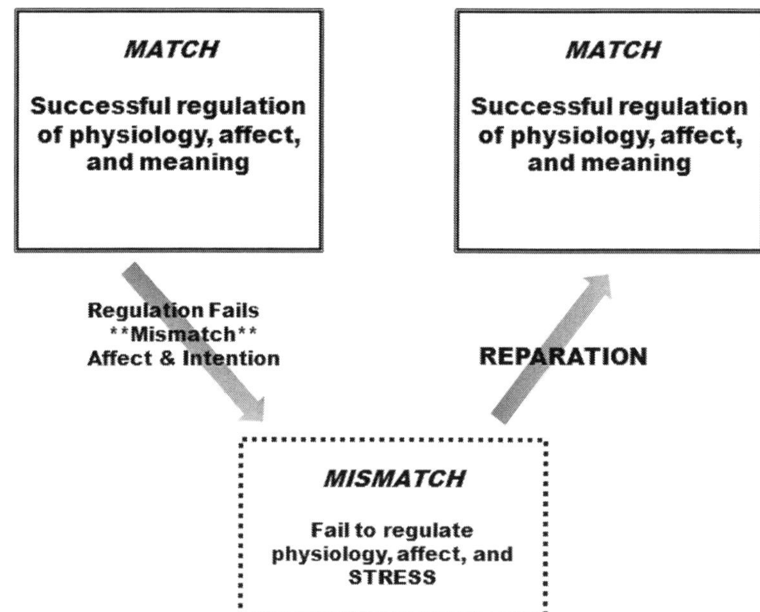

Fig. 4.1 Matching, mismatching, and reparation in the dyad. As the infant and mother transition from a matched state to a mismatched state, the stress level within the dyad increases (e.g., infant shows increased negative affect, dysregulated physiology). With reparation, the infant transitions back to a positive state (e.g., increased positive affect, regulated physiology). (Adapted from DiCorcia and Tronick 2011)

When regulation is even briefly disrupted, stress and negative affect are generated (Tronick 2006). Mismatched or miscoordinated states tend to be more the norm than the exception in face-to-face interactions, even with typical mother–infant dyads. In our studies, we have found that periods of mismatching in mother–infant dyads can make up as much as 70–80% of face-to-face interactive exchanges (Tronick and Cohn 1989).

A host of factors make mismatching inevitable. Signals are emitted at speeds as fast as 0.25 s per signal (Trevarthen and Schogller 2005) which demands responses at rates as fast as tenths of seconds (Beebe et al. 2010; Trevarthen et al. 2006; Condon and Sander 1974). Detecting and decoding signals at these speeds is very difficult. This process is further complicated by the occurrence of miscues or non-perfect signaling as well as the rapidly changing regulatory demands. Add to these reasons the fact that infants have limited and immature regulatory, behavioral, and attentional capacities, and the likelihood of mismatches become quite high. However, interactive disorganization is consequently repaired into a more organized state.

It is our view that the infant–caregiver dyad comprises a constantly interacting and evolving dynamic system. Dyadic interaction is nonlinear and goes through stochastic transformations into phases of sensitivity and insensitivity to the environment (Lewis et al. 1999). Dyadic interaction at any given moment is influenced by both the preceding communication history of the dyad as well as factors unique to that moment. Over time, the interactional patterns stabilize into a limited number of states representative of a particular infant–caregiver dyad (Granic and Hollenstein 2003). As the infant develops, the regulatory demands placed on the infant–caregiver dyadic system changes, in turn destabilizing the homeostasis of the system and leading to an increase in mismatched states. However, over time through moment-by-moment mismatches and reparations, the dyadic system eventually stabilizes into a new equilibrium. This new equilibrium is not a single point of balance but a new organization of hierarchically organized states with greater complexity and coherence for the dyad as well as the infant. The infant emerges better equipped to self-regulate and cope with the demands made by the environment.

Although typical interactions fluctuate between instances of coordination and miscoordination, a key point is that reparations do occur. Missteps are corrected. Thinking in these terms expands our notion of stressors from intense, perhaps traumatic, stressors to everyday stressors to micro-stressors. Without reparation and regulation, even micro-stressors have the potential to accumulate and disrupt development. The process of mutual regulation, in particular stress and its reparation, has been most carefully studied using an experimental stress-induction procedure, the Face-to-Face Still-Face paradigm (Mesman et al. 2009; Tronick et al. 1978). The Face-to-Face Still-Face paradigm highlights the match–mismatch–reparation process at a simulated macro-temporal level which allows for detailed measurement of infants' and caregivers' reactions (see Fig. 4.2a). The paradigm consists of three episodes: (1) an episode of typical infant–caregiver face-to-face play, (2) the still-face episode where the caregiver stops interacting with her infant and holds a still, expressionless face, and (3) a reunion episode where the caregiver resumes interacting with her infant.

Most infants enjoy and come to depend on the reciprocal nature of social interactions with their caregiver (e.g., reciprocal smiling, playful touching), and the violation of this expectation of reciprocity during the still-face is stressful. Affective and behavioral responses are striking and include decreases in positive affect, increases in negative affect, and infant behaviors that are aimed at changing the mothers' behavior or reducing stress, such as increases in protest, gaze aversion and turning away, back arching, and postural collapse (Mesman et al. 2009; Adamson and Frick 2003). Infants also show signs of physiologic activation with increases in heart rate (Bazhenova et al. 2001; Moore and Calkins 2004; Weinberg and Tronick 1996) and skin conductance (Ham and Tronick 2008), and a suppression of respiratory sinus arrhythmia (RSA) (Bazhenova et al. 2001; Moore and Calkins 2004; Weinberg and Tronick 1996; Ham and Tronick 2006; Moore 2009). Hypothalamic–pituitary–adrenal axis activation, as measured by increases in salivary cortisol, has also been observed in infants during the still-face (Ham and Tronick 2006; Feldman et al. 2010; Haley and Stansbury 2003; Montirosso et al. 2011). During the reparation or the reunion episode, mothers once again interact with their infants and attempt to reestablish dyadic regulation. In return infants gaze more toward their mothers and express more positive affect. Negative affect and stress-reduction behaviors also decrease though infants may still express higher levels of anger (Weinberg and Tronick 1996).

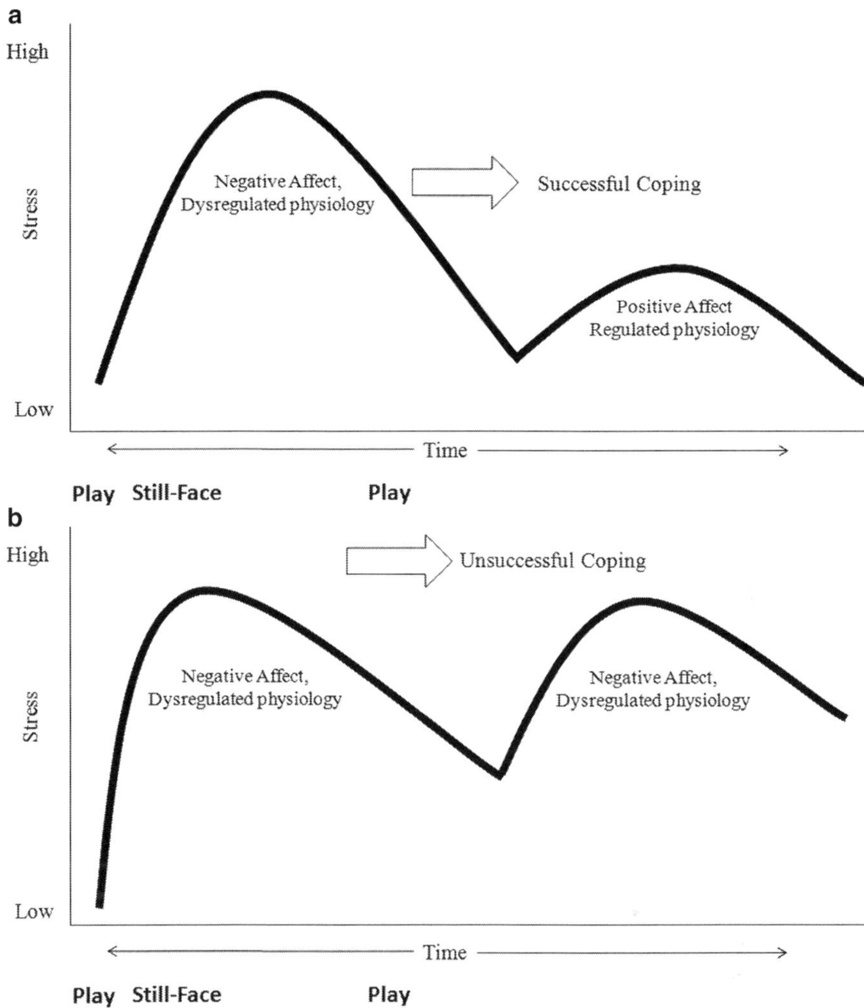

Fig. 4.2 (**a**, **b**) Consequence of a mismatched state in the dyad can be either (**a**). Successful reparation of stress and the discontinuation of stress or (**b**). Unsuccessful reparation and the continuation of stress. (Adapted from DiCorcia and Tronick 2011)

Cardiac measures recover (Bazhenova et al. 2001; Moore and Calkins 2004; Weinberg and Tronick 1996; Feldman et al. 2010; Haley and Stansbury 2003), although Ham and Tronick (2006) found that skin conductance remained high during the reunion episode.

Reparation is a dyadic process of matching regulatory input to regulatory need in order to provide the scaffolding for infants' intrinsic regulatory capacities. The ESR Hypothesis sees reparation as central to the development of regulatory capacities. As already noted, with development the regulatory task becomes increasingly self-organized, and new ways of regulating distress (e.g., language, executive functioning, inhibitory control, and emotion display rules) begin to emerge. However, it is not until later in childhood that these capacities begin to take their mature form (Carlson and Wang 2007; Casey 1993; Cole et al. 2004; Eisenberg et al. 2007; Saarni 1979; Stegge and Meerum Terwogt 2007; Thompson 1994; Saarni et al. 1998). Although emerging regulatory capacities are internalized by infants, their development is critically dependent on the successful provision of external regulation by the caregiver (Calkins and Hill 2007; Kopp 1989; Bernier et al. 2010). External regulation serves to foster the development of infants' self-regulatory capacities to cope with everyday stressors, and it is this development that propels and boosts their resilience when under greater duress. When deprived of regulatory support, infants, as well as the young of other species, show deficits in their regulatory capacities (Fogel 1993, 2000; Blandon et al. 2008; Champagne and Curley 2009; Meaney 2010; Tronick and Reck 2009; Weaver et al. 2004). They are chronically dysregulated and constantly recruit their resources to self-regulate which, in turn, undermines and disrupts their engagement with the world (see Fig. 4.2b). Just by its nature, severe chronic stress taxes reparatory capacity and subsequently reduces reparatory success. Consequently, infants' ability to self-regulate and the quality and form of the mutual regulation relationship between infants and their mothers, often referred to as "maternal sensitivity," are both important to the development of regulatory capacity in infants, as well as in their overall development (Beebe et al. 2010; Ainsworth et al. 1974; Beeghly et al. 2011).

4.5 The Three Tenets of the Everyday Stress Resilience Hypothesis

The ESR Hypothesis contextualizes the development of resilience in the exposure to and successful reparation of typical everyday stressors. In order for the hypothesis to be valid, there are several necessary conditions that need to be met that involve the infant, the nature of the stressor, and the effectiveness of the infant's external regulator, their mother.

4.5.1 Tenet 1: The Development of Self-Regulation

The first tenet of the ESR Hypothesis requires that the building blocks of self-regulation be in place early in development. Studies have demonstrated individual differences in behavioral and physiological reactivity early in development which have been linked to difficulties in self-regulation and later emotion regulation capacities (Bridgett et al. 2009; Calkins 1997; Gottman and Katz 2002; Hill-Soderlund and Braungart-Rieker 2008; Kagan and Snidman 2004; Kagan et al. 1992, 1998; Santucci et al. 2008). For example, using typical measures of heart rate variability, infants who have less variability at rest and who showed greater suppression during challenges have less temperamental difficulties, show more regulatory behaviors, and are typically more attentive (Fox 1989; Fox and Porges 1985; Porges et al. 1974, 1994; Porter 2003; Stifter and Corey 2001; Stifter and Fox 1990; Stifter and Jain 1996). This is compared to infants who have greater heart rate variability at rest and who also showed less suppression during challenges. Infants who have this pattern of variability were more likely to have difficult temperament styles and, when under duress, took longer to recover (Bazhenova et al. 2001; Moore and Calkins 2004; Fox 1989; Porges et al. 1994; Stifter and Fox 1990; Beauchaine 2001; Field and Diego 2008). In the context of the dyad, an infant who is highly reactive is more likely to take longer to recover, something that has the potential to make the reparation more difficult for the mother. The infant's high level of arousal makes it difficult to attend to the regulatory support provided by their mother which is likely to lead to an increase in her frustration and a continued disorganized state within the dyad. One can easily see how the cycle can then become self-amplifying and self-perpetuating.

Gains in self-regulation cascade into more advanced regulatory capacities that correlate with cognitive development including effortful control and executive function. Both effortful control and executive function possess elements of inhibition, planning, error detection, and problem solving (Best and Miller 2010; Posner and Rothbart 2000; Rothbart 2007). Gains in these two capacities are thought to mirror the flexibility that is required for successful emotion regulation (Dennis et al. 2010) and, with development, individual differences may influence one's ability to successfully use emotion regulation strategies when under duress (Opitz et al. 2012; Urry and Gross 2010).

4.5.2 Tenet 2: Practice Involves Exposure to Typical Stressors

Thinking about the capacity at which an individual is able to regulate and ultimately recover from stressors, the second tenet of the ESR Hypothesis emphasizes that stressors experienced early in life must be typical and not extreme or chronic. Numerous studies and reviews have demonstrated the toxic effect of chronic, high levels of stress on development (de Bellis et al. 1999a, b; Kaltsas and Chrousos 2007; Lupien et al. 2009; Nelson and Carver 1998). Chronic, intense stress overwhelms the regulatory capacities of the mother, her infant, and the dyad. When mothers need to cope with this type of stress, their ability to function and to parent are both compromised (Blandon et al. 2008; Hoffman et al. 2006; Lupien et al. 2000; Nicholson et al. 2011). They are inattentive and the timing and appropriateness of their responses are disrupted. As caregivers they have deficits in or lack the capacity to successfully regulate their own emotional state, let alone disruptions in their infant's emotional state. And if chronic and high levels of stress affecting both the mother and her infant were not enough, in combination this level of stress disrupts the regulatory operation of the dyad leading to rigid, inflexible patterns for resolving the stress and problematic interactive patterns (Reck et al. 2004; Warren et al. 2003). Therefore, not only are these infants experiencing an extreme amount of stress from their environment but they also lack the external support or scaffolding needed to successfully regulate (Bridgett et al. 2009; Hoffman et al. 2006; Bosquet Enlow et al. 2011; Calkins et al. 2008). In a way, this scenario is not unlike the marathon runner who continually overtrains and does not provide their body with an opportunity to recover. In the end, their body progressively weakens.

4.5.3 Tenet 3: The Mother as an Effective Regulator

The third tenet of the ESR Hypothesis stipulates that moderate levels of maternal sensitivity lead to the most successful regulatory development and, consequently, the greatest resilience. Following a classic inverted U-shaped distribution, both extremely low and extremely high levels of "sensitivity" may contribute to suboptimal regulation skills or vulnerabilities in the developing infant. This idea is based on evidence associating moderate levels of maternal sensitivity and secure attachment (Beebe and Lachmann 1994; Isabella and Belsky 1991). Sensitivity has and is an omnipresent concept in psychology with developmental effects that are viewed as wide reaching. From Freud (1974) to Bowlby (1980), the quality of maternal sensitivity has been seen as influencing infants' development of relationships with others over the life span. Higher levels of maternal sensitivity in infancy are associated with more effective regulation including physiological regulation (Calkins et al. 1998; Conradt and Ablow 2010; Moore et al. 2009) and stress management (Bugental et al. 1993; Conway and McDonough 2006; Waters et al. 2010). Higher levels of maternal sensitivity are also associated with later secure attachment (Ainsworth et al. 1974; Bigelow et al. 2010; Isabella et al. 1989; McElwain and Booth-Laforce 2006), greater sociability (Hobson et al. 2004), lower levels of aggression (Crockenberg et al. 2008; Leerkes et al. 2009), and gains in both cognitive (Bernier et al. 2010; Tamis-LeMonda et al. 1996) and socio-emotional development (Leerkes et al. 2009).

However, at face value maternal sensitivity is a multidimensional, complex psychological construct that can be measured in many different ways under many different circumstances. It can be measured by observing synchrony within the dyad or the matching of affect during times of distress, non-distress, or both, as well as during heightened states of positive arousal. When thinking about the mother's regulatory role within the dyad, we think it is more fitting to tease apart the construct of "maternal sensitivity" and limit our consideration to what we define as "reparatory sensitivity." Reparatory sensitivity refers to the quality and form of the mutual regulation relationship between the infant and mother during times when infants' regulatory strategies are overtaxed, and they cannot self-regulate their states, be the states negative or positive. Reparatory sensitivity occurs at multiple stress levels including the micro-temporal level where the mother provides regulatory scaffolding that leads to interactive reparation of the micro-stress that travels with short-lived rapidly occurring mismatches.

The idea of reparatory sensitivity can be conceptualized in terms of Hans Selye's (1998) classic General Adaptation Syndrome to stress theory. Selye stated that depending upon the individuals' regulatory resilience in the face of a stressor, they may progress through three biobehavioral states—the alarm state, the resistance state, and the exhaustion state. The alarm state prepares the individual for the stressor which is then followed by the resistance state where the individual may use emotion regulation or stress-behavior modification techniques to self-regulate. If those attempts fail, the individual succumbs to the stressor and moves into the exhaustion state where they are now vulnerable to stress-related diseases.

In Selye's model, how the individual adapts during the resilience stage is central to any understanding of how they cope with stress, but more recent thinking points to factors not considered in the model. He, of course, could not have considered the phenomena of plasticity, sensitive periods, and how experience can modify development through learning and changes in gene expression as researched in the emerging field of epigenetics (Champagne and Curley 2009; Meaney 2010; Weaver et al. 2004; Barry et al. 2008; Champagne 2010). But Selye's original model also did not consider development and the changes that occur in the regulatory systems ontogenetically. In this context, Selye saw adaptations as intrinsic to the individual organism, rather than considering the idea we are advancing that successful regulation for the infant is a dyadic process and that dyadic failure leads to stress; that it is not intrinsic and fixed but an experiential developmental process. For example, mismatched affect within the dyad is a key factor for behavioral and physiological disorganization in infants (Tronick et al. 1986).

Nonetheless these developmental and dyadic ideas can be readily incorporated into Selye's stress theory where the mother acts as a constant external regulator, a fail-safe, not only knowing when her infant's regulatory tolerance level has been exceeded and when to step in to intervene, but also when to let her infant self-regulate. With this organization of regulatory sensitivity, the mother allows her infant to experience a certain amount of stress or discomfort, a level that she knows her infant can cope with. Furthermore, through her scaffolding during the process of dyadic mismatches, matches, and reparation, the mother helps her infant build a self-soothing repertoire. More specifically, we believe that reparatory sensitivity to typical interactive macro- and micro-stressors leads to individual differences in infants' regulatory capacities and, consequently, the growth of resilience.

This brings us back to the marathon example. The marathon runner runs a series of shorter, but progressively longer, less-traumatic distances everyday in order to prepare for the actual shock of the marathon. It is this practice that prepares the runner for the longer distance. Monitoring their own training coupled with the insight of coaches and training mates helps prevent overtraining and subsequent damage that is not easily repaired, and at the same time allows for a level of training stress that can be repaired, thus resulting in a growth of capacity. Similarly, the mother's reparatory sensitivity to the match–mismatch process monitors infant's stress within the dyad. The mother acts to prevent stress that would overwhelm the infant's resources while allowing for appropriate levels of capacity increasing stress. Thus the infant does not necessarily need to be in a distressed state during a mismatch.

Instead, mismatches can be small and occur quite frequently in everyday social encounters. What is important is how the infant copes during mismatches. This everyday coping, fostered by the mother's reparatory sensitivity, is what leads to increased everyday resilience. Therefore it is important to consider, given infants' limited regulatory repertoire, the subsequent effects of maternal sensitivity on infant development.

4.6 The Effects of Sensitivity on Development

Research suggests that the quality of maternal sensitivity remains consistent across non-stressful and stressful contexts (Conradt and Ablow 2010; Moore et al. 2009; McElwain and Booth-Laforce 2006; Leerkes et al. 2009; Mills-Koonce et al. 2009). The stability of maternal sensitivity as broadly characterized in the literature, what we would prefer to see as "reparatory sensitivity," fits well with the ESR Hypothesis in its emphasis on chronic on-going events; that is a chronic progressive exposure to reparable levels of stress. Infants of mothers who showed greater maternal sensitivity at 6 months were less likely to show externalizing and internalizing behavioral problems at 24 and 36 months, problems reflecting regulatory issues (Leerkes et al. 2009). Calkins and colleagues (1998) found that greater maternal sensitivity to toddler's negative emotions coupled with a flexible parenting style increases young children's physiologic regulation across multiple stressors. Adding to this, recent epigenetic research emphasizes the protective nature of maternal sensitivity. Propper et al. (2008) found that infants with a genetic vulnerability for physiological dysregulation during stressors were likely to show signs of successful physiological coping (e.g., RSA withdrawal) at 12 months if their mothers were rated as more sensitive at 3 and 6 months. Maternal sensitivity has also been associated with decreases in infants' reactivity to fear (Braungart-Rieker et al. 2010) and later gains in executive functioning abilities that foster self-regulating capacities (Bernier et al. 2010). In contrast, a lack of maternal sensitivity, especially during distress, has been shown to predict later behavioral and emotion regulation problems (McElwain and Booth-Laforce 2006; Leerkes et al. 2009; Crockenberg and Leerkes 2006; Pauli-Pott et al. 2004). For example, Crockenberg and Leerkes (2006) found that greater reactivity to novelty in infancy was associated with later anxious behavior in toddlerhood among infants whose mothers were less sensitive. These findings highlight the important role of sensitivity in the development of the infants' self-regulation capabilities.

The development of successful coping and emotion regulation strategies fostered by maternal sensitivity is also associated with later secure attachment (McElwain and Booth-Laforce 2006; Bakermans-Kranenburg et al. 2003; Braungart-Rieker et al. 2001; Cassidy 1994; de Wolff and van Ijzendoorn 1997; Hill-Soderlund et al. 2008). In the attachment literature, secure attachment develops from infants' expectation that their needs and affective signals will be attended to (Cassidy 1994; Ainsworth et al. 1970, 1978). In a study by McElwain and Booth-Laforce (2006), infants whose mothers showed greater sensitivity at 6 months during a free-play session were more likely to be classified as secure at 15 months. This general pattern was replicated in a study by Fuertes and colleagues who found that mothers higher in sensitivity during play interactions at 9 months were more likely to have securely attached infants at 1 year (Fuertes et al. 2009). Sensitive parenting fosters a secure and trusting relationship whereas insensitive parenting leads to mistrust and insecurity. Trust in the mother and in oneself helps the infant cope with the stress of building new relationships and when exploring the environment (Ainsworth et al. 1970). Infants in secure relationships are also more attentive to their mothers, and it is this increase in attention that provides the mother with more opportunities to help her infant regulate during and after stressful experiences (Beebe et al. 2010; Crockenberg and Leerkes 2006; Evans and Porter 2009; Koulomzin et al. 2002). By contrast, insecure attachment styles are related to less adaptive regulatory capacities, and infants who are avoidant are more likely to disregard their mothers' regulatory attempts. Under duress, 1-year-old infants classified as insecure–avoidant did not show the expected RSA withdrawal response during the socially stressful Ainsworth Strange Situation Paradigm. Adding to this finding, these infants also had higher levels of salivary alpha-amylase compared to securely attached infants. This pattern suggests the insecure–avoidant infants had less of a parasympathetic response to the social stressor and were generally overaroused regardless of the presence of an external stressor when compared to securely attached infants (Hill-Soderlund et al. 2008).

Perhaps as a result of learning or imitation, infants' expression of emotions and regulation comes to resemble their mothers' expression and regulation. For example, mothers of avoidant infants show a narrower range of emotional expressions (Ainsworth et al. 1978). Likewise, avoidant infants have been shown to have a heightened physiologic reaction when under duress, even though they appear less distressed when solely observing their expressive behavior (Spangler and Grossmann 1993). Along this same line of thought, mothers with disorganized attachment have been shown to be biased in their attention. Atkinson et al. (2009) found that disorganized mothers responded slower during an emotional Stroop task involving negative attachment and negative emotion stimuli. As their reaction time increased, so did the likelihood of having the dyad classified as disorganized. Therefore disorganized mothers may have difficulty attending and, consequently, reacting to negative situations. Such attentional deficits have the potential to affect the timing and quality of the mother's intervention thereby disrupting the level of trust within the dyad and the stability of the mother's relationship with her developing infant.

Several additional studies highlight the important link between the timing and efficiency of reparation and infant recovery (Moore and Calkins 2004; Thompson 1994; Porter 2003; McElwain and Booth-Laforce 2006; Leerkes et al. 2009; Kogan and Carter 1996). Timing and efficiency are often related to levels of synchrony within the dyad. Synchrony within the dyad is the result of infants' increased levels of distress and the need for the dyad to achieve regulatory stability (Tronick and Cohn 1989). For example, in dyads who experienced more unilateral patterns of communication during play, where only one part of the dyad is engaged in some sort of communication while the other member is inattentive (e.g., the infant is looking away), infants showed lower vagal tone, a cardiac marker for poor physiological regulation (Porter 2003). Due to the lack of attention, infants in less synchronous dyads had less of an opportunity to experience their mothers' attempts to change their emotional state and to redirect their attention compared to infants who were part of more synchronous dyads. Additionally, in synchronous dyads infants exhibited more positive behaviors and greater vagal tone indicating greater physiological regulation. This suggests the important role of dyadic synchrony in situations where reparation is necessary (Porter 2003).

All in all, in our terms parents who intervene both on-time and efficiently are high in reparatory sensitivity. They recognize when their infant needs help and provide the appropriate level of attention and intervention. Intervening too soon may lead infants to seldom experience regulation on their own and preclude opportunities to develop regulatory capacities. As a consequence, when faced with a stressor in the absence of the caregiver, the infant may be unable to cope. Likewise intervening too late may result in an inconsolable infant who is unable to utilize the caregiver's soothing input. Observations by Beebe and Lachmann (1994) and Isabella and Belsky (1991) found that sensitivity in the mid-range, rather than at the low or high end, typifies normal interactions. Given this work and work by Tronick and colleagues (Tronick and Gianino 1986; Tronick and Cohn 1989), it is hypothesized that mid-range sensitivity is also characterized by mismatches and repairs, yielding strong, but not perfect, behavioral measures of synchrony and matching. This is compared to interactions where the mother is never sensitive (i.e., high mismatch/low synchrony and no reparation) or always overly sensitive (i.e., low mismatch/low synchrony and no reparation).

Research suggests that a lack of sensitivity, characterized by frequent maternal emotional withdrawal, affects infants' later developing physiological stress responses. Specifically infants with emotionally absent mothers (e.g., mothers suffering from postpartum depression) are more likely to have elevated resting cortisol levels (Bugental et al. 2003), a maladaptive response indicative of a hyperresponsive hypothalamic–pituitary–adrenal axis (Dickerson and Kemeny 2004; Hellhammer et al. 2009; Lovallo and Thomas 2000; Sapolsky 1996). Elevated basal cortisol levels, including elevated levels during reactivity, may help infants cope in the here and now, but chronically elevated levels have the potential for negative effects later in life including problems regulating future stressors, suppressed immune function, and the development of stress-related disorders (Bugental et al. 2003; Cacioppo 2000; Uchino et al. 2007; Feldman et al. 2009). These effects may emerge because mothers suffering from postpartum depression are typically unable to balance their own need for regulation with their infants' need for external regulation (Cohn et al. 1990; Goodman 2007). As a consequence their timing is off. They respond slower and are overall less responsive to their infants' attentional bids (Zlochower and Cohn 1996). In effect infants of depressed mothers tend to withdraw and must rely on their own, albeit insufficient, regulatory strategies (Manian et al. 2009). In the end, these infants are left to regulate on their own and, when faced with higher levels of distress, are likely unable to regain homeostatic balance.

Paradoxically too much sensitivity, where an infant is rarely allowed to experience reparation, can also lead to negative consequences. Hypervigilant parents try to buffer their infants from experiencing any stress. Infants of mothers who are over-involved or intrusive are less likely to develop a secure attachment style (Isabella and Belsky 1991; Isabella et al. 1989). Over-involved or intrusive styles of maternal interaction have the potential to limit infants' regulatory growth owing to fewer opportunities to regulate typical, everyday stress (Graziano et al. 2010). This lack of experience with stress and reparation can also translate to differences in the functioning of neurophysiological stress-response systems (Warren et al. 2003) with the final result being an arousal overload for infants when faced with a stressor in the absence of their mothers. Both in cases of under- and oversensitive mothers, it is our hypothesis that infants are at a disadvantage for developing coping mechanisms used to regulate their physiological, behavioral, and emotional stress responses. Reparatory sensitivity is intertwined with the mother's ability to regulate her own psychological state in addition to her infant's. In sum, the maternal sensitivity findings make it clear that the development of infants' resilience, as seen in the development of self-soothing and regulatory strategies, is both maintained and expanded by sensitive reductions in physiological arousal.

4.7 Conclusion

The concept of resilience is usually associated with coping and regulation under extreme, chronic stress. For that reason, examples of resilient behavior tend to focus more on the against-all-odds types of stories such as the inner-city youth who grew up in poverty and lost both parents to violence or the physical abuse survivor, both of whom managed to become influential leaders in society. In this chapter, we put forth a hypothesis, the ESR Hypothesis, to present the argument that resilience emerges from a process of regulating and coping with everyday life stressors. The more experience one has with successfully regulating everyday life stressors, the more prepared one is to cope with greater challenges. For infants, this can consist of coping with micro-stressors, the ubiquitous disruptions in the typical flow of communication within the mother–infant dyad. *Ubiquitous* is emphasized here since experience with severe, chronic stressors can compromise development by overwhelming the regulatory capacity of the dyad and affecting infants' early self-regulation and later emotion regulation capacities. But infants' coping experience is not solely dependent on their own capacities. They are part of a larger dyadic regulatory system, and their experience with the reparation of mismatches within the dyadic system is critical to successful regulation of stress in the short run and to the enhancement of infants' regulatory resilience in the long run (Tronick 2006; Thompson 1994; McElwain and Booth-Laforce 2006; Leerkes et al. 2009). Thus it is the quality of behavioral and biological reparation within the mother–infant dyad that serves a protective regulatory function—a function of preparing, expanding, and developing infants' regulatory capacity. The positive feedback loop within the dyad creates a greater propensity to regulate, in turn enhancing the process of mutual regulation.

Future research is needed to validate the implications brought forth by our hypothesis. A primary area of interest involves stress as a factor inducing regulatory capacity. How much stress is appropriate? How much is too little or too much? What markers can we develop to evaluate and predict outcomes? Are there sensitive periods when stress is *required* and does the stress have to be of a specific type to induce a positive effect? If there are sensitive periods, can they be overcome later in development? To explore these questions, more research involving microanalytic coding methods should be conducted at various ages in order to determine the dynamics of the interplay of behavior and physiology including their effects on reparation. Another area of research involves infants who do not fit the hypothesis because they function well despite the extremes of chronic, toxic stress, and trauma. Are these infants at the extreme of the continuum of individual differences in regulatory capacity? Do they have deficits that go unnoticed? Even under situations of extreme stress where there is slim possibility that the infant comes out unscathed, they may show deficits in some areas of functioning compared to others. Was there some fail-safe in their daily experience that protected them and allowed for normal development?

The ESR Hypothesis represents a critical first step in building a more comprehensive theory of resilience. The hypothesis demands a detailed knowing of an individual's experience, in particular the details of their relational experience, rather than a global characterization of events in their lives. Finally, how do differences in types of stress and types of infants fit into the picture? Is regulatory capacity specific to context and stimulus or is it unbounded? Does a sense of regulatory sensitivity extend to all domains (e.g., social, cognitive, and executive functioning)? To what extent do temperamental differences predetermine infants' regulatory capacity regardless of the mothers' reparatory sensitivity? Carefully modifying these variables in future studies opens the door to understanding the scope of everyday resiliency.

Small stressors are ubiquitous in everyday experience, even for infants. Unfortunately, traumatic, chronic stress is also far too common. As argued here, it is our view that understanding resilience in the face of extreme stress requires broadening our perspective and focusing on the developmental processes that lead to resilience—the reparation of the stress of simply being in and engaging the world.

References

Adamson LB, Frick JE (2003) The still face: a history of a shared experimental paradigm. Infancy 4(4):451–473. Available at http://doi.wiley.com/10.1207/S15327078IN0404_01. Accessed 29 Feb 2012

Adolph KE (1997) Learning in the development of infant locomotion. Monogr Soc Res Child Dev 62(3):I–VI, 1–158. Available at http://www.ncbi.nlm.nih.gov/pubmed/9394468. Accessed 29 Feb 2012

Ainsworth MD, Bell SM (1970) Attachment, exploration, and separation: illustrated by the behavior of one-year-olds in a strange situation. Child Dev 41(1):49–67. Available at http://www.ncbi.nlm.nih.gov/pubmed/5490680. Accessed 29 Feb 2012

Ainsworth MD, Bell SM, Strayton DJ (1974) Infant-mother attachment and social development: socialization as a product of reciprocal responsiveness to signals. In: Richards MPM (ed) The integration of a child into a social world. Cambridge University Press, New York, pp 99–119

Ainsworth MD, Blehar MC, Waters E, Wall S (1978) Patterns of attachment: a psychological study of the strange situation. Lawrence Erlbaum, Oxford, p 391

Atkinson L, Leung E, Goldberg S et al (2009) Attachment and selective attention: disorganization and emotional Stroop reaction time. Dev Psychopathol 21(1):99–126. Available at http://www.ncbi.nlm.nih.gov/pubmed/19144225. Accessed 29 Feb 2012

Bakermans-Kranenburg MJ, van IJzendoorn MH, Juffer F (2003) Less is more: meta-analyses of sensitivity and attachment interventions in early childhood. Psychol Bull 129(2):195–215. Available at http://doi.apa.org/getdoi.cfm?doi=10.1037/0033-2909.129.2.195. Accessed 29 Feb 2012

Barry RA, Kochanska G, Philibert RA (2008) G × E interaction in the organization of attachment: mothers' responsiveness as a moderator of children's genotypes. J Child Psychol Psychiatry 49(12):1313–1320. Available at http://www.ncbi.nlm.nih.gov/pubmed/19120710. Accessed 29 July 2011

Bazhenova OV, Plonskaia O, Porges SW (2001) Vagal reactivity and affective adjustment in infants during interaction challenges. Child Dev 72(5):1314–1326. Available at http://www.ncbi.nlm.nih.gov/pubmed/11699673. Accessed 29 Feb 2012

Beauchaine T (2001). Vagal tone, development, and Gray's motivational theory: toward an integrated model of autonomic nervous system functioning in psychopathology. Dev Psychopathol 13(2):183–214. Available at http://www.ncbi.nlm.nih.gov/pubmed/11393643. Accessed 29 Feb 2012

Beebe B, Lachmann FM (1994) Representation and internalization in infancy: three principles of salience. Psychoanal Psychol 11(2):127–165. Available at http://doi.apa.org/getdoi.cfm?doi=10.1037/h0079530. Accessed 29 Feb 2012

Beebe B, Jaffe J, Markese S et al (2010) The origins of 12-month attachment: a microanalysis of 4-month mother-infant interaction. Attach Hum Dev 12(1–2):3–141. Available at http://www.ncbi.nlm.nih.gov/pubmed/20390524. Accessed 29 Feb 2012

Beeghly M, Fuertes M, Liu C, Delonis MS, Tronick EZ (2011) Maternal sensitivity in dyadic context: mutual regulation, meaning-making, and reparation. In: Davis DW, Logsdon MC (eds) Maternal sensitivity: a scientific foundation for practice. Nova Science, Hauppauge, pp 45–69

Bernier A, Carlson SM, Whipple N (2010) From external regulation to self-regulation: early parenting precursors of young children's executive functioning. Child Dev 81(1):326–339. Available at http://www.ncbi.nlm.nih.gov/pubmed/20331670. Accessed 26 Sep 2011

Best JR, Miller PH (2010) A developmental perspective on executive function. Child Dev 81(6):1641–1660. Available at http://www.pubmedcentral.nih.gov/articlerender.fcgi?artid=3058827&tool=pmcentrez&rendertype=abstract. Accessed 29 Feb 2012

Bigelow AE, MacLean K, Proctor J et al (2010) Maternal sensitivity throughout infancy: continuity and relation to attachment security. Infant Behav Dev 33(1):50–60. Available at http://www.ncbi.nlm.nih.gov/pubmed/20004022. Accessed 29 July 2011

Blandon AY, Calkins SD, Keane SP, O'Brien M (2008) Individual differences in trajectories of emotion regulation processes: the effects of maternal depressive symptomatology and children's physiological regulation. Dev Psychol 44(4):1110–1123. Available at http://www.pubmedcentral.nih.gov/articlerender.fcgi?artid=2630713&tool=pmcentrez&rendertype=abstract. Accessed 29 Feb 2012

Bosquet Enlow M, Kitts RL, Blood E et al (2011) Maternal posttraumatic stress symptoms and infant emotional reactivity and emotion regulation. Infant Behav Dev 34(4):487–503. Available at http://www.pubmedcentral.nih.gov/articlerender.fcgi?artid=3180882&tool=pmcentrez&rendertype=abstract. Accessed 29 Feb 2012

Bowlby J (1980) Attachment and loss, vol 1, Attachment. Basic Books, New York

Braungart-Rieker JM, Garwood MM, Powers BP, Wang X (2001) Parental sensitivity, infant affect, and affect regulation: predictors of later attachment. Child Dev 72(1):252–270. Available at http://www.ncbi.nlm.nih.gov/pubmed/11280483. Accessed 29 Feb 2012

Braungart-Rieker JM, Hill-Soderlund AL, Karrass J (2010) Fear and anger reactivity trajectories from 4 to 16 months: the roles of temperament, regulation, and maternal sensitivity. Dev Psychol 46(4):791–804. Available at http://www.ncbi.nlm.nih.gov/pubmed/20604602. Accessed 29 Feb 2012

Brazelton TB (1992) Touchpoints: emotional and behavioral development. Addison-Wesley, Reading

Brazelton TB, Koslowski B, Main M (1974) The origins of reciprocity: the early mother-infant interaction. In: Lewis M, Rosenblum LA (eds) The effect of the infant on its caregiver. Wiley-Interscience, Oxford, pp 49–76

Bridgett DJ, Gartstein MA, Putnam SP et al (2009) Maternal and contextual influences and the effect of temperament development during infancy on parenting in toddlerhood. Infant Behav Dev 32(1):103–116. Available at http://www.ncbi.nlm.nih.gov/pubmed/19111913. Accessed 29 July 2011

Bugental DB, Blue J, Cortez V et al (1993) Social cognitions as organizers of autonomic and affective responses to social challenge. J Pers Soc Psychol 64(1):94–103. Available at http://www.ncbi.nlm.nih.gov/pubmed/8421254. Accessed 29 Feb 2012

Bugental DB, Martorell GA, Barraza V (2003) The hormonal costs of subtle forms of infant maltreatment. Horm Behav 43(1):237–244. Available at http://linkinghub.elsevier.com/retrieve/pii/S0018506X02000089. Accessed 23 Feb 2012

Cacioppo JT (2000) Autonomic, neuroendocrine, and immune responses to psychological stress. Psychol Beltrage 42:4–23

Calkins SD (1997) Cardiac vagal tone indices of temperamental reactivity and behavioral regulation in young children. Dev Psychobiol 31(2):125–135. Available at http://www.ncbi.nlm.nih.gov/pubmed/9298638. Accessed 29 Feb 2012

Calkins SD, Hill A (2007) Caregiver influences on emerging emotion regulation. In: Gross JJ (ed) Handbook of emotion regulation. Guilford, New York, pp 229–248

Calkins SD, Smith CL, Gill KL, Johnson MC (1998) Maternal interactive style across contexts: relations to emotional, behavioral and physiological regulation during toddlerhood. Social Dev 7(3):350–369. Available at http://doi.wiley.com/10.1111/1467-9507.00072. Accessed 28 Feb 2012

Calkins SD, Graziano PA, Berdan LE, Keane SP, Degnan KA (2008) Predicting cardiac vagal regulation in early childhood from maternal-child relationship quality during toddlerhood. Dev Psychobiol 50(8):751–766. Available at http://www.ncbi.nlm.nih.gov/pubmed/18814182. Accessed 29 Feb 2012

Carlson SM, Wang TS (2007) Inhibitory control and emotion regulation in preschool children. Cognitive Dev 22(4):489–510. Available at http://linkinghub.elsevier.com/retrieve/pii/S088520140700055X. Accessed 29 Feb 2012

Casey RJ (1993) Children's emotional experience: relations among expression, self-report, and understanding. Dev Psychol 29(1):119–129

Cassidy J (1994) Emotion regulation: influences of attachment relationships. Monogr Soc Res Child Dev 59(2–3):228–249. Available at http://www.ncbi.nlm.nih.gov/pubmed/7984163. Accessed 28 Feb 2012

Champagne FA (2010) Epigenetic influence of social experiences across the lifespan. Dev Psychobiol 52(4):299–311. Available at http://www.ncbi.nlm.nih.gov/pubmed/20175106. Accessed 22 July 2011

Champagne FA, Curley JP (2009) Epigenetic mechanisms mediating the long-term effects of maternal care on development. Neurosci Biobehav Rev 33(4):593–600. Available at http://www.ncbi.nlm.nih.gov/pubmed/18430469. Accessed 19 July 2011

Cicchetti D, Rogosch FA, Lynch M, Holt KD (1993) Resilience in maltreated children: processes leading to adaptive outcome. Dev Psychopathol 5(04):629–647. Available at http://www.journals.cambridge.org/abstract_S0954579400006209. Accessed 29 Feb 2012

Cohn JF, Campbell SB, Matias R, Hopkins J (1990). Face-to-face interactions of postpartum depressed and nondepressed mother-infant pairs at 2 months. Dev Psychol 26(1):15–23. Available at http://doi.apa.org/getdoi.cfm?doi=10.1037/0012-1649.26.1.15. Accessed 29 Feb 2012

Cole PM, Martin SE, Dennis TA (2004) Emotion regulation as a scientific construct: methodological challenges and directions for child development research. Child Dev 75(2):317–333. Available at http://www.ncbi.nlm.nih.gov/pubmed/15056186

Condon WS, Sander LW (1974) Neonate movement is synchronized with adult speech: interactional participation and language acquisition. Science 183(4120):99–101. Available at http://www.ncbi.nlm.nih.gov/pubmed/4808791. Accessed 29 Feb 2012

Conradt E, Ablow J (2010) Infant physiological response to the still-face paradigm: contributions of maternal sensitivity and infants' early regulatory behavior. Infant Behav Dev 33(3):251–265. Available at http://dx.doi.org/10.1016/j.infbeh.2010.01.001. Accessed 29 Feb 2012

Conway AM, McDonough SC (2006) Emotional resilience in early childhood: developmental antecedents and relations to behavior problems. Ann N Y Acad Sci 1094:272–277. Available at http://www.ncbi.nlm.nih.gov/pubmed/17347360. Accessed 29 Feb 2012

Crockenberg SC, Leerkes EM (2006) Infant and maternal behavior moderate reactivity to novelty to predict anxious behavior at 2.5 years. Dev Psychopathol 18(1):17–34. Available at http://www.ncbi.nlm.nih.gov/pubmed/16478550. Accessed 29 Feb 2012

Crockenberg SC, Leerkes EM, Bárrig Jó PS (2008) Predicting aggressive behavior in the third year from infant reactivity and regulation as moderated by maternal behavior. Dev Psychopathol 20(1):37–54. Available at http://www.ncbi.nlm.nih.gov/pubmed/18211727. Accessed 29 Feb 2012

de Bellis MD, Baum AS, Birmaher B et al (1999a) Developmental traumatology part I: biological stress systems. Biol Psychiatry 45(10):1259–1270. Available at http://linkinghub.elsevier.com/retrieve/pii/S000632239900044X. Accessed 29 Feb 2012

de Wolff MS, van Ijzendoorn MH (1997) Sensitivity and attachment: a meta-analysis on parental antecedents of infant attachment. Child Dev 68(4):571–591. Available at http://www.ncbi.nlm.nih.gov/pubmed/9306636. Accessed 29 Feb 2012

de Bellis MD, Keshavan MS, Clark DB et al (1999b) Developmental traumatology part II: brain development. Biol Psychiatry 45(10):1271–1284. Available at http://linkinghub.elsevier.com/retrieve/pii/S0006322399000451. Accessed 29 Feb 2012

Dennis TA, Hong M, Solomon B (2010) Do the associations between exuberance and emotion regulation depend on effortful control? Int J Behav Dev 34(5):462–472. Available at http://jbd.sagepub.com/cgi/doi/10.1177/0165025409355514. Accessed 29 Feb 2012

Dickerson SS, Kemeny ME (2004) Acute stressors and cortisol responses: a theoretical integration and synthesis of laboratory research. Psychol Bull 130(3):355–391. Available at http://www.ncbi.nlm.nih.gov/pubmed/15122924. Accessed 29 Feb 2012

Egeland B, Carlson E, Sroufe LA (1993) Resilience as process. Dev Psychopathol 5(4):517. Available at http://www.journals.cambridge.org/abstract_S0954579400006131. Accessed 29 Feb 2012

Eisenberg N, Hofer C, Vaughan J (2007) Effortful control and its socioemotional consequences. In: Gross JJ (ed) Handbook of emotion regulation. Guilford, New York, pp 287–306

Evans CA, Porter CL (2009) The emergence of mother-infant co-regulation during the first year: links to infants' developmental status and attachment. Infant Behav Dev 32(2):147–158. Available at http://www.ncbi.nlm.nih.gov/pubmed/19200603. Accessed 27 Sep 2011

Feldman R (2007) Parent-infant synchrony and the construction of shared timing; physiological precursors, developmental outcomes, and risk conditions. J Child Psychol Psychiatry 48(3–4):329–354. Available at http://www.ncbi.nlm.nih.gov/pubmed/17355401. Accessed 18 July 2011

Feldman R, Granat A, Pariente C et al (2009) Maternal depression and anxiety across the postpartum year and infant social engagement, fear regulation, and stress reactivity. J Am Acad Child Psychiatry 48(9):919–927. Available at http://dx.doi.org/10.1097/CHI.0b013e3181b21651. Accessed 3 Aug 2011

Feldman R, Singer M, Zagoory O (2010) Touch attenuates infants' physiological reactivity to stress. Dev Sci 13(2):271–278. Available at http://www.ncbi.nlm.nih.gov/pubmed/20136923. Accessed 29 Feb 2012

Field T (1994) The effects of mother's physical and emotional unavailability on emotion regulation. Monogr Soc Res Child Dev 59(2–3):208–227. Available at http://www.ncbi.nlm.nih.gov/pubmed/7984162. Accessed 29 Feb 2012

Field T, Diego M (2008) Vagal activity, early growth and emotional development. Infant Behav Dev 31(3):361–373. Available at http://www.ncbi.nlm.nih.gov/pubmed/18295898. Accessed 29 Feb 2012

Fogel A (1993) Developing through relationships: origins of communication, self, and culture. University of Chicago Press, Chicago, p 230

Fogel A (2000) Developmental pathways in close relationships. Child Dev 71(5):1150–1151. Available at http://www.blackwell-synergy.com/links/doi/10.1111%2F1467-8624.00217. Accessed 29 Feb 2012

Fox NA (1989) Psychophysiological correlates of emotional reactivity during the first year of life. Dev Psychol 25(3):364–372. Available at http://doi.apa.org/getdoi.cfm?doi=10.1037/0012-1649.25.3.364. Accessed 29 Feb 2012

Fox NA, Porges SW (1985) The relation between neonatal heart period patterns and developmental outcome. Child Dev 56(1):28–37. Available at http://www.ncbi.nlm.nih.gov/pubmed/3987407. Accessed 29 Feb 2012

Freud A (1974) Normality and pathology in childhood: assessments of development 1965, vol 6. International Universities Press, New York

Fuertes M, Lopes-dos-Santos P, Beeghly M, Tronick EZ (2009) Infant coping and maternal interactive behavior predict attachment in a Portuguese sample of healthy preterm infants. Euro Psychol 14:320–331

Gianino A, Tronick EZ (1988) The mutual regulation model: the infant's self and interactive regulation and coping and defensive capacities. In: Field TM, McCabe PM, Schneiderman N (eds) Stress and coping across development. Lawrence Erlbaum, Hillsdale, pp 47–68

Goodman SH (2007). Depression in mothers. Ann Rev Clin Psychol 3:107–135. Available at http://www.ncbi.nlm.nih.gov/pubmed/17716050. Accessed 29 Feb 2012

Gottman JM, Katz LF (2002) Children's emotional reactions to stressful parent-child interactions. Marriage Family Rev 34(3–4):265–283. Available at http://www.informaworld.com/openurl?genre=article&doi=10.1300/J002v34n03_04&magic=crossref||D404A21C5BB053405B1A640AFFD44AE3. Accessed 29 Feb 2012

Granic I, Hollenstein T (2003) Dynamic systems methods for models of developmental psychopathology. Dev Psychopathol 15(3):641–669. Available at http://www.ncbi.nlm.nih.gov/pubmed/14582935. Accessed 29 Feb 2012

Graziano PA, Keane SP, Calkins SD (2010) Maternal behavior and children's early emotion regulation skills differentially predict development of children's reactive control and later effortful control. Infant Child Dev 19(4):333–353. Available at http://www.pubmedcentral.nih.gov/articlerender.fcgi?artid=3034150&tool=pmcentrez&rendertype=abstract. Accessed 29 Feb 2012

Haglund ME, Nestadt PS, Cooper NS, Southwick SM, Charney DS (2007) Psychobiological mechanisms of resilience: relevance to prevention and treatment of stress-related psychopathology. Dev Psychopathol 19(3):889–920. Available at http://www.ncbi.nlm.nih.gov/pubmed/17705907. Accessed 29 Feb 2012

Haley DW, Stansbury K (2003) Infant stress and parent responsiveness: regulation of physiology and behavior during still-face and reunion. Child Dev 74(5):1534–1546. Available at http://www.ncbi.nlm.nih.gov/pubmed/14552412. Accessed 29 Feb 2012

Ham J, Tronick E (2006) Infant resilience to the stress of the still-face: infant and maternal psychophysiology are related. Ann N Y Acad Sci 1094:297–302. Available at http://www.ncbi.nlm.nih.gov/pubmed/17347365. Accessed 29 Feb 2012

Ham J, Tronick E (2008) A procedure for the measurement of infant skin conductance and its initial validation using clap induced startle. Dev Psychobiol 50(6):626–631. Available at http://www.ncbi.nlm.nih.gov/pubmed/18683186. Accessed 29 Feb 2012

Harrist AW, Waugh RM (2002) Dyadic synchrony: its structure and function in children's development. Dev Rev 22(4):555–592. Available at http://linkinghub.elsevier.com/retrieve/pii/S0273229702005002. Accessed 31 Dec 2011

Hellhammer DH, Wüst S, Kudielka BM (2009) Salivary cortisol as a biomarker in stress research. Psychoneuroendocrinology 34(2):163–171. Available at http://www.ncbi.nlm.nih.gov/pubmed/19095358. Accessed 20 July 2011

Hill-Soderlund AL, Braungart-Rieker JM (2008) Early individual differences in temperamental reactivity and regulation: implications for effortful control in early childhood. Infant Behav Dev 31(3):386–397. Available at http://www.ncbi.nlm.nih.gov/pubmed/18279967. Accessed 29 Feb 2012

Hill-Soderlund AL, Mills-Koonce WR, Propper C et al (2008) Parasympathetic and sympathetic responses to the strange situation in infants and mothers from avoidant and securely attached dyads. Dev Psychobiol 50(4):361–376. Available at http://www.ncbi.nlm.nih.gov/pubmed/18393278. Accessed 29 Feb 2012

Hobson RP, Patrick MPH, Crandell LE, García Pérez RM, Lee A (2004) Maternal sensitivity and infant triadic communication. J Child Psychol Psychiatry 45(3):470–480. Available at http://www.ncbi.nlm.nih.gov/pubmed/15055367. Accessed 29 Feb 2012

Hofer MA (1987) Early social relationships: a psychobiologist's view. Child Dev 58(3):633–647. Available at http://www.ncbi.nlm.nih.gov/pubmed/3608643. Accessed 29 Feb 2012

Hofer MA (1994) Hidden regulators in attachment, separation, and loss. Monogr Soc Res Child Dev 59(2–3):192–207. Available at http://www.ncbi.nlm.nih.gov/pubmed/7984161. Accessed 29 Feb 2012

Hofer MA (2006) Evolutionary basis of adaptation in resilience and vulnerability: response to Cicchetti and Blender. Ann N Y Acad Sci 1094:259–262. Available at http://www.ncbi.nlm.nih.gov/pubmed/17347357. Accessed 7 Sept 2011

Hoffman C, Crnic KA, Baker JK (2006) Maternal depression and parenting: implications for children's emergent emotion regulation and behavioral functioning. Parenting 6(4):271–295. Available at http://www.tandfonline.com/doi/abs/10.1207/s15327922par0604_1. Accessed 29 Feb 2012

Isabella RA, Belsky J (1991) Interactional synchrony and the origins of infant-mother attachment: a replication study. Child Dev 62(2):373–384. Available at http://www.ncbi.nlm.nih.gov/pubmed/2055128. Accessed 29 Feb 2012

Isabella RA, Belsky J, von Eye A (1989) Origins of infant-mother attachment: an examination of interactional synchrony during the infant's first year. Dev Psychol 25(1):12–21. Available at http://doi.apa.org/getdoi.cfm?doi=10.1037/0012-1649.25.1.12. Accessed 29 Feb 2012

Kagan J, Snidman NC (2004) The long shadow of temperament. Harvard University Press, Cambridge, p 292

Kagan J, Snidman N, Arcus DM (1992) Initial reactions to unfamiliarity. Curr Dir Psychol Sci 1(6):171–174. Available at http://cdp.sagepub.com/lookup/doi/10.1111/1467-9566.ep10770010. Accessed 29 Feb 2012

Kagan J, Snidman N, Arcus D (1998) Childhood derivatives of high and low reactivity in infancy. Child Dev 69(6):1483–1493. Available at http://doi.wiley.com/10.1111/j.1467-8624.1998.tb06171.x. Accessed 29 Feb 2012

Kaltsas GA, Chrousos GP (2007) The neuroendocrinology of stress. In: Cacioppo JT, Tassinary LG, Berntson GG (eds) Handbook of psychophysiology, 3rd edn. Cambridge University Press, New York, pp 303–318

Kogan N, Carter AS (1996) Mother-infant reengagement following the still-face: the role of maternal emotional availability an infant affect regulation. Infant Behav Dev 19(3):359–370. Available at http://linkinghub.elsevier.com/retrieve/pii/S016363839690034X. Accessed 29 Feb 2012

Kopp CB (1989) Regulation of distress and negative emotions: a developmental view. Dev Psychol 25(3):343–354. Available at http://doi.apa.org/getdoi.cfm?doi=10.1037/0012-1649.25.3.343. Accessed 24 Jan 2012

Koulomzin M, Beebe B, Anderson S et al (2002) Infant gaze, head, face and self-touch at 4 months differentiate secure vs. avoidant attachment at 1 year: a microanalytic approach.

Attach Hum Dev 4(1):3–24. Available at http://www.ncbi.nlm.nih.gov/pubmed/12065027. Accessed 29 Feb 2012

Leerkes EM, Nayena Blankson A, O'Brien M (2009) Differential effects of maternal sensitivity to infant distress and nondistress on social-emotional functioning. Child Dev 80(3):762–775. Available at http://www.pubmedcentral.nih.gov/articlerender.fcgi?artid=2854550&tool=pmcentrez&rendertype=abstract. Accessed 29 Feb 2012

Lewis MD, Lamey AV, Douglas L (1999) A new dynamic systems method for the analysis of early socioemotional development. Dev Sci 2(4):457–475. Available at http://doi.wiley.com/10.1111/1467-7687.00090. Accessed 29 Feb 2012

Lovallo WR, Thomas TL (2000) Stress hormones in psychophysiological research. In: Cacioppo JT, Tassinary LG, Berntson GG (eds) Handbook of psychophysiology, 2nd edn. Oxford University Press, New York, pp 342–367

Lupien SJ, King S, Meaney MJ, McEwen BS (2000) Child's stress hormone levels correlate with mother's socioeconomic status and depressive state. Biol Psychiatry 48(10):976–980. Available at http://linkinghub.elsevier.com/retrieve/pii/S0006322300009653. Accessed 29 Feb 2012

Lupien SJ, McEwen BS, Gunnar MR, Heim C (2009) Effects of stress throughout the lifespan on the brain, behaviour and cognition. Nat Rev Neurosci 10(6):434–445. Available at http://www.nature.com/doifinder/10.1038/nrn2639. Accessed 19 July 2011

Luthar SS, Cicchetti D, Becker B (2000) The construct of resilience: a critical evaluation and guidelines for future work. Child Dev 71(3):543–562. Available at http://doi.wiley.com/10.1111/1467-8624.00164. Accessed 4 Nov 2011

Manian N, Bornstein MH (2009) Dynamics of emotion regulation in infants of clinically depressed and nondepressed mothers. J Child Psychol Psychiatry 50(11):1410–1418. Available at http://www.pubmedcentral.nih.gov/articlerender.fcgi?artid=2844731&tool=pmcentrez&rendertype=abstract. Accessed 29 Feb 2012.

McElwain NL, Booth-Laforce C (2006) Maternal sensitivity to infant distress and nondistress as predictors of infant-mother attachment security. J Fam Psychol 20(2):247–255. Available at http://www.ncbi.nlm.nih.gov/pubmed/16756400. Accessed 29 Feb 2012

Meaney MJ (2010) Epigenetics and the biological definition of gene × environment interactions. Child Dev 81(1):41–79. Available at http://www.ncbi.nlm.nih.gov/pubmed/20331654. Accessed 29 Feb 2012

Mesman J, van IJzendoorn MH, Bakermans-Kranenburg MJ (2009) The many faces of the Still-Face Paradigm: a review and meta-analysis. Dev Rev 29(2):120–162. Available at http://linkinghub.elsevier.com/retrieve/pii/S0273229709000021. Accessed 29 Feb 2012

Mills-Koonce WR, Propper C, Gariepy JL et al (2009) Psychophysiological correlates of parenting behavior in mothers of young children. Dev Psychobiol 51(8):650–661. Available at http://www.ncbi.nlm.nih.gov/pubmed/19739135. Accessed 28 Feb 2012

Montirosso R, Cozzi P, Morandi F, Ciceri, F, Provenzi L, Borgatti R, Riccardi B, Tronick EZ (2011). Four-month-old infant's memory for a stressful social event measured by cortisol reactivity. Paper presented at the meeting of the Society for Research in Child Development, Montreal, QC

Moore GA (2009) Infants' and mothers' vagal reactivity in response to anger. J Child Psychol Psychiatry 50(11):1392–1400. Available at http://www.pubmedcentral.nih.gov/articlerender.fcgi?artid=2891761&tool=pmcentrez&rendertype=abstract. Accessed 29 Feb 2012

Moore GA, Calkins SD (2004) Infants' vagal regulation in the still-face paradigm is related to dyadic coordination of mother-infant interaction. Dev Psychol 40(6):1068–1080. Available at http://www.ncbi.nlm.nih.gov/pubmed/15535757. Accessed 29 Feb 2012

Moore GA, Hill-Soderlund AL, Propper CB et al (2009) Mother-infant vagal regulation in the face-to-face still-face paradigm is moderated by maternal sensitivity. Child Dev 80(1):209–223. Available at http://www.ncbi.nlm.nih.gov/pubmed/19236402. Accessed 29 Feb 2012

Nelson CA, Carver LJ (1998) The effects of stress and trauma on brain and memory: a view from developmental cognitive neuroscience. Dev Psychopathol 10(4):793–809. Available at http://www.ncbi.nlm.nih.gov/pubmed/9886227. Accessed 29 Feb 2012

Nicholson JS, Deboeck PR, Farris JR, Boker SM, Borkowski JG (2011) Maternal depressive symptomatology and child behavior: transactional relationship with simultaneous bidirectional coupling. Dev Psychol 47(5):1312–1323. Available at http://www.pubmedcentral.nih.gov/articlerender.fcgi?artid=3168584&tool=pmcentrez&rendertype=abstract. Accessed 29 Feb 2012

Nomura Y, Chemtob CM, Fifer WP, Newcorn JH, Brooks-Gunn J (2006) Additive interaction of child abuse and perinatal risk as signs of resiliency in adulthood. Ann N Y Acad Sci 1094:330–334. Available at http://www.ncbi.nlm.nih.gov/pubmed/17347371. Accessed 29 Feb 2012

Opitz PC, Gross JJ, Urry HL (2012) Selection, optimization, and compensation in the domain of emotion regulation: applications to adolescence, older age, and major depressive disorder. Soc Personal Psychol Compass 6(2):142–155. Available at http://doi.wiley.com/10.1111/j.1751-9004.2011.00413.x. Accessed 29 Feb 2012

Parker KJ, Buckmaster CL, Sundlass K, Schatzberg AF, Lyons DM (2006) Maternal mediation, stress inoculation, and the development of neuroendocrine stress resistance in primates. Proc Nat Acad Sci U S A 103(8):3000–3005. Available at http://www.pubmedcentral.nih.gov/articlerender.fcgi?artid=1413772&tool=pmcentrez&rendertype=abstract. Accessed 29 Feb 2012

Pauli-Pott U, Mertesacker B, Beckmann D (2004) Predicting the development of infant emotionality from maternal characteristics. Dev Psychopathol 16(1):19–42. Available at http://www.ncbi.nlm.nih.gov/pubmed/15115063. Accessed 29 Feb 2012

Porges SW, Stamps LE, Walter GF (1974) Heart rate variability and newborn heart rate responses to illumination changes. Dev Psychol 10(4):507–513. Available at http://content.apa.org/journals/dev/10/4/507. Accessed 29 Feb 2012

Porges SW, Doussard-Roosevelt JA, Maiti AK (1994) Vagal tone and the physiological regulation of emotion. Monogr Soc Res Child Dev 59(2/3):167–186

Porter CL (2003) Coregulation in mother-infant dyads: links to infants' cardiac vagal tone. Psychol Rep 92(1):307–319. Available at http://www.ncbi.nlm.nih.gov/pubmed/12674298. Accessed 29 Feb 2012

Posner MI, Rothbart MK (2000) Developing mechanisms of self-regulation. Dev Psychopathol 12:427–441

Prigogine I, Stengers I (1984) Order out of chaos: man's new dialogue with nature. Bantam New Age Books, Toronto, p 349

Propper C, Moore GA, Mills-Koonce WR et al (2008) Gene-environment contributions to the development of infant vagal reactivity: the interaction of dopamine and maternal sensitivity. Child Dev 79(5):1377–1394. Available at http://www.ncbi.nlm.nih.gov/pubmed/18826531. Accessed 29 Feb 2012

Reck C, Hunt A, Fuchs T et al (2004) Interactive regulation of affect in postpartum depressed mothers and their infants: an overview. Psychopathology 37(6):272–280. Available at http://www.ncbi.nlm.nih.gov/pubmed/15539778. Accessed 29 Feb 2012

Rothbart MK (2007) Temperament, development, and personality. Curr Dir Psychol Sci 16(4):207–212. Available at http://cdp.sagepub.com/lookup/doi/10.1111/j.1467-8721.2007.00505.x. Accessed 29 Feb 2012

Saarni C (1979) Children's understanding of display rules for expressive behavior. Dev Psychol 15(4):424–429. Available at http://content.apa.org/journals/dev/15/4/424. Accessed 29 Feb 2012

Saarni C, Mumme DL, Campos JJ (1998) Emotional development: action, communication, and understanding. In: Damon W, Eisenberg N (eds) Handbook of child psychology, social, emotional, and personality development, vol 3, 5th edn. Wiley, Hoboken, pp 237–309

Santucci AK, Silk JS, Shaw DS et al (2008) Vagal tone and temperament as predictors of emotion regulation strategies in young children. Dev Psychobiol 50(3):205–216. Available at http://www.ncbi.nlm.nih.gov/pubmed/18335488. Accessed 29 Feb 2012

Sapolsky RM (1996) Why stress is bad for your brain. Science 273(5276):749–750. Available at http://www.ncbi.nlm.nih.gov/pubmed/8701325. Accessed 29 Feb 2012

Selye H (1998) A syndrome produced by diverse nocuous agents. J Neuropsychiatry Clin Neurosci 10(2):230–231. Available at http://www.ncbi.nlm.nih.gov/pubmed/9722327. Accessed 29 Feb 2012

Spangler G, Grossmann KE (1993) Biobehavioral organization in securely and insecurely attached infants. Child Dev 64(5):1439–1450. Available at http://www.ncbi.nlm.nih.gov/pubmed/8222882. Accessed 29 Feb 2012

Stegge H, Meerum Terwogt M (2007) Awareness and regulation of emotion in typical and atypical development. In: Gross JJ (ed) Handbook of emotion regulation. Guilford, New York, pp 269–286

Stifter CA, Corey JM (2001) Vagal regulation and observed social behavior in infancy. Soc Dev 10(2):189–201. Available at http://doi.wiley.com/10.1111/1467-9507.00158. Accessed 29 Feb 2012

Stifter CA, Fox NA (1990) Infant reactivity: physiological correlates of newborn and 5-month temperament. Dev Psychol 26(4):582–588. Available at http://doi.apa.org/getdoi.cfm?doi=10.1037/0012-1649.26.4.582. Accessed 29 Feb 2012

Stifter CA, Jain A (1996) Psychophysiological correlates of infant temperament: stability of behavior and autonomic patterning from 5 to 18 months. Dev Psychobiol 29(4):379–391. Available at http://www.ncbi.nlm.nih.gov/pubmed/8732809. Accessed 29 Feb 2012

Tamis-LeMonda CS, Bornstein MH, Baumwell L, Melstein Damast A (1996) Responsive parenting in the second year: specific influences on children's language and play. Early Dev Parenting 5(4):173–183. Available at http://doi.wiley.com/10.1002/%28SICI%291099-0917%28199612%295%3A4%3C173%3A%3AAID-EDP131%3E3.0.CO%3B2-V. Accessed 29 Feb 2012

Thompson RA (1994) Emotion regulation: a theme in search of definition. Monogr Soc Res Child Dev 59(2–3):25–52. Available at http://www.ncbi.nlm.nih.gov/pubmed/7984164. Accessed 29 Feb 2012

Trevarthen C (1982) Basic patterns of psychogenetic change in infancy. In: Bever TG (ed) Regression in mental development: basic phenomena and theories. Erlbaum, Hillsdale, pp 7–46

Trevarthen C, Aitken KJ (2001) Infant intersubjectivity: research, theory, and clinical applications. J Child Psychol Psychiatry 42(1):3–48. Available at http://www.ncbi.nlm.nih.gov/pubmed/11205623. Accessed 29 Feb 2012

Trevarthen C, Schogller B (2005) Musicality and the creation of meaning: infants' voices and jazz duets show us how, not what, music means. In: Grund CM (ed) Cross disciplinary studies in music and meaning. Indiana University Press, Indianapolis

Trevarthen C, Aitken KJ, Vandekerckhove MMP, Delafield-Butt J, Nagy E (2006) Collaborative regulations of vitality in early childhood: stress in intimate relationships and postnatal psychopathology. In: Cicchetti D, Cohen DJ (eds) Developmental psychopathology, vol 2, 2nd edn. Wiley, New York, pp 65–126

Tronick EZ (1989) Emotions and emotional communication in infants. Am Psychol 44(2):112–119. Available at http://www.ncbi.nlm.nih.gov/pubmed/2653124. Accessed 30 Dec 2011

Tronick E (2006) The inherent stress of normal daily life and social interaction leads to the development of coping and resilience, and variation in resilience in infants and young children: comments on the papers of Suomi and Klebanov & Brooks-Gunn. Ann N Y Acad Sci 1094:83–104. Available at http://www.ncbi.nlm.nih.gov/pubmed/17347343. Accessed 29 Feb 2012

Tronick E, Beeghly M (2011) Infants' meaning-making and the development of mental health problems. Am Psychol 66(2):107–119. Available at http://www.pubmedcentral.nih.gov/articlerender.fcgi?artid=3135310&tool=pmcentrez&rendertype=abstract. Accessed 29 Feb 2012

Tronick EZ, Cohn JF (1989) Infant-mother face-to-face interaction: age and gender differences in coordination and the occurrence of miscoordination. Child Dev 60(1):85–92. Available at http://www.ncbi.nlm.nih.gov/pubmed/2702877. Accessed 29 Feb 2012

Tronick EZ, Gianino A (1986) Interactive mismatch and repair: challenges to the coping infant. Zero to Three 6(3):1–6

Tronick E, Reck C (2009) Infants of depressed mothers. Harvard Rev Psychiatry 17(2):147–156. Available at http://www.ncbi.nlm.nih.gov/pubmed/19373622. Accessed 29 Feb 2012

Tronick E, Als H, Adamson L, Wise S, Brazelton TB (1978) The infant's response to entrapment between contradictory messages in face-to-face interaction. J Am Acad Child Psychiatry 17(1):1–13. Available at http://www.ncbi.nlm.nih.gov/pubmed/632477. Accessed 29 Feb 2012

Tronick EZ, Cohn JF, Shea E (1986) The transfer of affect between mothers and infants. In: Brazelton TB, Yogman MW (eds) Affect development in infancy. Ablex, Westport, pp 11–25

Uchino BN, Smith TW, Holt-Lunstad J, Campo R, Reblin M (2007) Stress and illness. In: Cacioppo JT, Tassinary LG, Berntson GG (eds) Handbook of psychophysiology, 3rd edn. Cambridge University Press, New York, pp 608–632

Urry HL, Gross JJ (2010) Emotion regulation in older age. Curr Dir Psychol Sci 19(6):352–357. Available at http://cdp.sagepub.com/lookup/doi/10.1177/0963721410388395. Accessed 29 Feb 2012

van de Rijt-Plooij HH, Plooij F (1992) Infantile regressions: disorganization and onset of transition periods. J Reprod Infant Psychol 10:129–149

van de Rijt-Plooij HH, Plooij FX (1993) Distinct periods of mother-infant conflict in normal development: sources of progress and germs of pathology. J Child Psychol Psychiatry 34(2):229–245. Available at http://www.ncbi.nlm.nih.gov/pubmed/8444994. Accessed 29 Feb 2012

Warren SL, Gunnar MR, Kagan J et al (2003) Maternal panic disorder: infant temperament, neurophysiology, and parenting behaviors. J Am Acad Child Adolesc Psychiatry 42(7):814–825. Available at http://www.ncbi.nlm.nih.gov/pubmed/12819441. Accessed 24 Feb 2012

Waters SF, Virmani EA, Thompson RA et al (2010) Emotion regulation and attachment: unpacking two constructs and their association. J Psychopathol Behav Assess 32(1):37–47. Available at http://www.pubmedcentral.nih.gov/articlerender.fcgi?artid=2821505&tool=pmcentrez&rendertype=abstract. Accessed 9 Aug 2011

Weaver IC, Cervoni N, Champagne FA et al (2004) Epigenetic programming by maternal behavior. Nat Neurosci 7(8):847–854. Available at http://www.ncbi.nlm.nih.gov/pubmed/15220929. Accessed 16 July 2011

Weinberg MK, Tronick EZ (1996) Infant affective reactions to the resumption of maternal interaction after the still-face. Child Dev 67(3):905–914. Available at http://www.ncbi.nlm.nih.gov/pubmed/8706534. Accessed 29 Feb 2012

Weinberg MK, Tronick EZ, Cohn JF, Olson KL (1999) Gender differences in emotional expressivity and self-regulation during early infancy. Dev Psychol 35(1):175–188. Available at http://www.ncbi.nlm.nih.gov/pubmed/9923473. Accessed 29 Feb 2012

Zlochower AJ, Cohn JF (1996). Vocal timing in face-to-face interaction of clinically depressed and nondepressed mothers and their 4-month-old infants. Infant Behav Dev 19(3):371–374. Available at http://linkinghub.elsevier.com/retrieve/pii/S0163638396900351. Accessed 29 Feb 2012

Chapter 5
Ontogeny of Stress Reactivity in the Human Child: Phenotypic Flexibility, Trade-Offs, and Pathology

Mark V. Flinn, Davide Ponzi, Pablo Nepomnaschy, and Robert Noone

Abstract We humans are highly sensitive to our social environments. Our brains have special abilities such as empathy and social foresight that allow us to understand each other's feelings and communicate in ways that are unique among all living organisms. Our extraordinary social minds, however, come with some significant strings attached. Our emotional states can be strongly influenced by what others say and do. Our hearts can soar, but they also can be broken. Our bodies use internal chemical messengers—hormones and neurotransmitters—to help guide responses to our social worlds. From romantic daydreams to jealous rage, from orgasm to lactation and parent–child bonding, the powerful molecules produced and released by tiny and otherwise seemingly insignificant cells and glands help orchestrate our thoughts and actions. Understanding this chemical language is important for many research questions in human health. Here we focus on the question of why social relationships can affect health—why it is that words can hurt children. Stress hormones appear to play important roles in this puzzle.

The hypothalamic–pituitary–adrenal axis (HPAA) is highly responsive to traumatic experiences including social challenges. For the past 23 years we have conducted a field study of child stress and family environment in a rural community in Dominica. The primary objective is to document hormonal responses of children to

M.V. Flinn (✉)
Department of Anthropology, University of Missouri, 107 Swallow Hall,
Columbia, MO 65211, USA
e-mail: Flinnm@missouri.edu

D. Ponzi
Division of Biological Sciences, University of Missouri, Columbia, MO, USA

P. Nepomnaschy
Simon Fraser University, Burnaby, BC, Canada

R. Noone
Center for Family Consultation, Evanston, IL, USA

G. Laviola and S. Macrì (eds.), *Adaptive and Maladaptive Aspects of Developmental Stress*, Current Topics in Neurotoxicity 3, DOI 10.1007/978-1-4614-5605-6_5,

everyday interactions with their parents and other care providers, concomitant with longitudinal assessment of developmental and health outcomes. Results indicate that difficult family environments and traumatic social events are associated with temporal elevations of cortisol and elevated morbidity. The long-term effects of traumatic early experiences on cortisol profiles are complex and indicate domain-specific effects, with normal recovery from physical stressors, but some heightened response to negative-affect social challenges. These results are consistent with the hypothesis that developmental programming of the HPAA and other neuroendocrine systems associated with stress responses may facilitate cognitive targeting to salient social challenges in specific environmental contexts.

5.1 Introduction

Living organisms are flexible; they can respond to changing conditions with a variety of morphological, physiological, and behavioral mechanisms. The processes that organisms use to change and respond to environmental challenges are posited to be evolved adaptations (West-Eberhard 2003). Flexibility involves both immediate, temporary responses and longer-term developmental changes. A classic example of adaptive developmental response is exhibited by the water flea, Daphnia ambigua, which grows a "helmet" of protective spikes if exposed during juvenile stages to chemicals released by one of its common predators, Chaoborus flavicans (Hanazato 1990). In the absence of the chemical signal of the Chaoborus predator, the juvenile Daphnia economize and allocate the resources that would have been used to grow the protective helmet into other somatic functions (Agrawal et al. 1999).

The human child does not grow spikes or other such dramatic morphological options. He or she nonetheless exhibits a most complex form of phenotypic plasticity. For the child must master the dynamics of social networks and the near-infinitely shifting sands of culture, supported by the extraordinary information-processing capacities of the human brain (Flinn 2006c; Rilling and Sanfey 2011) and its socio-cognitive and linguistic programs. Instead of growing a helmet, the human child can acquire the information necessary to build one out of animal skins or even duct tape.

We are especially interested in the unusual sensitivity of the fetus and child to the social environment—interpersonal relationships—and the consequent changes that occur in physiological stress systems. Our curiosity is piqued by both the paradoxical nature of this phenomenon—for physiological stress response has attendant somatic costs—and its importance for human health. Adverse conditions in utero appear to have a wide range of negative effects on child development and well-being with grave health consequences throughout the life span. Maternal stress during gestation has been reported to affect child development and postnatal behavior, mood, and mental and physical health (Seckl and Meaney 2004; Van den Bergh et al. 2008; Glover 2011; Brand et al. 2011; Hatzinger et al. 2012). For example, maternal depression and high levels of social anxiety during pregnancy are associated with low birth weight, elevated stress reactivity, and subsequent disease risk for offspring (Gluckman and Hanson 2006; Weinstock 2005). The processes that underlay this biological embedding of information from the social environment in humans remain obscure. Our objective here is to use evolutionary theory to evaluate possible mechanisms and developmental trajectories that link early life events to physiological stress response, psychological development, and health outcomes.

Here we use an evolutionary framework to discuss the role of stress as a critical adaptation. First, we briefly review the general evolutionary logic of what is commonly termed "developmental programming" and discuss the effects of social environment on the ontogeny of neuroendocrine stress response and subsequent health outcomes. We evaluate our ideas with a brief overview of our field study of child stress and family environment in Bwa Mawego, Dominica. In our discussion we translate our interpretations of the results from this basic research into concepts with potential clinical application. We conclude with some evolutionary perspectives on the rather unusual life history and ontogeny of the human fetus and child.

5.2 Alternative Phenotypes

Waddington (1956) termed the mysteries of the translation of information from genetic materials during the development of the phenotype as the "great gap in biology." Ontogeny is an astonishingly complex process. Multi cellular organisms such as humans have significant portions of their genomes that perform developmental "regulatory" functions—in effect, switches that turn some genes off and others on during development. To complete their development takes humans nearly three decades. Physical and reproductive maturation take roughly two decades, while the brain continues its development for up to another decade (Bogin 1999; Fox et al. 2010). Throughout these periods, environmental exposures including social, energetic, and immune challenges can alter an individual's developmental trajectory in a variety of ways (Fig. 5.1a). Social, environmental, and physical challenges have been associated with mental health outcomes, academic performance, memory capabilities, sexual behavior, energy metabolism, immune response, aging patterns, and life-span length (Felitti et al. 1998; Harkonmaki et al. 2007; Thomas et al. 2008) (Fig. 5.1e). Consequently, the quality of the environment during development may affect an individual's health and well-being across their life span (Glover 2011; Flinn et al. 2011; Nepomnaschy and Flinn 2009; Swain et al. 2007; Talge et al. 2007). Importantly, the effects of those environmental exposures on ontogenesis appear to be more intense and extensive earlier during development that the exposure takes place (Van den Bergh and Marcoen 2004; Weinstock 2008; Laplante et al. 2004). This inverse relationship between the timing of an exposure and the intensity and breadth of its effects may potentially be explained by the mediating mechanisms involved.

The underlying mechanisms through which environmental challenges "get under the skin" and influence physical and mental health and behavior are the subject of much scientific scrutiny. Epigenetic processes have emerged as important candidate

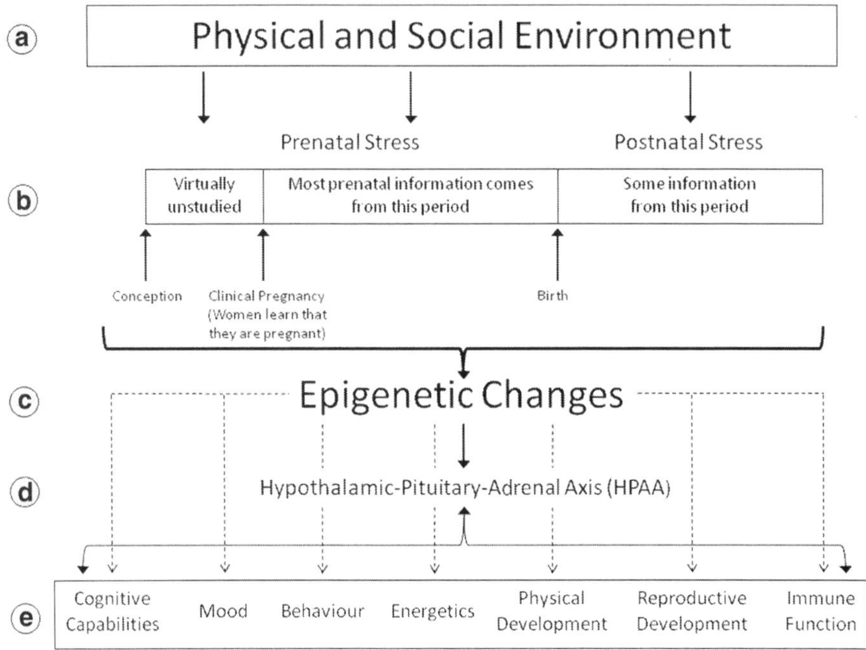

Fig. 5.1 The "great gap" in understanding the ontogeny of the human child

mechanisms. The expression of every gene depends on the individual's epigenome. The epigenome consists of methylations, histone acetylations, and other chemical "marks" that influence gene expression by affecting the extent to which specific sequences of DNA are accessible for transcription. The differential transcription of particular genes and the translation of their encoded proteins ultimately contribute to the observable traits in an individual (phenotype). DNA methylation, the addition of a methyl group to a cytosine base, is considered to be one of the more important and representative epigenetic modifications. Environmental challenges can affect the enzymatic pathways that regulate DNA methylation patterns, which can lead to specific changes in endocrine and metabolic regulation. These changes can in turn modify a wide range of critical outcomes, including growth trajectory, energy metabolism, reactivity, developmental pace, cancer risk, behavioral profiles, long-term memory capabilities, and neurological function (Fig. 5.1).

Although not without cost, these regulatory epigenetic switches are expected to result in phenotypic modifications that aid an organism to survive environmental challenges (West-Eberhard 2003). For example, food availability and crowding determine development of wingless grasshopper Phaulacridium vittatum nymphs into two distinct adult morphs. Scarce food and overcrowding result in a migratory, winged morph, whereas the opposite conditions result in the strikingly different wingless morph. Each morph appears well suited to their respective environmental challenges. Plants also exhibit strategic modifications of phenotype in response to cues of future environmental conditions such as light availability and herbivore density. For example, thale cress (Arabidopsis thaliana) increases production of chemicals that deter caterpillars and other insects in response to loss of foliage (van Hulten et al. 2006; for general review, see Agrawal 2001, 2007).

Neurological processes provide for more rapid temporal adjustments of phenotype. For example, a young rat rapidly elevates pulse rate when it sees a cat to prepare for the energetic demands of a potential chase. Frequent exposure to cats results in more permanent changes to parts of the rat's brain—in particular the amygdala—that enhances predator anxiety and wariness in the future (Ademec et al. 2005; see also Amaral 2003; Sabatini et al. 2007). Human studies focused on epigenetic modifications are still scarce. Essex and colleagues recently published the results of a retrospective study where they report to have found an association between DNA methylation patterns in adolescents and adversity levels during infancy and preschool. These effects appear to depend on whether the stressor was maternal or paternal and were modulated by the individual's gender (Essex et al. 2012). Importantly, the ultimate effects of life events on the stress response system and their reversibility seem to be linked to the specific timing of their occurrence during ontogeny.

The general objective of developmental plasticity is to modify the phenotype so as to better meet future challenges. The key problem is predictability. How reliable are the cues that are used to assess future contingencies? The difficulty of prediction increases with distances in time and space. Dark clouds, thunder, and lightning are good indicators of oncoming rain in the immediate future. Cues for next week's or next year's weather conditions are less certain. Nonetheless we build roofs on our houses to protect us from future rains. Early preparation allows for more specialization and economy of development. The sooner one can adjust development to fit future environments, the better; in this way, resources need not be wasted covering other options. The balance between predictability and specialization during development influences the ontogenetic trajectories of phenotypic plasticity. "Critical periods" for environmental input involve this inherent temporal trade-off between the reliability of cues and the advantages of earlier specialization.

5.3 What Is Special About the Fetal and Infant Environments for Ontogeny?

The fetus has a unique source of information upon which to guide development of its phenotype: via the placenta, it can monitor its mother's neuroendocrine systems, including levels of cortisol, oxytocin, epinephrine, and other bioactive molecules. This hormonal data can provide cues to maternal condition and how the mother responds to different environmental stimuli, invaluable knowledge for decisions about ontogenetic trajectories. After birth the offspring loses this direct link to maternal neuroendocrine response and must instead rely upon breast milk and external cues—such as behavior and olfactory signals—to monitor levels of maternal hormones such as cortisol.

Fetal and infant development, therefore, present somewhat of a paradox: on the one hand, it is very difficult to predict what conditions will be like in 10 or 20 years. On the other hand, this period offers a wealth of information about the mother's internal states that can be used to adaptively modify ontogenetic trajectories. In preparation for his or her future postnatal environment, it may be advantageous for the fetus to adjust the baseline functioning of its own HPA axis to that of its mother. Following a similar logic, sudden alterations to the mother's HPA baseline functioning (acute stress) could also indicate relevant changes in the conditions to be faced postnatally, and the fetus should benefit from adjusting its stress response to those changes as well. A simple example of this type of scenario could concern the loss of a contributing partner/father taking place during gestation. Such an event could trigger modifications in the mother's HPA functioning and also affect the prenatal and postnatal environments of development for the fetus. Neurophysiological changes that help the developing fetus to survive a fatherless gestation, first, and a fatherless childhood, second, should be positively selected. Whether the resulting adult phenotype enjoys his or her life or whether peers deem the individual a carrier of one pathology or another is irrelevant to the process of natural selection. Some of the undesirable outcomes associated with stress may represent the unavoidable costs of adaptations that allowed the individual to survive exogenous challenges at some point during development or to be better adapted for the current environment. Labeling all stress outcomes as pathologies ignores the adaptive role of stress function, reduces our ability to achieve a complete understanding of the role stress plays in the unusual life history and ontogeny of the human fetus and child, and fails to help us curtail environments that lead to those undesirable outcomes.

5.4 Mother–Infant Synchrony

One interesting aspect of infant–mother relations involves synchrony or entrainment (Feldman et al. 2011, 2012). Some mother–infant pairs have activity and hormone levels that are highly correlated (Fig. 5.2a). Other pairs are more discordant (Fig. 5.2b).

Disruptions in child development as identified by delays in the Denver developmental assessment are associated with synchrony of cortisol response in mother–infant pairs (Fig. 5.3) and with reported and observed attachment behavior (Fig. 5.4).

The reasons why synchrony of cortisol levels between mother and infant is associated with positive outcomes for child development are uncertain. We speculate, based largely on 23 years of ethnographic study of this community, that mothers with fewer constraints on child care (such as separation from infant due to work and other outside demands) are more likely to be engaged in mutual activities (e.g., sleeping together, eating together, regular breastfeeding schedule, and playing together) and therefore have more similar cortisol levels. Additionally, as cortisol levels are lower in infants than in adults and there is much variation in the timing at which the circadian pattern of cortisol secretion is established during infancy (Hucklebridge et al. 2005; Laakso et al. 1994; Shimada et al. 1995; Touitou and Haus 2000; Touitou et al. 1983), it is possible that the observed synchrony between

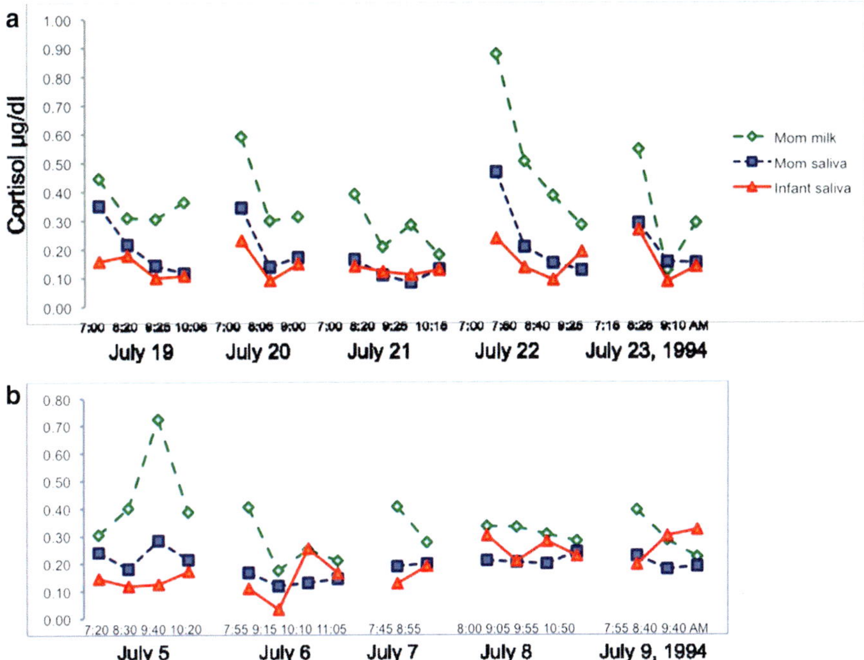

Fig. 5.2 Synchrony of mothers and their 8-month-old infants cortisol levels. Case #5 (**a**) exhibits moderate relation between a mother's salivary and breast milk cortisol levels and the salivary cortisol levels of her infant ($r=0.743$ for saliva). Case #12 (**b**) ($r=-0.182$ for saliva) exhibits lower levels of relation (Data from Flinn et al. 2011. Biological samples collected by Mark Turner)

mothers and their nursing infants is just reflecting the mothers' circadian profiles as maternal cortisol is transmitted through mothers' milk (Brummelte et al. 2010). Hence the synchrony relation may be an incidental (noncausal) marker of the intensity and quality of parental care.

The information that is conveyed by neuroendocrine response between mothers and infants is an understudied research area. The shift from the enormous amount of neuroendocrine information available in utero to postpartum may represent an important life history transition. At a minimum the postpartum transition presents new communicative challenges for both mothers and infants.

5.5 Why Is the Human Child so Sensitive to the Social Environment?

Human behavioral plasticity is unusual in its potential for novelty and for cumulative directional changes occurring over lifetimes and multiple generations. Natural selection equipped humans with rather special abilities to adjust and respond to the current environment and, in some circumstances, to struggle to improve upon the current

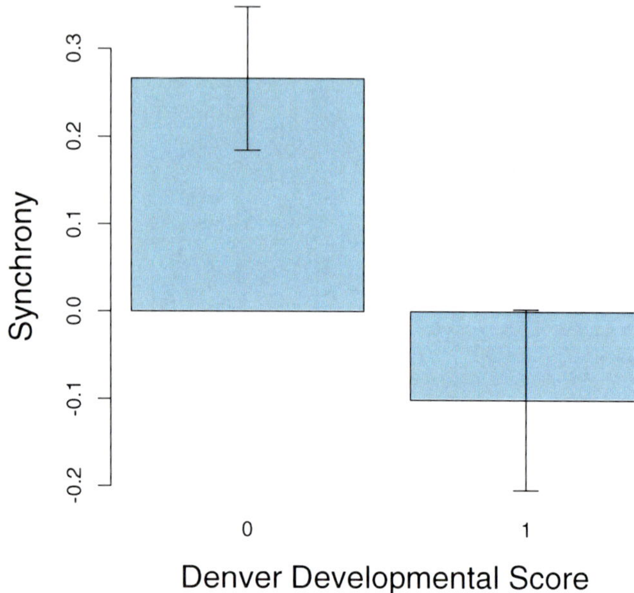

Fig. 5.3 Mother–infant cortisol synchrony (r values as in Fig. 5.1) and Denver developmental scores (0 = no risks or delays; 1 = one or more risks or delays). $N=22$ and 9, respectively

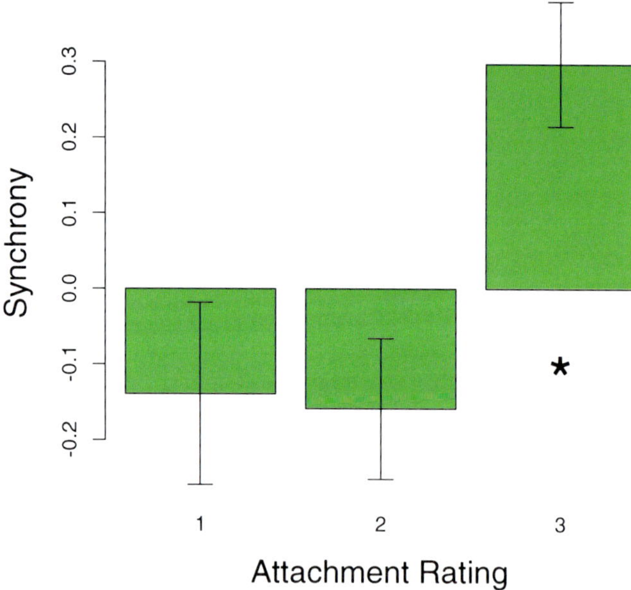

Fig. 5.4 Mother–infant cortisol synchrony (r values as in Fig. 5.1) and attachment (physical and emotional closeness: composite measure of reported and observed affection and care, including shared attention, nuzzling, and response to distress). $N=6$, 12, and 13, respectively. Mother–infant pairs with attachment rating 3 had higher synchrony than those with attachment ratings of 1 or 2

strategies, that is, to be creative, to develop a way of doing something that is different, and potentially better, than the way that it is currently being done by others.

Novelty is a risky business in the evolutionary game. Mutation is organic evolution's source of variability, new genetic information that may alter the phenotype in beneficial or, much more likely, deleterious ways. Hence organisms have evolved mechanisms to protect and correct their DNA if it has been damaged. Changes, especially random ones, to a machine/organism that have been finely honed over millions of generations are not a good bet. And yet the human mind does so regularly and persistently with the information bits that it processes. It seems to have a system of filters and checks that improve the odds of hitting upon a good idea, a better mousetrap.

One cognitive area in which humans are truly extraordinary is social relationships. Humans are able to mentally represent the feelings and thoughts of others. Humans have unusually well-developed mechanisms for theory of mind (Amodio and Frith 2006; Gallese et al. 2004; Leslie et al. 2004) and associated specific pathologies in this domain (Baron-Cohen 1995; Gilbert 2001, 2005). We have exceptional linguistic abilities for transferring information from one brain to another (Pinker 1994), enabling complex social learning. Social and linguistic competencies are roughly equivalent in both males and females, although human mothers appear to have especially important roles in the development of their offspring's socio-cognitive development (Deater-Deckard et al. 2004; Simons et al. 2001). In apparent contrast with chimpanzees and gorillas, human females have substantial social influence or power, based not only on modeling a behavior but on the use of information transmitted via language (e.g., Hess and Hagen 2006).

The human child is an extraordinarily social creature, motivated by and highly sensitive to interpersonal relationships (Gopnik et al. 1999). The life history stage of human childhood enables the development of necessary social skills (Muehlenbein and Flinn 2011), including emotional regulation. Learning, practice, and experience are imperative for social success. The information-processing capacity used in human social competition and cooperation is considerable and perhaps significantly greater than that involved with foraging skills (Roth and Dicke 2005). An extended human childhood may be attributed to the selection for development and necessity of a social brain that requires a lengthy ontogeny to master complex dynamic tasks such as learning the personalities, social biases, and so forth of peers and adults in the local community and developing appropriate emotional responses to these challenges (Battaglia et al. 2004; Bugental 2000). The learning environments that facilitate and channel these astonishing aspects of human mental phenotypic plasticity appear to take on a special importance.

Parents and other kin may be especially important for the child's mental development of social and cultural maps due to their level of emotional involvement/attachment and because they may be relied upon as landmarks who provide relatively honest information. From this perspective, the evolutionary significance of the human family in regard to child development is viewed as a nest from which social skills may be acquired and emotional regulation developed (Flinn et al. 2005a, b, 2007; Flinn and Coe 2007), in addition to its importance as an economic unit centered on the sexual division of labor. The links among psychosocial stimuli, emotions, and physiological stress response may guide both the acute and long-term neurological plasticity necessary for adapting to the dynamic aspects of human sociality. Adjustments to developmental trajectories begin with the zygote's adaptation to the maternal uterine environment and continue throughout pregnancy, infancy, and childhood. The family appears to play a key role in this adaptive process.

5.6 Social Brain and the Human Family

One of the most remarkable developments in the evolution of life has been the dramatic growth of brain size and complexity found among primates over the past several million years. The adaptive advantage in this spurt in brain growth may be related to the complex and highly integrated social lives found among primates (Alexander 1990; Byrne and Suomi 2002; Flinn and Ward 2005). The size of the neocortex relative to the rest of the brain is correlated with the size of the social group in which a particular primate species lives (Dunbar 1997, 2007; Dunbar and Schultz 2007). The primary selective pressures that led to the rapid evolution of the human brain may have involved competition in the social world of our hominid ancestors and the cognitive skills required to successfully manage the kin- and reciprocity-based coalitions (Alexander 1990; Flinn and Alexander 2007).

The selection for brain size and complexity found among primates also occurred in tandem with the selection for prolonged parental care (Allman 1999; Geary and Flinn 2001; Kaplan et al. 2003; Flinn et al. 2005a, b). The coevolution of the brain and the family permitted both a protective environment in which brain growth after birth could occur and a structure in which information beyond that encoded in DNA could be transmitted over the generations. A lengthened period of parental care also allowed the developing brain to adapt to a more complex social environment, with the family providing a social realm in which a child could learn the competitive and cooperative relationship skills needed to successfully navigate in the larger world (Flinn 2004; Kaplan et al. 2003).

The prolonged period of protective development necessary for the growth of the human brain, then, entails more than the mother–child dyad. In Evolving Brains (1999) Allman states, "Without the extended family, big brains would not have evolved in hominids. … The human evolutionary success story depends on two great buffers against misfortune, large brains and extended families, with each supporting and enhancing the adaptive value of the other" (pp. 202–203; see also Hrdy 2005; Walker et al. 2010, 2011).

The brain/family coevolution also entailed the selection of the powerful neuroendocrine mechanisms required for the attachments involved in human mating, parenting, and extended family bonds (Carter 2005; Flinn 2011; Panksepp 1998). For not only were highly sophisticated cognitive skills required to observe, remember, analyze, and respond to complex social scenarios, the emotional forces forging intricate and enduring bonds among family and larger social groups were also required. Although the family provided the protective environment in which a complex brain can develop, the attachments entailed also exerted pressures on family members to which they must have had to continuously respond. The strong bonds, required for prolonged pair bonding, parental investment, and connections in the larger family and social systems, can generate intense emotional states. Since such bonds are vital to individual and family well-being, disturbances in important relationships can be perceived as threatening and activate the fear or stress response. The evolution of the "social mind" required the integration of increasingly complex cognitive and emotional functions, allowing the individual not only to attend to the external social world but to regulate the intense emotional reactions stimulated by events in the relationship environment.

5.7 Parental Care and Stress Reactivity

One aspect of how well individuals respond and adapt to challenges over the course of their lives is related to the effectiveness of their stress response systems. This entails how accurately they are able to perceive social stressors as well as how effectively they respond to them (Del Giudice 2012; Huether 1996, 1998; McEwen and Seeman 1999). Over the past several decades significant knowledge has been gained about the mammalian neuroendocrine stress response system and the influence parental care has on its development.

Much research has focused on the limbic–hypothalamic–pituitary–adrenal (L-HPA) system and the release of "stress hormones" into the bloodstream, which help to regulate the mind–body's response to challenge. The stress response is designed to release and channel energy to allow individuals to adapt to changing conditions and threats. This mobilization is aimed at facilitating cognitive, emotional, physiological, and behavioral responses. Those responses are mediated in part by the release of the catecholamines (epinephrine and norepinephrine) and the stress hormones corticotropin-releasing hormone (CRH) and cortisol. The "stress hormones" help to coordinate a system-wide response, involving the brain, cardiovascular, immune, digestive, and reproductive systems, in order to respond to the challenge or threat at hand. This automatic response is vital to survival. However, while the short-term elevation of cortisol in the bloodstream results in adaptive responses chronic elevations can lead to impairment in one or more of biological systems (Huether 1996, 1998; LeDoux 1996; Lupien and McEwen 1997; McEwen 1995, 1998; Sapolsky 1992). Prolonged heightened L-HPA responses to stress have been found to be associated with autoimmune and cardiovascular illnesses as well as anxiety, depressive, and addictive disorders (Heim and Nemeroff 2001; McEwen 1998; McEwen et al. 1997).

Individuals differ in how they respond to stress including their behavioral, emotional, and physiological resilience in the face of life pressures. The L-HPA system of individuals is not equally influenced by all stressors, but primarily by those that are seen as uncontrolled and as having a social-evaluative aspect (Dickerson and Kemeny 2004). Entailed in the phenotypic plasticity of the human are the significant and enduring effects the parental–offspring relationship has on the offspring's responsiveness to stress throughout life. Animal studies have demonstrated that subjecting one generation to uncontrollable stress results in observable behavioral and physiological changes in their offspring.

5.7.1 Prenatal Stage

The parental influence on the development of an offspring's physical and behavioral responsiveness to stress begins even prior to birth. The stressing of pregnant mammals can result in changes among offspring that persist into adulthood. Among rodents such stressing has been found to influence their offspring's neurotransmitter functioning (Fride and Weinstock 1988; Moyer et al. 1978; Takahashi et al. 1992); opiate and benzodiazepine receptors (Insel et al. 1990; Fride et al. 1985); maternal and sexual behavior (Champagne and Meaney 2006; Fride et al. 1985; Kinsley and Bridges 1988); exploratory, cognitive, and aggressive behaviors (Grimm and Frieder 1987; Kinsley and Svare 1986); and reactivity to stressful situations (D'Amato et al. 1988; Pollard 1984; Takahashi 1992). In a review, O'Regan et al. (2001) cite further findings of prenatal exposure to exogenous or endogenous stress hormones associated with hypertension, hyperglycemia, hyperinsulinemia, as well as altered behavior and neuroendocrine responses into adulthood. Cross-fostering and adoption studies have demonstrated that some of the effects of prenatal stressing can be moderated during postnatal development (Maccari et al. 1995; Weinstock 1997).

Similar findings have been found in studies with nonhuman primates (Clarke et al. 1994; Clarke and Schneider 1993; Schneider 1992; Schneider and Coe 1993). Though most of the primate studies focused on the behavioral and physiological functioning of infants, Clarke and Schneider (1993, 1994) studied the long-term effects of prenatal stressing of rhesus monkeys and found that the offspring continued to demonstrate more anxious social behavior and more elevated ACTH and cortisol levels when stressed than did controls.

Human studies are consistent with the prenatal stress findings of the nonhuman animal studies. Glover and O'Connor (2002), for example, reported in an epidemiological study that a strong relationship existed between maternal anxiety during the third trimester and behavioral and emotional problems in those children at age 4. In a study of mothers who had been exposed to the World Trade Center collapse while pregnant, Yehuda et al. (2005) found that both the mothers who developed PTSD and their infants at age 1 had reduced cortisol levels as compared to the mothers who did not develop PTSD and their infants. Reduced cortisol levels had previously been found among individuals with PTSD.

5.7.2 Postnatal Stage

The study of the early postnatal period has also demonstrated that the parent–offspring relationship can have a lifelong influence in regulating an offspring's responsiveness to stress. In both human and nonhuman studies, the disruption of the maternal–offspring relationship as well as overprotective, restrictive maternal behavior have been found to be associated with long-term cognitive, physiological, and behavioral effects on offspring (Anderson et al. 1999; Byrne and Suomi 2002; Essex et al. 2002; Fairbanks 1989; Fairbanks and McGuire 1988; Francis and Meaney 2002; Mirescu et al. 2004; Suomi 2002; Suomi et al. 1983).

The influence of the early parental relationship on an offspring's later responsiveness to stress appears to be shaped by the actual "programming" of the neuroendocrine stress response system. Investigators have demonstrated that variation in particular maternal behaviors (licking and grooming, arched-backed nursing) among rodents results in stable individual differences in the neural systems that mediate fearfulness in their offspring which persist into adulthood (Caldji et al. 1998; Francis et al. 1999; Liu et al. 1997). More specifically, higher or lower levels of these maternal behaviors influence gene expression of the stress hormone, CRH, in the hypothalamus and amygdala, which are involved in activating the HPA stress response system. These maternal behaviors also influence gene expression for glucocorticoid (cortisol in primates; corticosterone in rodents) receptors in the hippocampus, which is instrumental in down regulating the HPA stress response. Maternal behaviors, then, can shape individual differences among offspring in their reactivity to stress. Elegant cross-fostering methods determined that stress reactivity was influenced by nursing mothers, but not by biological mothers (Caldji et al. 1998; Francis et al. 1999; Liu et al. 1997).

Human studies also suggest that the early environment and parental care influence individual differences in the stress reactivity of children. Most of the studies have been retrospective and usually investigate the impact of either early life stressors and trauma or maternal depression. Several studies, for example, have found that early trauma or loss is associated with increased stress reactivity (Heim et al. 2002; Kaufman et al. 2000; Nicolson 2004). The experience of early life stress or maternal stress during childhood is related to the later development of depression or anxiety disorders (Brunson et al. 2001; Grossman et al. 2003; Heim and Nemeroff 2001), both of which may be related to dysregulation of the stress response system. Further, there is some evidence that the stress reactivity of children is influenced by depression in their mothers (Ashman et al. 2002; Harkness and Monroe 2002; Lupien et al. 2000). Maternal depression in a child's first 2 years of life, for example, was the best predictor of elevated cortisol levels at age 7 (Ashman et al. 2002).

Given the lengthy period of development and plasticity of the human brain (Sapolsky 2003), a number of studies highlight the importance of long-term developmental studies of children in order to determine the influence of early adverse experiences on the later functioning of the L-HPA stress response system. Gunnar and Donzella (2002) found that a child's L-HPA stress response system is to some degree regulated by parental care, thus mediating the influence of stressful events. In a longitudinal study, Flinn (2006a, 2009) found that the cortisol reactivity of children to stressful events was influenced by the family environment. Another longitudinal study (Essex et al. 2002) found higher cortisol levels among children exposed both during infancy and at 4.5 years, but not among children exposed only in infancy or at 4.5 years of age. Socioeconomic status and maternal depression were associated with cortisol reactivity in a study of Quebec children, but SES effects emerged over time (Lupien et al. 2000). These studies suggest that while there are sensitive periods in development, during which the stress response system is more open to influence, in humans such periods may occur over a significant part of development.

The link between parental care and stress axis function is likely to be a primitive one. McGowan and colleagues compared methylation patterns between suicide victims with and without a history of childhood abuse and found that the postmortem hippocampus obtained from those who suffered abuse presented lower levels of glucocorticoid receptor mRNA and its transcripts (McGowan et al. 2009). These results are consistent with similar findings in nonhuman models. An ancient origin for the role of parental care on the epigenetic regulation of the functioning of the L-HPA stress response system, together with the obvious costs that some of the resulting behaviors have for the carriers in some contexts, strengthens the idea that under dire circumstances, traits such as anxiety and depression should provide a selective advantage.

The critical question is, what are those circumstances? And why can early programming of the L-HPA stress response system be reversed by later experience (Francis et al. 2002)? To answer those questions, it is necessary to have thorough understanding of the complexity of human society and its social units—communities, groups, and especially the human family—in which the mother–child relationship is embedded.

5.8 Family Influence on Children's Stress Reactivity

There is, then, a significant body of research which demonstrates that the quality of maternal care can influence the development of individual differences in stress reactivity among offspring and, as a result, their health and behavior (Caldji et al. 1998; Essex et al. 2002; Fairbanks and McGuire 1988; Flinn 1999, 2006b; Suomi 2003). The evidence suggests such differences occur due in part to the influence maternal care can have on the development of an offspring's L-HPA and CRH stress response systems (Brunson et al. 2001; Francis and Meaney 2002; Heim and Nemeroff 2001; Ladd et al. 2000; McEwen and Magarinos 2001; Weaver et al. 2004).

Most studies on the interaction between maternal behavior and stress reactivity have focused on one or more of four areas: (1) maternal separation or early trauma (Anderson et al. 1999; Byrne and Suomi 2002; Heim and Nemeroff 2001; Huot et al. 2004; Kaufman et al. 2000; Mirescu et al. 2004), (2) maternal behavior (Ashman et al. 2002; Caldji et al. 1998; Fairbanks and McGuire 1988; Fleming et al. 1999; Halligan et al. 2004; Kraemer 1992), (3) maternal social and physical environment (Clarke and Schneider 1993; Coplan et al. 1998; Flinn 1999; Lupien et al. 2000; O'Connor et al. 2001), and (4) influence of the previous generation on parental care (Berman 1996; Champagne and Meaney 2001, 2006; Fairbanks 1989; Fleming 2005; Francis et al. 1999).

While the maternal–offspring dyad has been the principle focus in most of these studies, the latter two areas suggest that the broader relationship environment, particularly that of the family, plays a vital role in the development of an individual's responsiveness to stress. This broader relationship environment includes not only fathers and siblings but the extended families as well. From this perspective, the mother–child dyad represents a link in a larger, highly integrated family system, which is seen as playing a significant role in the shaping and regulation of the individual's neuroendocrine stress response system (Noone 2008).

The mother–child relationship is embedded in and influenced by the larger family relationship system that can serve as a mediator of environmental stress as well as a potential source of stress for its members (Carter 2005; Flinn 2006a; Lupien et al. 2000; Nicolson 2004; Uvnas-Moberg 1998). The human brain is highly sensitized to relationships and social behavior (Adolphs 2003), and this sensitivity is strongly influenced by the family relationship network.

5.9 Multigenerational Transmission of Reactivity to Stress

In addition to the influence of family interactions on maternal care and the development of an offspring's stress reactivity, there is evidence that the maternal caretaker's experience of being parented influences her own maternal care and the stress reactivity of her offspring. The previous generation, then, can be seen to influence the development of individual differences in stress reactivity. Rodent (Champagne and Meaney 2001; Francis and Meaney 2002), nonhuman primate (Fairbanks 1989; Suomi 2002), and human (Fleming 2005) studies provide evidence that maternal care can influence not only the stress reactivity of offspring but the quality of maternal care provided in the next generation as well. Maternal care is seen as influencing the development of L-HPA stress reactivity, which in turn affects neural systems involved in parental care.

A series of studies indicate that parental care influencing individual differences in the stress response systems of offspring can be transmitted nongenomically over the generations (Caldji et al. 1998; Francis et al. 1999). Champagne and Meaney (2001) state, "These findings suggest that for neurobiologists, the function of the family is an important level of analysis and the critical question is that of how environmental events regulate neural systems that mediate the expression of parental care."

The plasticity of the neural systems involved in stress reactivity and the evidence for epigenetic programming of maternal behaviors has been discussed in a number of recent studies (Champagne 2008; Francis et al. 1999, 2002; Lupien et al. 2000; Weaver et al. 2004; Zhang and Meaney 2010). Fleming (2005) describes a number of experiential factors during childhood and adolescence, such as alloparenting and maternal deprivation that may alter psychobiological mechanisms mediating parental behavior. Family instability prior to age 12, for example, leads to higher cortisol levels and long-lasting effects on maternal behavior (Krpan et al. 2005). While much remains to be learned, the existing animal and human research provide evidence that the development of the neuroendocrine stress response systems in mammals is shaped by the parent–offspring relationship and that this influence can continue to have an effect in the next generation.

It has been suggested that the intergenerational regulation of parental care and stress reactivity may be adaptive (Champagne and Meaney 2001; Zhang et al. 2006). A phenomenon known as maternal or parental effects may evolve when there is some predictability between the parental and offspring environments. "Adaptive trans generational phenotypic plasticity" is observed when the maternal phenotype response to the environment stimulates phenotypic plasticity in offspring (Mousseau and Fox 1998). For example, parental reactivity to a threatening environment might stimulate offspring to respond with greater wariness and thereby enhance adaptiveness in a threatening environment. From this perspective, adverse environmental conditions may result in maternal behavior that "programs" gene expression for neuroendocrine systems more responsive to such conditions (Zhang et al. 2006).

This does not, however, account for within family variability of stress reactivity or the range of stress reactivity found in both highly stable and unstable environmental conditions. Increased stress reactivity increases vulnerability to stress-related illnesses, and higher levels of reactivity are associated with a decrease in the more recently evolved cognitive capacity to regulate social and emotional responsiveness to the environment. If prolonged, heightened reactivity of the L-HPA system can impair cognitive and other systems central to survival (McEwen 1998; McEwen and Seeman 1999) and so the adaptive value may be limited.

Another view is that while such programming may be detrimental to one or more offspring, it may have adaptive value to the other siblings and the family. While one child is likely to be the object of anxious parental involvement, the other children will be less involved, less constrained, and as a result more adaptive or resilient in their lives. More intense involvement with one child can take the form of either a positive or anxious emotional responsiveness, but each is seen as leading to the child's heightened responsiveness to other relationships. The containment of much of the parental anxiety in one relationship is seen as having a buffering effect for the other children. Among the factors observed to contribute to such positional effects on the functioning of children are birth order, nuclear family behavioral patterns, and family stressors occurring during pregnancy, early development, and adolescence (Kerr and Bowen 1988).

Differential parental involvement and other family interactions lead to individual differences in responsiveness to stress; such differences can influence mate selection and be transmitted into the next generation, hence affecting family interactional patterns resulting in a wide range of individual differences in functioning over the generations. From this perspective the "trans generational phenotypic plasticity" found in the human family would enhance the adaptiveness of some offspring, but be maladaptive for others.

Given the plasticity of the human brain, its extended development, and the complexity of the family environment, the degree to which individual differences in stress reactivity in the human population are shaped and transmitted by the family will require the long-term study of families over several generations.

As Fleming (2005) has remarked, "the role of epigenetics in the regulation of parental behavior and of the effects of experience on individual differences in development is totally uncharted territory in nonhuman mammals or humans" (2005, p. 162).

5.10 Family and Stress Reactivity Among Children in a Caribbean Village

One of the few studies that have included family relationships as factors in maternal care and stress reactivity is a 23-year longitudinal study of children, their families, and health in a rural village on the Caribbean island of Dominica (Flinn 2006a; Flinn et al. 2005a, b). This study of stress reactivity and health includes over 300 children and their families and has entailed the collection of extensive family genealogies, medical histories, growth measures, household compositions, and detailed observations of behavior and daily routines. Utilizing noninvasive saliva immunoassay techniques, the physiological stress responses of children in the naturalistic settings of their families were obtained. Saliva is relatively easy to collect and store. Concomitant monitoring of a child's daily activities, stress hormones, and psychological conditions provides a powerful research design for investigating the effects of naturally occurring psychosocial events in the family environment.

In this study community, family household composition is related to the stress hormone levels of children (Flinn and England 1997). Children who lived with both parents or with single mothers who had grandparental or other kin support had lower mean cortisol levels than those children who lived with single parents without kin support or who lived in the households of more distant relatives or stepparents with half-siblings.

High-stress events (cortisol increases from 100 to 2,000%) were most commonly found to involve trauma from family conflict or change (Flinn and England 2003; Flinn et al. 1996). Punishment, quarreling, and residence change substantially increased cortisol levels, whereas calm affectionate contact was associated with diminished (−10 to−50%) cortisol levels. Of all cortisol values that were more than two standard deviations above mean levels (i.e., indicative of substantial stress), 19.2% were temporally associated with traumatic family events (residence change of child or parent/caretaker, punishment, "shame," serious quarreling, and/or fighting) within a 24-h period. Of all recorded traumatic family events, 42.1% were temporally associated with substantially elevated cortisol (i.e., at least one of the saliva samples collected within 24 h was >2 S.D. above mean levels).

It is important to note that there was considerable variability among children in cortisol response to family disturbances. Not all individuals had detectable changes in cortisol levels associated with family trauma. Some children had significantly elevated cortisol levels during some episodes of family trauma but not during others. Cortisol response is not a simple or uniform phenomenon. Numerous factors, including preceding events, habituation, specific individual histories, context, and temperament, might affect how children respond to particular situations.

Nonetheless, traumatic family events were associated with elevated cortisol levels for all ages of children more than any other factor examined. These results suggest that family interactions were a critical psychosocial stressor in most children's lives, although the sample collection during periods of intense family interaction (early morning and late afternoon) may have exaggerated this association.

Although elevated cortisol levels are associated with traumatic events such as family conflict, long-term stress may result in diminished cortisol response. In some cases chronically stressed children had blunted response to physical activities that normally evoked cortisol elevation. Comparison of cortisol levels during "non-stressful" periods (no reported or observed: crying, punishment, anxiety, residence change, family conflict, or health problem during 24-h period before saliva collection) indicates a striking reduction and, in many cases, reversal of the family environment–stress association Flinn 2007. Chronically stressed children sometimes had subnormal cortisol levels when they were not in stressful situations. For example, cortisol levels immediately after school (walking home from school) and during noncompetitive play were lower among some chronically stressed children (cf. Long et al. 1993). Some chronically stressed children appeared socially "tough" or withdrawn and exhibited little or no arousal to the novelty of the first few days of the saliva collection procedure.

Relations between family environment and cortisol stress response appear to result from a combination of factors. These include frequency of traumatic events, frequency of positive "affectionate" interactions, frequency of negative interactions such as irrational punishment, frequency of residence change, security of "attachment," development of coping abilities, and availability or intensity of caretaking attention. Probably the most important correlate of household composition that affects childhood stress is maternal care.

Mothers in socially "secure" households (i.e., permanent amiable co-residence with mate and/or other kin) appeared more able and more motivated to provide physical, social, and psychological care for their children (Fig. 5.4). Mothers without mate or kin support were likely to exert effort attracting potential mates and may have viewed dependent children as impediments to this. The mothers in such stable households may also be less socially isolated and thus less anxious. Hence co-residence of father may provide not only direct benefits from paternal care, but also affect maternal care (Konner 2010). Young mothers without mate support usually relied extensively upon their parents or other kin for help with child care (Quinlan and Flinn 2005; Quinlan et al. 2005).

While children who experienced early trauma prior to age 6 had significantly more elevated cortisol levels at age 10 than those children who did not, this did not hold true for those children whose mothers had high levels of maternal support (Flinn 2006a). Children experiencing early trauma whose mothers had social support did not differ significantly from children who had not experienced early trauma. The stress reactivity of children was also found to be mediated by significant contact with caretakers other than their mothers. Children with extensive alloparental care, especially by grandparents, were found to recover normal HPA function more quickly following social trauma than did those children who had fewer caretakers (Flinn and Leone 2006, 2009).

5.11 Discussion

The human brain represents one of the most extraordinary developments in the evolution of life on earth. It is extraordinary in terms of its complexity and rapid evolution. The brain and family coevolved and are intimately interconnected as the latter provides the protective environment required for the prolonged development of the brain. The prolonged postnatal development allows for the learning required to respond to and navigate through the complexity of the human social environment. A central element in the human's responsiveness to the social environment is the L-HPA stress response system, which heightens an individual's cognitive, emotional, and physiological responses to the challenges at hand. It plays a vital role in an individual's capacity to adapt throughout life.

The family not only provides a safe haven in which the infant and child's brain can observe and learn to respond to a changing social environment, it also plays a significant role in shaping this responsiveness. The neuroendocrine stress response systems, along with other neural systems, are shaped by the family relationship environment during both the prenatal and postnatal development. The study of the influence of the family on the developing brain and individual differences in responding to the complex social world of the human will continue to be enhanced by both animal and human studies. The study of mother–child dyadic relationships will also contribute to the knowledge of the family and individual development, but prospective, longitudinal studies of the larger family network in naturalistic settings will be required to observe the rich and complex role the larger family plays in mother–child relationships and individual development. A family system and an evolutionary theoretical framework will play a vital role in both the integration of emerging knowledge and the development of researchable questions in the growing field of social neuroscience.

Acknowledgements The authors thank the editors and reviewers for helpful advice. We also thank Dr. Katrina Salvanteand for aid in the literature review and development of the figures. This chapter has been possible, thanks to the financial support provided by the National Science Foundation (BCS-SBE #0640442, SBR-0136023, SBR-9205373, and BNS-8920569 to MVF) and a CIHR IGH Operating Grant (CIHR #106705), a Simon Fraser University (SFU) President's Start-up Grant, and a seed grant from the Human Evolutionary Program at SFU (funded by the Simon Fraser University Community Trust Endowment Fund) to PAN.

References

Ademec RE, Blundell J, Burton P (2005) Neural circuit changes mediating lasting brain and behavioral response to predator stress. Neurosci Biobehav Rev 29:1225–1241

Adolphs R (2003) Cognitive neuroscience of human social behavior. Nat Rev Neurosci 4:165–179

Agrawal AA (2001) Phenotypic plasticity in the interactions and evolution of species. Science 294:321–326

Agrawal AA (2007) Macroevolution of plant defense strategies. Trends Ecol Evol 22(2):103–109. doi:10.1016/j.tree.2006.10.012

Agrawal AA, Laforsch C, Tollrian R (1999) Transgenerational induction of defences in animals and plants. Nature 401(6748):60–63

Alexander RD (1990) How did humans evolve? Special Publication No. 1. Museum of Zoology, The University of Michigan, Ann Arbor

Allman J (1999) Evolving brains. Scientific American Library, New York

Amaral DG (2003) The amygdala, social behavior, and danger detection. Ann N Y Acad Sci 1000:337–347

Amodio DM, Frith CD (2006) Meeting of minds: the medial frontal cortex and social cognition. Nat Rev Neurosci 7(4):268–277

Anderson S, Lyss P, Dumont PN, Teicher M (1999) Enduring neurochemical effects of early maternal separation on limbic structures. Ann N Y Acad Sci 877:756–759

Ashman S, Dawson G, Panagiotides H, Yamada E, Wilkinson C (2002) Stress hormone levels of children of depressed mothers. Dev Psychopathol 14:333–349

Baron-Cohen S (1995) Mindblindness: an essay on autism and theory of mind. MIT Press/Bradford Books, Boston

Battaglia M, Ogliari A, Zanoni A, Villa F, Citterio A, Binaghi F et al (2004) Children's discrimination of expressions of emotion: relationship with indices of social anxiety and shyness. J Am Acad Child Adolesc Psychiatry 43:358–365

Berman CM (1996) An animal model for the intergenerational transmission of maternal style? Fam Syst 3:125–140

Bogin B (1999) Patterns of human growth, 2nd edn. Cambridge University Press, Cambridge

Brand S, Wilhelm FH, Kossowsky J, Holsboer-Trachsler E, Schneider S (2011) Children suffering from separation anxiety disorder (SAD) show increased HPA axis activity compared to healthy controls. J Psychiatr Res 45:452–459

Brummelte S, Galea LA (2010) Depression during pregnancy and postpartum: contribution of stress and ovarian hormones. Prog Neuropsychopharmacol Biol Psychiatry 34:766–776

Brunson K, Sarit Avishai-Eliner S, Hatalski C, Baram T (2001) Neurobiology of the stress response early in life: evolution of a concept and the role of corticotropin releasing hormone. Mol Psychiatry 6:647–656

Bugental DB (2000) Acquisition of the algorithms of social life: a domain-based approach. Psychol Bull 126:187–219

Byrne G, Suomi S (2002) Cortisol reactivity and its relation to homecage behavior and personality ratings in tufted capuchin (Cebus paella) juveniles from birth to six years of age. Psychoneuroendocrinology 27:139–154

Caldji C, Tannenbaum B, Sharma S, Francis D, Plotsky P, Meaney MJ (1998) Maternal care during infancy regulates the development of neural systems mediating the expression of fearfulness in the rat. Neurobiology 95:5335–5340

Carter CS (2005) Biological perspectives on social attachment and bonding. In: Carter CS et al (eds) Attachment and bonding: a new synthesis. The MIT Press, Cambridge

Champagne F (2008) Epigenetic mechanisms and the transgenerational effects of maternal care. Front Neuroendocrinol 29:386–397

Champagne F, Meaney MJ (2001) Like mother, like daughter: evidence for non-genomic transmission of parental behavior and stress responsivity. Prog Brain Res 133:287–302

Champagne F, Meaney MJ (2006) Stress during gestation alters postpartum maternal care and the development of the offspring in a rodent model. Biol Psychiatry 59:1227–1235

Clarke AS, Schneider M (1993) Prenatal stress has long-term effects on behavioral responses to stress in juvenile rhesus monkeys. Dev Psychobiol 26:293–304

Clarke AS, Wittwer D, Abbott D, Schneider M (1994) Long-term effects of prenatal stress on HPA activity in juvenile rhesus monkeys. Dev Psychobiol 27:257–269

Coplan J, Trost R, Owens M, Cooper T, Gorman J, Nemeroff C, Rosenblum L (1998) Cerebrospinal fluid concentrations of somatostatin and biogenic amines in grown primates reared by mothers exposed to manipulated foraging conditions. Arch Gen Psychiatry 55:473–477

D'Amato FR, Ammassari-Teule M, Oliverio A (1988) Prenatal antagonism of stress by naltexone administration: early and long-lasting effects on emotional behaviors in mice. Dev Psychobiol 21:283–292

Deater-Deckard K, Atzaba-Poria N, Pike A (2004) Mother- and father-child mutuality in Anglo and Indian British families: a link with lower externalizing behaviors. J Abnorm Child Psychol 32(6):609–620

Del Giudice M (2012) Fetal programming by maternal stress: insights from a conflict perspective. *Psychoneuroendocrinology* 37:1614–1629

Dickerson SS, Kemeny ME (2004) Acute stressors and cortisol responses: a theoretical integration and synthesis of laboratory research. Psychol Bull 130:355–391

Dunbar RIM (1997) Gossip, grooming, and evolution of language. Harvard University Press, Cambridge, MA

Dunbar RIM, Schultz S (2007) Evolution in the social brain. Science 317:1344–1347

Essex M, Klein M, Cho E, Kalin N (2002) Maternal stress beginning in infancy may sensitize children to later stress exposure: effects on cortisol and behavior. Biol Psychiatry 52:776–784

Essex MJ, Boyce WT, Hertzman C, Lam LL, Armstrong JM, Neumann SMA, Kobor MS (2012) Epigenetic vestiges of early developmental adversity: childhood stress exposure and DNA methylation in adolescence. Child Dev, published online: 1 SEP 2011. doi:10.1111/j.1467-8624.2011.01641

Fairbanks L (1989) Early experience and cross-generational continuity of mother-infant contact in vervet monkeys. Dev Psychobiol 22:669–681

Fairbanks L, McGuire M (1988) Long-term effects of early mothering behavior on responsiveness to the environment in vervet monkeys. Dev Psychobiol 21:711–724

Feldman R, Gordon I, Zagoory-Sharon O (2011) Maternal and paternal plasma, salivary, and urinary oxytocin and parent–infant synchrony: considering stress and affiliation components of human bonding. Dev Sci 14(4):752–761. doi:10.1111/j.1467-7687.2010.01021.x

Feldman R, Zagoory-Sharon O, Weisman O, Schneiderman I, Gordon I, Maoz R, Shalev I, Ebstein RP (2012) Sensitive parenting is associated with plasma oxytocin and polymorphisms in the *OXTR* and *CD38* genes. Biol Psychiatry 72:175–181. doi:10.1016/j.biopsych.2011.12.025

Felitti VJ, Anda RF, Nordenberg D, Williamson DF, Spitz AM et al (1998) Relationship of childhood abuse and household dysfunction to many of the leading causes of death in adults—The Adverse Childhood Experiences (ACE) Study. Am J Prev Med 14:245–258

Fleming A (2005) Plasticity of innate behavior: experiences throughout life affect maternal behavior and its neurobiology. In: Carter CS et al (eds) Attachment and bonding: a new synthesis. The MIT Press, Cambridge

Fleming A, O'Day D, Kraemer GW (1999) Neurobiology of mother-infant interactions: experience and central nervous system plasticity across development and generations. Neurosci Biobehav Rev 23:673–685

Flinn MV (1999) Family environment, stress, and health during childhood. In: Panter-Brick C, Worthman C (eds) Health, hormones, and behavior. Cambridge University Press, Cambridge

Flinn MV (2004) Culture and developmental plasticity: evolution of the social brain. In: MacDonald K, Burgess RL (eds) Evolutionary perspectives on child development. Sage, Thousand Oaks, CA

Flinn MV (2006a) Evolution and ontogeny of stress response to social challenge in the human child. Dev Rev 26:138–174

Flinn MV (2006b) Evolution of stress response to social-evaluative threat. In: Barrett L, Dunbar R (eds) Oxford handbook of evolutionary psychology. Oxford University Press, Oxford

Flinn MV (2006c) Cross-cultural universals and variations: the evolutionary paradox of informational novelty. Psychol Inq 17(2):118–123

Flinn MV (2007) Social relationships and health. In: Trevathan W, Smith EO, McKenna J (eds) Evolutionary medicine & health. Oxford University Press, Oxford

Flinn MV (2011) Evolutionary anthropology of the human family. In: Salmon C, Shackleford T (eds) Oxford handbook of evolutionary family psychology, (chapter 2). Oxford University Press, Oxford, pp 12–32

Flinn MV, Alexander RD (2007) Runaway social selection. In: Gangestad SW, Simpson JA (eds) The evolution of mind. Guilford Press, New York

Flinn MV, Coe MK (2007) The linked red queens of human cognition, reciprocity, and culture. In: Gangestad SW, Simpson JA (eds) The evolution of mind. Guilford Press, New York
Flinn MV, England B (1997) Social economics of childhood glucocorticoid stress response and health. Am J Phys Anthropol 102:33–53
Flinn MV, England BG (2003) Childhood stress: endocrine and immune responses to psychosocial events. In: Wilce JM (ed) Social & cultural lives of immune systems. Routledge Press, London, pp 107–147
Flinn MV, Leone DV (2006) Early trauma and the ontogeny of glucocorticoid stress response in the human child: grandmother as a secure base. J Dev Processes 1(1):31–68
Flinn MV, Leone DV (2009) Alloparental care and the ontogeny of glucocorticoid stress response among stepchildren. In: Bentley G, Mace R (eds) Alloparental care in human societies. Biosocial Society Symposium Series. Berghahn Books, Oxford
Flinn MV, Ward CV (2005) Evolution of the social child. In: Ellis B, Bjorklund D (eds) Origins of the social mind: evolutionary psychology and child development. Guilford Press, London
Flinn MV, Quinlan RJ, Turner MT, Decker SA, England BG (1996) Male–female differences in effects of parental absence on glucocorticoid stress response. Hum Nat 7:125–162
Flinn MV, Geary D, Ward CV (2005a) Ecological dominance, social competition, and coalitionary arms races: why humans evolved extraordinary intelligence. Evol Hum Behav 26:10–46
Flinn MV, Ward CV, Noone R (2005b) Hormones and the human family. In: Buss D (ed) Handbook of evolutionary psychology. Wiley, New York
Flinn MV, Quinlan R, Ward CV, Coe MK (2007) Evolution of the human family: cooperative males, long social childhoods, smart mothers, and extended kin networks. In: Salmon C, Shackelford T (eds) Evolutionary family psychology. Oxford University Press, Oxford
Flinn MV, Nepomnaschy PA, Muehlenbein MP, Ponzi D (2011) Evolutionary functions of early social modulation of hypothalamic-pituitary-adrenal axis development in humans. Neurosci Biobehav Rev 35:1611–1629
Flinn MV, Duncan C, Quinlan RL, Leone DV, Decker SA, Ponzi D (2012a) Hormones in the wild: monitoring the endocrinology of family relationships. Parent Sci Pract 12(2):124–133. doi:10.1080/15295192.2012.683338
Flinn MV, Ponzi D, Muehlenbein MP (2012b) Hormonal mechanisms for regulation of aggression in human coalitions. Hum Nat 22(1):68–88. doi:10.1007/s12110-012-9135-y
Fox S, Levitt P, Nelson CA (2010) How the timing and quality of early experiences influence the development of brain architecture. Child Dev 81:28–40
Francis D, Meaney M (2002) Maternal care and the development of stress responses. In: Cacioppo J et al (eds) Foundations in social neuroscience. The MIT Press, Cambridge, MA
Francis D, Diorio J, Liu D, Meaney MJ (1999) Nongenomic transmission across generations of maternal behavior and stress responses in the rat. Science 286:1155–1158
Francis D, Diorio J, Plotsky P, Meaney M (2002) Environmental enrichment reverses the effects of maternal separation on stress reactivity. J Neurosci 22:7840–7843
Fride E, Weinstock M (1988) Prenatal stress increases anxiety-related behavior and alters cerebral laterization of dopamine activity. Life Sci 42:1059–1065
Fride E, Dan Y, Gavish M, Weinstock M (1985) Prenatal stress impairs maternal behavior in a conflict situation and reduces hippocampal benzodiazepine receptors. Life Sci 36:2103–2109
Gallese V, Keysers C, Rizzolatti G (2004) A unifying view of the basis of social cognition. Trends Cogn Sci 8:396–403
Geary D, Flinn MV (2001) Evolution of human parental behavior and the human family. Parent Pract Sci 1:5–61
Gilbert P (2001) Evolutionary approaches to psychopathology: the role of natural defences. Aust N Z J Psychiatry 35:17–27
Gilbert P (2005) Social mentalities: a biopsychosocial and evolutionary approach to social relationships. In: Baldwin MW (ed) Interpersonal cognition. Guilford, New York, pp 299–333
Glover V (2011) Prenatal stress and the origins of psychopathology: an evolutionary perspective. J Child Psychol Psychiatry 52:356–367

Glover V, O'Connor T (2002) Effects of antenatal stress and anxiety: implications for development and psychiatry. Br J Psychiatry 180:389–391

Gluckman PD, Hanson MA (2006) Developmental origins of health and disease. Cambridge University Press, Cambridge

Gopnik A, Meltzoff AN, Kuhl PK (1999) The scientist in the crib: minds, brains, and how children learn. William Morrow & Co., New York

Grimm V, Frieder B (1987) The effects of mild maternal stress during pregnancy on the behavior of rat pups. Int J Neurosci 35:65–72

Grossman A, Churchill J, McKinney B, Kodish I, Otte S, Greenough W (2003) Experience effects on brain development: possible contributions to psychopathology. J Child Psychol Psychiatry 44:33–63

Gunnar M, Donzella B (2002) Social regulation of the cortisol levels in early human development. Psychoneuroendocrinology 27:199–220

Halligan S, Herbert J, Goodyer I, Murray L (2004) Exposure to postnatal depression predicts elevated cortisol in adolescent offspring. Biol Psychiatry 55:376–381

Hanazato T (1990) Induction of helmet development by a Chaoborus factor in Daphnia ambigua during juvenile stages. J Plankton Res 12:1287–1294

Harkness K, Monroe S (2002) Childhood adversity and the endogenous versus nonendogenous distinction in women with major depression. Am J Psychiatry 159:387–393

Harkonmaki K, Korkeila K, Vahtera J, Kivimaki M, Suominen S et al (2007) Childhood adversities as a predictor of disability retirement. J Epidemiol Community Health 61:479–484

Hatzinger M, Brand S, Perren S, von Wyl A, Standelmann S et al (2012) Pre-schoolers suffering from psychiatric disorders show increased cortisol secretion and poor sleep compared to healthy controls. J Psychiatr Res 46(5):590–599

Heim C, Nemeroff C (2001) The role of childhood trauma in the neurobiology of mood and anxiety disorders: preclinical and clinical studies. Biol Psychiatry 49:1023–1039

Heim C, Newport D, Wagner D, Wilcox M, Miller A, Nemeroff C (2002) The role of early adverse experience and adulthood stress in the prediction of neuroendocrine stress reactivity in women: a multiple regression analysis. Depress Anxiety 15:117–125

Hess NH, Hagen EH (2006) Sex differences in indirect aggression: psychological evidence from young adults. Evol Hum Behav 27:231–245

Hrdy SB (2005) Evolutionary context of human development: the cooperative breeding model. In: Carter CS, Ahnert L (Eds.) *Attachment and bonding: a new synthesis*. Dahlem Workshop 92. MIT Press, Cambridge, MA

Hucklebridge F, Hussain T, Evans P, Clow A (2005) The diurnal patterns of the adrenal steroids cortisol and dehydroepiandrosterone (DHEA) in relation to awakening. Psychoneuroendocrinology 30:51–57

Huether G (1996) The central adaptation syndrome: psychosocial stress as a trigger for adaptive modifications of brain structure and brain function. Prog Neurobiol 48:569–612

Huether G (1998) Stress and the adaptive self organization of neuronal connectivity during early childhood. Int J Dev Neurosci 16:297–306

Huot R, Gonzalez C, Ladd C, Thrivikraman K, Plotsky P (2004) Foster litters prevent hypothalamic-pituitary-adrenal axis sensitization mediated by neonatal maternal separation. Psychoneuroendocrinology 29:279–289

Insel T, Kinsley C, Mann P, Bridges R (1990) Prenatal stress has long-term effects on brain opiate receptors. Brain Res 511:93–97

Kaplan H, Lancaster J, Robson A (2003) Embodied capital and the evolutionary economics of the human life span. Population Dev Rev 29:152–182

Kaufman J, Plotsky P, Nemeroff C, Charney D (2000) Effects of early adverse experiences on brain structure and function: clinical implications. Biol Psychiatry 48:778–790

Kerr M, Bowen M (1988) Family evaluation. W.W. Norton, New York

Kinsley C, Bridges R (1988) Prenatal stress and maternal behavior in intact virgin rats: response latencies are decreased in males and increased in females. Horm Behav 22:76–89

Kinsley C, Svare B (1986) Prenatal stress reduces intermale aggression in mice. Physiol Behav 36:783–786

Konner M (2010) The evolution of childhood: relationships, emotion, mind. Harvard University Press, Cambridge, MA

Kraemer G (1992) A psychobiological theory of attachment. Behav Brain Sci 15:493–511

Krpan K, Coombs R, Zinga D, Steiner M, Fleming A (2005) Experiential and hormonal correlates of maternal behavior in teen and adult mothers. Horm Behav 47:112–122

Laakso M, Porkka-Heiskanen T, Alila A, Stenberg O, Johansson G (1994) 24-hour rhythms in relation to the natural photoperiod: a field study in humans. J Biol Rhythms 9:283–293

Ladd C, Huot R, Thrivikraman K, Nemeroff C, Meaney M, Plotsky P (2000) Long-term behavioral and neuroendocrine adaptations to adverse early experience. In: Mayer E, Super C (eds) Progress in brain research. Elsevier Science, Amsterdam, pp 81–103

Laplante DP, Barr RG, Brunet A, Galbaud du Fort G, Meaney MJ et al (2004) Stress during pregnancy affects general intellectual and leanguage functioning in human toddlers. Pediatr Res 56:400–410

LeDoux J (1996) The emotional brain. Simon and Schuster, New York

Leslie AM, Friedmann O, German TP (2004) Core mechanisms in 'theory of mind'. Trends Cogn Sci 8:529–533

Liu D, Diorio J, Tannenbaum B, Caldji C, Francis D, Freedman A, Sharma S, Pearson D, Plotsky P, Meaney MJ (1997) Maternal care, hippocampal glucocorticoid receptors and HPA responses to stress. Science 277:1659–1662

Long B, Ungpakorn G, Harrison GA (1993) Home-school differences in stress hormone levels in a group of Oxford primary school children. J Biosoc Sci 25:73–78

Lupien S, McEwen BS (1997) The acute effects of corticosteroids on cognition: integration of animal and human model studies. Brain Res Rev 24:1–27

Lupien S, King S, Meaney M, McEwen BS (2000) Child's stress hormone levels correlate with mother's socioeconomic status and depressive state. Biol Psychiatry 48:976–980

Maccari S, Piazza P, Kabbaj M, Barbazanges A, Simon H, Le Moal M (1995) Adoption reverses the long-term impairment in glucocorticoid feedback induced by prenatal stress. J Neurosci 15:110–115

McEwen BS (1998) Protective and damaging effects of stress mediators. N Engl J Med 338:171–179

McEwen BS, Magarinos AM (2001) Stress and hippocampal plasticity: implications for the pathophysiology of affective disorders. Hum Psychopharmacol 16:S7–S19

McEwen BS (1995) Stressful experience, brain, and emotions: developmental, genetic, and hormonal influences. In: Gazzaniga MS (ed) The cognitive neurosciences. MIT, Cambridge, MA, pp 1117–1135

McEwen BS, Seeman T (1999) Protective and damaging effects of mediators of stress: elaborating and testing the concepts of allostasis and allostatic load. Ann N Y Acad Sci 896:30–47

McEwen B, Biron C, Brunson K, Bulloch K, Chambers W, Dhabhar F, Goldfarb R, Kitson R, Miller A, Spencer R, Weiss J (1997) The role of adrenocorticoids as modulators of immune function in health and disease: neural, endocrine and immune interactions. Brain Res Rev 23:79–133

McGowan PO, Sasaki A, D'Alessio AC, Dymov S, Labonté B, Szyf M, Turecki G, Meaney MJ (2009) Epigenetic regulation of the glucocorticoid receptor in human brain associates with childhood abuse. Nat Neurosci 12:342–348

Mirescu C, Peters J, Gould E (2004) Early life experience alters response of adult neurogenesis to stress. Nat Neurosci 7:841–846

Mousseau TA, Fox CW (1998) Maternal effects as adaptations. Oxford University Press, New York

Moyer J, Herrenkohl L, Jacobowitz D (1978) Stress during pregnancy: effect on catecholamines in discrete brain regions of offspring as adults. Brain Res 144:173–178

Muehlenbein MP, Flinn MV (2011) Patterns and processes of human life history evolution. In: Flatt T, Heyland A (eds) Mechanisms of life history evolution. Oxford University Press, New York, pp 153–168

Nepomnaschy P, Flinn MV (2009) Early life influences on the ontogeny of neuroendocrine stress response in the human child. In: Gray P, Ellison P (eds) The endocrinology of social relationships (Chapter 16). Harvard University Press, Cambridge, MA
Nicolson N (2004) Childhood parental loss and cortisol levels in adult men. Psychoneuroendocrinology 29:1012–1018
Noone R (2008) The multigenerational transmission process and the neurobiology of attachment and stress reactivity. Fam Syst 8:21–34
O'Connor T, Dunn J, Jenkins J, Pickering K, Rasbash J (2001) Family settings and children's adjustment: differential adjustment within and across families. Br J Psychiatry 179:110–115
O'Regan D, Welberg L, Holmes M, Seckl J (2001) Glucocorticoid programming of pituitary-adrenal function: mechanisms and physiological consequences. Semin Neonatol 6:319–329
Panksepp J (1998) Affective neuroscience. Oxford University Press, New York
Pinker S (1994) *The language instinct*. William Morrow, New York
Pollard I (1984) Effects of stress administered during pregnancy on reproductive capacity and subsequent development of the offspring of rats: prolonged effects on the litters of a second pregnancy. J Endocrinol 100:301–306
Quinlan R, Flinn MV (2005) Kinship, sex and fitness in a Caribbean community. Hum Nat 16(1):32–57
Quinlan R, Quinlan MB, Flinn MV (2005) Local resource enhancement & sex-biased breastfeeding in a Caribbean community. Curr Anthropol 46(3):471–480
Rilling JK, Sanfey AG (2011) The neuroscience of social decision-making. Annu Rev Psychol 62:23–48
Roth G, Dicke U (2005) Evolution of the brain and intelligence. Trends Cogn Sci 9:250–257
Sabatini MJ, Ebert P, Lewis DA, Levitt P, Cameron JL, Mirnics K (2007) Amygdala gene expression correlates of social behavior in monkeys experiencing maternal separation. J Neurosci 27(12):3295–3304
Sapolsky R (1992) Stress, the aging brain, and the mechanisms of neuron death. A Bradford Book, The MIT Press, Cambridge, Massachusetts
Sapolsky R (2003) Stress and plasticity in the limbic system. Neurochem Res 28:1735–1742
Schneider M (1992) Prenatal stress exposure alters postnatal behavioral expression under conditions of novelty challenge in rhesus monkey infants. Dev Psychobiol 25:529–540
Schneider M, Coe C (1993) Repeated social stress during pregnancy impairs neuromotor development of the primate infant. J Dev Behav Pediatr 14:81–87
Seckl JR, Meaney MJ (2004) Glucocorticoid programming. Ann N Y Acad Sci 1032:63–84
Shimada M, Takahashi K, Ohkawa T, Segawa M, Higurashi M (1995) Determination of salivary cortisol by ELISA and its application to the assessment of the circadian rhythm in children. Horm Res 44:213–217
Simons RL, Chao W, Conger RD, Elder GH (2001) Quality of parenting as mediator of the effect of childhood defiance on adolescent friendship choices and delinquency: a growth curve analysis. J Marriage Fam 63:63–79
Suomi S (2002) Attachment in rhesus monkeys. In: Cacioppo J et al (eds) Foundations in social neuroscience. The MIT Press, Cambridge, MA, pp 775–795
Suomi S (2003) Social and biological mechanisms underlying impulsive aggressiveness in rhesus monkeys. In: Lahey B, Moffit T, Caspi A (eds) Causes of conduct disorder and juvenile delinquency. Guilford Press, New York, p 2003
Suomi S, Mineka S, DeLizio RD (1983) Short- and long-term effects of repetitive mother-infant separations on social development in rhesus monkeys. Dev Psychobiol 19:770–786
Swain JE, Lorberbaum JP, Kose S, Strathearn L (2007) Brain basis of early parent–infant interactions: psychology, physiology, and in vivo functional neuroimaging studies. J Child Psychol Psychiatry 48:262–287
Takahashi LK, Turner JG, Kalin NH (1992) Prenatal stress alters brain catecholaminergic activity and potentiates stress-induced behavior in adult rats. Brain Res 574:131–137
Talge NM, Neal C, Glover V (2007) Antenatal maternal stress and long-term effects on child neurodevelopment: how and why? J Child Psychol Psychiatry 48:245–261

Thomas C, Hypponen E, Power C (2008) Obesity and type 2 diabetes risk in midadult life: the role of childhood adversity. Pediatrics 121:e1240–e1249

Touitou Y, Haus E (2000) Alterations with aging of the endocrine and neuroendocrine circadian system in humans. Chronobiol Int 17:369–390

Touitou Y, Sulon J, Bogdan A, Reinberg A, Sodoyez JC, Demey-Ponsart E (1983) Adrenocortical hormones, ageing and mental condition: seasonal and circadian rhythms of plasma 18-hydroxy-11-deoxycorticosterone, total and free cortisol and urinary corticosteroids. J Endocrinol 96:53–64

Uvnas-Moberg K (1998) Oxytocin may mediate the benefits of positive social interaction and emotions. Psychoneuroendocrinology 23:819–835

Van den Bergh BRH, Marcoen A (2004) High antenatal maternal anxiety is related to ADHD symptoms, externalising problems, and anviety in 8- and 9-year-olds. Child Dev 75:1085

Van den Bergh BRH, Van Calster B, Smits T, Van Huffel S, Lagae L (2008) Antenatal maternal anxiety is related to HPA-axis dusregulation and self-reported depressive symptioms in adolescence: a prospective study on the fetal origins of depressed mood. Neuropsychopharmacology 33:536–545

van Hulten M, Pelser M, van Loon LC, Pieterse CM, Ton J (2006) Costs and benefits of priming for defense in Arabidopsis. Proc Natl Acad Sci U S A 103(14):5602–5607

Waddington CH (1956) Principles of embryology. Macmillan, New York

Walker RS, Flinn MV, Hill K (2010) The evolutionary history of partible paternity in lowland South America. Proc Natl Acad Sci 107(45):19195–19200

Walker RS, Hill K, Flinn MV, Ellsworth R (2011) Evolutionary history of hunter-gatherer marriage practices. PLoS One 6(4):e19066. doi:10.1371/ journal.pone.0019066

Weaver I, Cervoni N, Champagne F, Alessio A, Sharma S, Seckl J, Dymov S, Szyl M, Meaney M (2004) Epigenetic programming by maternal behavior. Nat Neurosci 7:847–854

Weinstock M (1997) Does prenatal stress impair coping and regulation of hypothalamic-pituitary-adrenal axis? Neurosci Biobehav Rev 21:1–10

Weinstock M (2005) The potential influence of maternal stress hormones on development and mental health of the offspring. Brain Behav Immun 19:296–308

Weinstock M (2008) The long-term behavioural consequences of prenatal stress. Neurosci Biobehav Rev 32:1073–1086

West-Eberhard MJ (2003) Developmental plasticity and evolution. Oxford University Press, Oxford

Yehuda R, Mulherin Engel S, Brand S, Marcus S, Berkowitz G (2005) Transgenerational effects of posttraumatic stress disorder in babies of mothers exposed to the World Trade Center attacks during pregnancy. J Clin Endocrinol Metab 90:4115–4118

Zhang T-Y, Meaney MJ (2010) Epigenetics and the environmental regulation of the genome and its function. Annu Rev Psychol 61:439–466

Zhang T-Y, Bagot R, Parent C, Nesbitt C, Bredy TW, Caldji C, Fish E, Anisman H, Szyf M, Meaney MJ (2006) Maternal programming of defensive responses through sustained effects on gene expression. Biol Psychol 73:72–89

Chapter 6
Consequences of Developmental Stress in Humans: Prenatal Stress

Nadine Skoluda and Urs M. Nater

Abstract Stress is a phenomenon which can occur throughout the entire life span. There is ample evidence of prenatal programming, i.e., in utero experiences may have long-term consequences for the unborn child. In this chapter, human studies are reviewed regarding the impact of various sources of prenatal stress on various outcomes in infants. Recent findings suggest that prenatal stress is predominately associated with maladaptive consequences, such as negative birth outcomes, altered physiological stress responses, behavior problems, and impaired cognitive and motor development. There is some evidence for genetic predisposition and stress-buffering factors which protect the unborn child from negative effects of prenatal stress. A psychobiological stress model is proposed to integrate the findings.

6.1 Introduction

During the past decades, it has become increasingly evident that stress and its psychological and physiological concomitants are associated with sometimes dramatic negative health outcomes. Research using both animal and human models has shown that stress experienced early in life may be particularly harmful to health. Such early experiences seem to profoundly alter psychobiological

N. Skoluda • U.M. Nater, Ph.D. (✉)
Department of Psychology, Philipps University of Marburg, Gutenbergstrasse 18, Marburg 35032, Germany
e-mail: nater@uni-marburg.de

G. Laviola and S. Macrì (eds.), *Adaptive and Maladaptive Aspects of Developmental Stress*, Current Topics in Neurotoxicity 3, DOI 10.1007/978-1-4614-5605-6_6,

stress-responsive systems[1] in a critical period of development, ultimately leading to dysregulation in endocrine, autonomic, and immune processes in adulthood. In this chapter, it is important to first address the concept of stress as it has been described in the human literature. Although almost everybody has an implicit concept of what stress is, even scientists find it difficult to define this sometimes elusive phenomenon. In human research, the term stress usually describes (a) psychological or physiological demands (stressor, stressful stimulus), (b) the psychological response to a stressful stimulus (appraisal, emotions), and/or (c) the physiological response and long-term consequences (activation of stress-responsive systems, behavior). It is important to note that the definition in (a) equals the stimulus with the response. This is conceptually difficult, and we will thus adhere to the definitions used in (b) and (c) in the following sections. In this sense, psychological or physiological demands (i.e., stressors) appraised by an individual as challenge or threat lead to psychological and psychological responses which facilitate dealing with the challenge or threat at hand.

The acute activation of endocrine and autonomic stress axes (see Sect. 6.2) leads to temporary metabolic, immune, behavioral, and cognitive changes. Such acute changes may be initially adaptive because they help an individual to survive a threatening situation, to adapt to an adverse condition in the current environment, and to be prepared for similar situations in the future. However, excessive, repeated, or chronic stress, particularly during critical phases of development, may have detrimental long-term effects on bodily functions (e.g., alterations in neuronal circuits, inappropriate stress reactivity) and may thus increase the risk for developing diseases.

In this chapter, we focus on the impact of stress experienced during the prenatal period (also termed *prenatal stress*) on various health outcomes in humans. In the next chapter, we will then focus on early life stress experienced during childhood and adolescence. Considering the amount of literature and the limited space of a book chapter, a selected overview of the literature on prenatal stress will be presented here.[2] The overall goal of this chapter is to briefly summarize the literature and give an overview of the current state of knowledge regarding short- and long-term consequences of prenatal stress.

---

[1] *Stress-responsive systems* refer to various physiological systems that are activated in response to stress. The most prominent systems are the hypothalamus–pituitary–adrenal (HPA) axis, autonomic nervous system (ANS, particularly the sympatho–adrenal–medullary [SAM] system), the central nervous system, and the immune system. We introduce the term *stress axes*, which refer to the HPA axis and the ANS, since these well-studied stress-responsive systems are known to interact and partly regulate the other aforementioned stress-responsive systems.

[2] A systematic literature search was conducted on databases PubMed and Google Scholar using a combination of the following keywords: prenatal (perceived) stress infant, prenatal depression infant, prenatal anxiety (infant), pregnancy (synthetic) glucocorticoid human outcome, pregnancy betamethasone human outcome, and prenatal stress and social support. The list of abstracts was reviewed with regard to the type of stress and outcome (neuroendocrine, behavioral, psychological, or health-related consequences). Only English-language, peer-reviewed original research papers were included. Additionally, the references of the selected papers and those of published review articles have been checked for further articles meeting our search criteria. Although we have tried to consider as many publications as possible, we are aware that the presented literature is not comprehensive.

6.2 The Psychobiology of Stress

6.2.1 Psychobiological Stress Theories

Early stress theories, such as Cannon's *homeostasis*[3] concept (Cannon 1935) and Selye's concept of the *General Adaptation Syndrome*, GAS[4] (Selye and Fortier 1950), have emphasized the role of physiological processes during stress. Based on these concepts which have described a relative diffuse physiological activation due to stress, later stress theories have suggested possible specific biological mechanisms in order to explain the connection between dysregulation of the stress-responsive systems and the risk for disease (Chrousos 2009). Importantly, the terms *allostasis* and *allostatic load* have been introduced in the literature. Both refer to physiological processes of stress-responsive systems which are involved in maintaining homeostasis (allostasis) and "the wear and tear on the body and brain" as a cost of allodynamic adaption (allostatic load) (McEwen and Gianaros 2011). In current theories, the *brain* plays a key role in the processing of potentially threatening internal and external stimuli (stressors) and in the regulation of the ensuing behavioral and physiological stress responses. There is sufficient evidence that the central nervous system, hypothalamus–pituitary–adrenal (HPA) axis, autonomic nervous system (ANS), and the immune system communicate with each other via neural and chemical pathways, i.e., afferent nerves, as well as cytokines, endocrine peptides, catecholamines, and glucocorticoids, respectively (Ziemssen and Kern 2007). The bidirectional interactions highlight the ability of each system to influence the functioning of the other systems. Many of the aforementioned stress theories are applicable for animal and human research. However, the main difference between animals and humans is the human's ability for performing complex cognitive processes (anticipation and evaluation of a potential threat), although there is evidence that higher developed animals (e.g., primates) are capable of comparable cognitive performances. It is assumed that only in the case of a situation being appraised as threat, loss/harm, or challenge stress is experienced and physiological stress-responsive systems are activated via top–down processes. Thus, peripheral physiological stress responses originate from the central nervous system.

In summary, stress processes according to this psychobiological model have psychological (e.g., appraisal), physiological (e.g., the activation of stress-responsive systems), and behavioral (e.g., coping with stressful situation) components. The physiological component will now be described in more detail.

---

[3]Homoeostasis describes the body's ability to regulate internal states ("internal environment") to maintain stable within certain limits.

[4]GAS includes three stages: (a) the alarm reaction or shock phase (activation of the system, particularly autonomic nervous system and hypothalamus–pituitary–adrenal axis), (b) the stage of resistance (adaptation: most of the responses of the alarm reaction phase disappear), and (c) the stage of exhaustion (return of the responses which were present during the alarm reaction phase).

6.2.2 Physiology of Stress

Under conditions of stress, the central nervous system, particularly the limbic system, activates the two main stress axes, namely, the sympatho–adrenal–medullary (SAM) system and the HPA axis, which both play an important role in the physiological stress response and regulation of other stress-responsive systems. Since most of the studies described in this chapter involve the HPA axis, this stress axis will be described in greater detail in the following paragraphs.

The fast-responding SAM system, as part of the ANS, induces the adrenal medulla to release catecholamines (norepinephrine and epinephrine) which are responsible for a number of responses typical of "flight or fight" (e.g., increased heart rate, alertness, mobilization of energy stores, immunostimulation) (McEwen 2003). The slower responding HPA axis originates in the paraventricular nucleus (PVN) of the hypothalamus where vasopressin and corticotrophin-releasing hormone (CRH) are synthesized and secreted. CRH stimulates the release of adrenocorticotropic hormone (ACTH) from the anterior lobe of the pituitary gland into the bloodstream. When ACTH reaches the adrenal cortex, glucocorticoids (predominately cortisol in humans) are released in response to this stimulus. Cortisol itself inhibits the further secretion of CRH and ACTH and thus regulates HPA axis activity via a negative feedback loop. Cortisol has various effects on cognitive functions (e.g., memory), behavior (e.g., alertness, euphoria), metabolism (gluconeogenesis, lipo- and proteolysis), and the immune system (immunosuppression) due to its binding to glucocorticoid receptors (GR) both in the central nervous system (e.g., hippocampus, amygdala, prefrontal cortex) and in peripheral tissue (McEwen 2003). The pulsatile secretion of cortisol follows a diurnal rhythm characterized by high morning levels and a decrease of cortisol levels during the course of the day. The cortisol awakening response (CAR), which refers to the morning peak of cortisol levels observed about 30 min after awakening, constitutes an important marker of HPA axis regulation and has been related to a number of negative health outcomes (Fries et al. 2009; Kudielka and Wust 2010). Thus, studying cortisol release during the morning and/or during the day constitutes a well-researched approach to study HPA axis *activity*.

The exposure to a (experimental) stressor or pharmacological stimulant provides the possibility to study HPA axis *reactivity*. Some prominent and effective stressors in human research are the following: (a) the heel-stick blood draw (also known as "heel prick test") in neonates, (b) the performance of a cognitive task (giving a speech or/and solving a difficult arithmetic task) in the presence of a social-evaluating audience, or (c) the administration of pharmacological substances.

Psychobiological approaches enable scientists to study the underlying mechanisms which explain how stress may lead to possible negative physiological, psychological, behavioral, and developmental consequences.

6.3 Prenatal Stress

Exposure to stress may occur even before birth, i.e., during the time in the mother's womb.[5] Experiences of stress before birth may have tremendous effects on the as yet unborn child, since the maternal and fetal physiological systems are connected in the placenta by the umbilical cord. Maternal cortisol is able to pass the placental membrane and may thus enter into the fetal circulation. In addition to exposure to maternal cortisol, the placenta itself produces CRH (placental CRH or pCRH) which stimulates both the maternal and the fetal HPA axis. Under normal conditions, the excessive release of stress hormones is inhibited due to the negative feedback regulation of the HPA axis (see Sect. 6.2). Particularly, cortisol suppresses the secretion of CRH and ACTH and thus its own release. However, in the state of pregnancy, cortisol enhances the production of pCRH in the placenta, which then leads to gradual increases in ACTH and cortisol levels in the course of pregnancy, reaching a peak in the third trimester (Cottrell and Seckl 2009). This dramatic rise of cortisol levels observed in the final stages of pregnancy has a crucial function in fetal development, since glucocorticoids promote lung maturation. At the same time, the body has to deal with these vast amounts of cortisol. In order to protect the fetus from overexposure to cortisol and its potentially deleterious effects, the placental enzyme 11beta-hydroxysteroid dehydrogenase type 2 (11beta-HSD2) metabolizes cortisol into its biologically inactive form cortisone. As pregnancy progresses, 11beta-HSD2 levels increase initially but fall rapidly several weeks before parturition. This drop of 11beta-HSD2 before parturition results in increased cortisol levels that are essential for the aforementioned fetal lung maturation and preparation of delivery (Davis and Sandman 2010).

It is conceivable that there is a tolerable amount of cortisol which the fetus can deal with, not least due to the regulating effect of 11beta-HSD2 on cortisol production. However, additional secretion of cortisol due to prenatal maternal stress or administration of synthetic glucocorticoids in sensitive periods during pregnancy (e.g., in phases of decreased 11beta-HSD2 production) might not be sufficiently metabolized. Excessive prenatal exposure to cortisol may alter the sensitivity of the glucocorticoid receptors in the brain and peripheral tissue. Consequences of such overexposure may be immediate negative birth-related characteristics[6] (e.g., low birth weight), negative

[5]Since the human body undergoes extreme biological and hormonal changes throughout pregnancy, it is important to consider the pregnancy stage at which stress occurs. We will use the terms early pregnancy, mid-pregnancy, or late pregnancy which refer to the first trimester (week of gestation 1–13), second trimester (week of gestation 14–27), or third trimester (week of gestation 28 until parturition), respectively.

[6]Birth outcomes are widely used indicators in scientific research of prenatal stress. Especially low birth weight (LBW) as a prominent prenatal stress marker is discussed to be a risk factor for chronic diseases, such as cardiovascular diseases, hypertension, and diabetes (see review, Osmond and Barker 2000). There are several approaches to determine birth weight. The direct assessment of the actual weight immediately after delivery provides one possibility. Another possibility is the estimation of fetal weight by fetal ultrasound biometry measurements during pregnancy, predominately conducted in mid- and late gestation.

health outcomes, as well as to physiological, behavioral, and cognitive aspects of development throughout the life span. There is ample evidence for the so-called *fetal programming*, i.e., prenatal stress may have long-term consequences on fetal development and impact on later life (Seckl et al. 2000).

In the following, findings are summarized by the type of prenatal stressors. There are several approaches to operationalize prenatal stress. For this purpose, we consider following types of maternal stress during pregnancy:

- Maternal stress and increases in basal stress hormones during pregnancy (Sect. 6.3.1)
- Administration of synthetic glucocorticoids (Sect. 6.3.2)
- Perceived stress and daily hassles (Sect. 6.3.3)
- Work-related stress (Sect. 6.3.4)
- Traumas and major life events (Sect. 6.3.5)
- Anxiety and depression (Sect. 6.3.6)

Prenatal stress may have effects on various outcomes. For establishing a coherent structure within the subchapters, we summarize the findings in respect to the reported outcomes, whenever possible, in the following order: birth outcomes, biological outcomes, temperament and behavior, as well as mental and motor development. *Birth outcomes* constitute birth-related characteristics, such as birth weight or length of gestation, and may be risk factors for diseases in later life. *Biological outcomes* encompass measures of the metabolic, immune, endocrine, autonomic nervous, or central nervous system (e.g., insulin levels, cytokine levels, cortisol levels, heart rate, or activity of certain brain regions). These biological markers can be studied under baseline conditions (e.g., morning values) or under challenge. Examples for these "challenging conditions" are stimulation of immune cells with a pathogen to assess the resulting cytokine production or acute stress tests to provoke HPA axis responses. Prenatal stress may also affect the infant's *temperament and behavior* which both can be assessed by questionnaires or by behavioral observations. In such questionnaires, the parents are asked to rate their child's behavior which in turn is supposed to reflect its temperament. Prominent dimensions of temperament are affective reactivity to novelty, shy or aggressive behavior in social interactions, or attention/hyperactivity indicated by restless/disruptive behavior. Also, manifestation of behavior may indicate internalizing (e.g., anxiety, depression) or externalizing (e.g., aggression, delinquent behavior) behavioral problems. Behavioral observations provide a tool to assess an infant's temperament: an infant's behavioral responses (e.g., crying, fussing, or attention) can be studied in an experimental setting (e.g., reactivity to novelty) or in a stressful situation (e.g., physical examination or prick test). Finally, prenatal stress may partly determine an infant's *mental and motor development* which can be measured by various test batteries. Mental development means age-related cognitive abilities such as attention/concentration, classification abilities, verbal comprehension, reasoning, or memory performances. On the other hand, motor development includes body control as well as gross and fine motor skills. The Bayley Scales of Infant Development (BSID) enable to evaluate both mental and motor development of infants up to 4 years and thus are one of the most used instruments in the following studies.

Most of the studies investigating stress in human research focus on possible negative effects of stress. Reflecting this trend in the literature, the following findings predominately report maladaptive consequences of prenatal stress. However, to do justice to adaptive consequences of prenatal stress, we added a summary of the few findings available in the literature at the end of this section.

6.3.1 Maternal Stress and Increases in Basal Stress Hormones during Pregnancy

The assessment of basal stress hormones levels constitutes one possibility to determine maternal stress during pregnancy and requires a prospective longitudinal study design. An overview of selected findings of associations between basal stress hormones and birth outcomes, biological outcomes, temperament and behavior, as well as mental and motor development will be given.

Effects on birth outcomes. Current evidence points to the notion that stress hormones may affect birth-related outcomes. In one study, e.g., high basal CRH levels during mid- and late pregnancy were associated with shortening of gestation length (Mancuso et al. 2004). In another study, Diego et al. (2006) have found that maternal urinary cortisol and norepinephrine levels during pregnancy were negatively associated with fetal weight. Finally, an augmented maternal CAR, describing the morning increase of cortisol levels 30 min after awakening (see Sect. 6.2), in mid- and late pregnancy was associated with low birth weight (Bolten et al. 2011). Negative birth outcomes such as these may be considered as risk factors for illness in later life. Thus, elevated basal stress levels in pregnancy may predict the risk for such negative birth outcomes and, consequently, for negative health outcomes in the offspring's later life.

Effects on biological outcomes. One study has found that elevated early morning cortisol levels of mothers during mid-pregnancy were related to increased cortisol responses to a vaccination procedure in their children when they were 4–6 years old (Gutteling et al. 2004). Another study reported higher salivary cortisol concentrations in response to the heel prick test in 1–2-day-old neonates when their mothers showed elevated cortisol levels during mid- and late pregnancy (Davis et al. 2011a). These preliminary findings indicate that high basal maternal stress hormone levels in mid- and late pregnancy may be related to augmented endocrine stress responses in children.

Effects on temperament and behavior. In one study, elevated maternal cortisol levels in early pregnancy were associated with slower neonatal behavioral recovery after the heel prick test, i.e., these neonates exhibited increased levels of distress behavior within the 5 min period following the end of the procedure (Davis et al. 2011a). Two other studies have shown that elevated maternal CRH or cortisol levels during late pregnancy were related to mothers' reports that their 8-week-old infants show increased levels of fear and distress as responses to novel stimuli in daily life situations (Davis et al. 2005, 2007). However, another study has failed to find this relationship between prenatal maternal cortisol levels and children's affective reactivity to novel stimuli, indicated by affective behavior in a novelty experiment (Rothenberger et al. 2011). A longitudinal study showed that maternal cortisol levels during pregnancy did not predict behavioral problems in 27-month-old children (Gutteling et al. 2005a). The majority of these findings imply that enhanced basal maternal hormone concentrations may be related to increased infants' behavioral responses to novel and stressful situations.

Effects on mental and motor development. High maternal morning cortisol in late pregnancy has been found to be associated with impaired mental and motor development at the age of 3 and 8 months (Huizink et al. 2003). Another study reported that accelerated mental development in infants of 12 months of age was characterized by lower maternal cortisol levels in early pregnancy and higher cortisol levels in late pregnancy (Davis and Sandman 2010). In contrast to these findings, Gutteling et al. (2006) found no association between maternal diurnal cortisol levels during pregnancy and infant's attention/concentration and learning performance at the mean age of 6 years. Possible explanations for these inconsistent findings may be different outcome measurements that were used and the infant's age when developmental assessment has occurred.

In summary, elevated stress hormone levels seem to be related to adverse birth outcomes, such as shortened gestation length and low birth weight, as well as enhanced endocrine and behavioral stress responses. Some of the studies indicate that increased basal maternal stress hormone levels may negatively affect development of the child. This last finding is in line with the notion that prenatal overexposure to stress hormones may alter receptor sensitivity in brain regions that are related to cognitive processes (Welberg and Seckl 2001).

6.3.2 Administration of Synthetic Glucocorticoids During Pregnancy

As mentioned above, glucocorticoids play an essential role in lung maturation at the end of gestation. Therefore, in case of a high risk of preterm birth, pregnant women can be given doses of synthetic glucocorticoids, such as dexamethasone or betamethasone, several days before the expected delivery date in order to enhance survival changes of the newborn. As the effects of increased stress hormone concentrations might be considered a specific (artificial) form of prenatal stress, this might have potential health implications for the offspring. A selection of findings about associations between administration of synthetic glucocorticoids during pregnancy and biological outcomes, temperament and behavior, as well as mental and motor development will be discussed in the following paragraphs.

Effects on biological outcomes. Several studies examined the impact of prenatal glucocorticoid exposure on metabolic parameters. In adults whose mothers received betamethasone treatment during pregnancy, lower morning baseline levels of insulin (Finken et al. 2008) as well as higher insulin and lower glucose concentrations following a glucose tolerance test were found (Dalziel et al. 2005b). These preliminary findings indicate an altered metabolism which might be associated with enhanced risk for diseases later in life. In a study assessing autonomic functioning, prenatal exposure to synthetic glucocorticoids was unrelated to blood pressure in children (Dalziel et al. 2004) or adults (Dalziel et al. 2005b; Finken et al. 2008). Thus, prenatal treatment with synthetic glucocorticoids might increase the risk for certain diseases in adulthood, although this effect might be relevant for metabolic disorders rather than cardiovascular diseases. Assessing heart rates in stressful situation is a commonly used measure for reactivity of the ANS. Using this parameter, two studies found that 3 days and older preterm neonates who were prenatally exposed to betamethasone exhibited an increased heart rate in response to a heel prick test (Davis et al. 2004b, 2006), indicating an enhanced ANS reactivity. In both studies, salivary cortisol levels as a measure for HPA axis activity were additionally obtained. In contrast to preterm infants whose mothers received no betamethasone treatment during pregnancy, infants with such an exposure showed not the expected cortisol response to the heel prick test. In contrast to this aforementioned finding, full-term infants whose mothers were treated with betamethasone showed increased HPA axis reactivity, as indicated by a higher cortisol response to the heel prick test administered between 15 and 48 h after delivery, compared to nonexposed controls (Davis et al. 2011b). These contrary findings are difficult to interpret due to the different study designs: differences in the gestational age of the sample (preterm vs. full-term children) and the time of examination (within 2 days vs. 3 or more days after birth) might explain these diverging results.

Effect on temperament and behavior. To our knowledge, there is only one study investigating the effect of synthetic glucocorticoids on infants' temperament (Trautman et al. 1995). In this study, children (aged 2-3 years) who were prenatally exposed to dexamethasone showed increased shy behavior and internatizing problems compared to children withou such an exposure.

Effects on mental and motor development. Based on both a neurological evaluation by a neuropsychiatrist and the BSID, one study reported neurodevelopmental impairment (e.g., spastic diplegia, cognitive deficits) in 2-year-old children who were prenatally exposed to dexamethasone (Spinillo et al. 2004). The majority of studies, however, have found no associations between synthetic glucocorticoids treatment during pregnancy and an infant's mental and motor development, i.e., children whose mothers received synthetic glucocorticoids during pregnancy exhibited no impairment in domains such as memory, verbal skills, or motor skills (MacArthur et al. 1981; Meyer-Bahlburg et al. 2004; Peltoniemi et al. 2009; Trautman et al. 1995). In line with these findings in children, adults who were prenatally exposed to betamethasone did not differ from adults whose mothers received a placebo (cortisone acetate with a lower glucocorticoid potency) in regard to cognitive functioning or memory performances assessed by intelligence or memory tests (Dalziel et al. 2005a).

In conclusion, preliminary findings suggest that exposure to synthetic glucocorticoids at the end of pregnancy may negatively affect on infant's biochemical equilibrium and physiological stress response. One study found shy behavior and internalizing problems in children whose mothers received dexamethasone during pregnancy. It appears that synthetic glucocorticoid treatment during pregnancy may have no detrimental effect on the child's mental and motor development.

6.3.3 Perceived Stress and Daily Hassles During Pregnancy

Assessment of perceived stress and daily hassles during pregnancy reported by expectant mothers is another possibility to study maternal prenatal stress. Perceived stress is measured by questionnaires asking the subject to retrospectively evaluate the subjective stress level during a certain time period (e.g., last month). Daily hassles, on the other hand, are often assessed by lists of possible everyday stressful situations, e.g., being stuck in a traffic jam or family matters. Cutoff values provide a method to assign subjects to a group of either low or high levels of daily hassles. Alternatively, the frequency of daily stressors can be used to determine the mother's stress during pregnancy. The majority of the reported studies used a prospective longitudinal design, i.e., perceived stress or daily hassles were obtained during pregnancy, and the outcome of interest was measured after birth in newborns or children. In the following, a selection of studies with regard to the impact on birth outcomes, biological outcomes, temperament and behavior, as well as mental and motor development will be outlined.

Effects on birth outcomes. One prospective longitudinal study reported that the number of daily stressors in the first trimester of pregnancy was a predictor for low birth weight. This positive relationship disappeared, however, when using a less stringent definition of low birth weight (Paarlberg et al. 1999). In accordance with this finding, the majority of longitudinal and cross-sectional studies have found no associations between subjective prenatal maternal stress and pregnancy- or birth-related measurements (e.g., fetal weight, birth weight, or gestational age) (Bolten et al. 2011; Diego et al. 2006; Henrichs et al. 2010).

Effects on biological outcomes. Infants aged 4–6 years whose mothers reported more daily hassles during the early period of mid-pregnancy exhibited increased cortisol concentrations on a vaccination day compared to infants whose mothers reported less daily hassles during pregnancy; no such association was found for maternal perceived stress during pregnancy in this longitudinal study (Gutteling et al. 2004). In another prospective longitudinal study, maternal perceived stress during pregnancy was not associated with the neonate's cortisol response to the heel prick test (Davis et al. 2011a). Based on these two studies, it appears that perceived stress during pregnancy is not related to the infant's HPA axis reactivity. A possible explanation for this finding may be the lack of "psychoendocrine covariance," which describes the often-observed finding that perceived stress levels and stress hormone levels are only weakly correlated (Schlotz et al. 2008; Schommer et al. 2003). Indeed, Davis et al. (2011a), who assessed both perceived stress and cortisol levels during pregnancy in the aforementioned study, found no association between these two measurements. It can be assumed that despite the mother's high level of perceived stress, the mother and thus the unborn child exhibited "normal" cortisol secretion during pregnancy, which led to a normal HPA axis activity of the infant, possibly explaining this finding.

Effects on temperament and behavior. Neonates whose mothers perceived high levels of stress throughout pregnancy showed slower behavioral recovery after the heel prick test, i.e., they remained longer in a behavioral state of distress, indicated by fussiness, after this stressful event (Davis et al. 2011a). Mothers of highly reactive 5-month-old infants, defined as excessive affective reactivity to novel stimuli in a laboratory setting, reported that they had perceived more stress in mid- and late pregnancy than mothers whose children did not display such responses (Rothenberger et al. 2011). Another longitudinal study, however, has found that maternal perceived stress throughout pregnancy was not associated with negative reactivity of 8-week-old infants, in this case defined as the fear reaction to a novel or surprising stimulus in daily life reported by the infants' mothers (Davis et al. 2007). The inconsistency of the aforementioned findings may be explained by the different methods, i.e., the studies differ in the instruments for prenatal stress assessment, operationalization of affective reactivity, and age of the infant sample. In another study, prenatally perceived stress, but not the reported frequency of daily hassles during pregnancy, was associated with decreased attention regulation at the age of 3 months as well as with increased difficult behavior in daily life (indicated by the infant's negative mood, nonadaptive behavior to novelty, problems with daily routine such as sleeping) in infants at age of 3 and 8 months (Huizink et al. 2002). A prospective longitudinal study found that high levels of maternal prenatal perceived stress predicted low levels of restless/disruptive temperament as well as more behavior problems, especially externalizing behavioral problems (Gutteling et al. 2005b). These findings suggest that maternal perceived stress during pregnancy may be associated with the infant's difficulties to regulate negative emotions, particularly in novel situation, and behavioral problems.

Effects on mental and motor development. While one longitudinal study reported that high levels of daily hassles in early pregnancy were associated with impaired mental development at the age of 8 months (Huizink et al. 2003), another longitudinal study has found that levels of daily hassles during pregnancy were unrelated to the infants' attention and learning performance at the age of 6 years (Gutteling et al. 2006). Further, subjective maternal perceived stress during pregnancy was not associated with the infants' mental and psychomotor development at the age of 12 months (Davis and Sandman 2010). These few studies permit only preliminary conclusions.

The only study that has found a positive association between daily hassles and mental development was carried out in a relative young cohort 8-month-old infants. It is conceivable that maternal perceived stress during pregnancy may possibly sustain for a certain time period after delivery. Mothers' stress perception during pregnancy and shortly after the birth may moderate the mother–infant interaction which in turn influences the infants' early development. As time progresses, however, positive and negative experiences that may occur during childhood and adolescence become more important for the further development of the child.

It seems that perceived stress and daily hassles during pregnancy are unrelated to the majority of negative birth outcomes. However, there is some evidence for associations between perceived prenatal stress with elevated behavioral stress responses and behavior problems. On the basis of the presented studies, it is unclear whether and to what degree perceived stress and daily hassles are associated with the infant's mental development.

6.3.4 Work-Related Stress During Pregnancy

Since many pregnant women continue to work until several weeks or even days before expected delivery, the question arises whether job-related stress may have an impact on birth outcomes. In contrast to the previous subchapter in which we discussed the effects of the mother's general perceived stress on the unborn child, we want to focus here on work-related stress, one of the main stressors in modern societies. According to Karasek's job strain model (Karasek et al. 1998), which describes the dimensions "job demands" (d) and "decision latitude/control" (c), there are four job categories, i.e., passive (d–/c–), low-strain (d–/c+), high-strain (d+/c–), and active jobs (d+/c+). The presented findings only include studies that used this job strain model to characterize job strain. In the following, the impact of work-related stress on birth outcomes is discussed.

Effects on birth outcomes. There are a handful of longitudinal studies that found increased risk for low birth weight in women with passive (Lee et al. 2011) or high-strain jobs (Oths et al. 2001; Vrijkotte et al. 2009). However, one (cross-sectional) study reported that this association between retrospectively assessed high job strain and low birth weight was only present in pregnant women who did not want to remain in the job (Homer et al. 1990). A longitudinal study showed that passive and high-strain jobs were positively related to risk of preterm delivery; however, this association was not statistically significant (Henriksen et al. 1994).

All in all, maternal perceived stress at the workplace may increase the risk of negative birth outcomes, such as low birth weight, which are suggested as possible risk factors for diseases in later life. However, not many studies have investigated the impact of occupational stress during pregnancy on birth outcomes, and thus, interpretation should be considered with caution.

6.3.5 Traumas and Major Life Events During Pregnancy

According to the DSM-IV (APA 2000), a traumatic event requires the following criteria: "the person experienced, witnessed, or was confronted with an event or events that involved actual or threatened death or serious injury, or a threat to the physical integrity of self or others" and that "the person's response involved intense fear, helplessness, or horror." Experiencing a natural disaster, becoming a victim of a criminal act, or a serious accident are examples for such traumatic events. Major life events, on the other hand, are major changes or events that some people may evaluate as stressful. In contrast to traumatic events, major life events can be evaluated as negative (e.g., death of the partner, illness of someone close, car accident, severe financial problems) or as positive (e.g., marriage, change in residence/school). We focus on negative major life events in this chapter. It is also important to differentiate major life events from daily hassles. The main difference is that the former one does not occur in everyday life and that it is often associated with long-lasting consequences. Both traumatic events or major life events constitute natural stressors that may entail temporary psychobiological stress (e.g., subjective experience of stress, elevated stress hormones levels) in the mother and may have thus long-term consequences (e.g., development of psychiatric disorders, altered biochemical equilibrium) for both the mother and the yet unborn child. A selection of findings in the literature in regard to birth outcomes, biological outcomes, and mental and motor development is summarized in the following paragraphs.

Effects on birth outcomes. There is a set of studies that found shortening of gestation length or a decrease of the infants' birth weight of mothers who retrospectively reported a traumatic event during pregnancy, such as an ice storm (Dancause et al. 2011), a strong earthquake (Glynn et al. 2001), the hurricane Katrina in New Orleans in 2005 (Xiong et al. 2008), or the World Trade Center disaster at 9/11 (Eskenazi et al. 2007; Lederman et al. 2004). Based on Swedish population registers reflecting three decades (1973–2004), another study found an increased risk for shortened gestational age and low birth weight when the mother experienced the death of a close person (father of the child or first-degree relative of the mother) during the time of mid-pregnancy (Class et al. 2011). These studies suggest negative consequences for birth outcomes possibly due to adverse major life events during pregnancy.

Effects on biological outcomes. Young adults (average age: 25 years) whose mothers retrospectively reported negative major life events during pregnancy exhibited increased cytokine production in response to antigen stimulation (Entringer et al. 2008a) and higher insulin and C-peptide levels following an oral glucose tolerance test (Entringer et al. 2008b), suggesting alterations in immune processes and metabolism that may serve as risk factors for a variety of diseases. Adults whose mothers reported stressful major life events during pregnancy have shown shortened telomere length which is supposed to be a risk factor for an early onset of diseases and a biomarker for aging (Entringer et al. 2011). Additionally, adults of mothers who retrospectively reported major life events during pregnancy displayed low cortisol levels before a laboratory psychosocial stress test but enhanced cortisol increases as a response to the stress test (Entringer et al. 2009b). In the same sample, the two groups also differed in their response to an $ACTH_{1-24}$ stimulation test, with adults of mothers with major life events during pregnancy showing lower cortisol levels in response to the stimulation test compared to control subjects. The same study found no differences in diurnal cortisol levels between the two groups, indicating altered stress reactivity but no alterations in basal cortisol levels. In summary, these findings suggest that exposure to stressful and traumatic events before birth may negatively alter the infants' metabolic, immune, or endocrine processes that may contribute to an increased risk for diseases later in life.

Effects on mental and motor development. Compared to controls, adults of mothers with major life events during pregnancy had longer reaction times in a working memory performance task after the intake of a dose of hydrocortisone, which was found to influence memory performances (Entringer et al. 2009a). There is a model which proposes that the relation of activated mineralocorticoid receptor (MR) and glucocorticoid receptor (GR) in brain tissue (hippocampus) has an impact on cognitive performance, with a low MR/GR ratio being related to impaired cognitive functioning (de Kloet et al. 1999). Relying on this model, Entringer et al. (2009a) suggested that subjects whose mothers reported major life effects during pregnancy may have a lower GR density in the frontal cortex, which is involved in working memory. Consequently, hydrocortisone administration may lead to this low MR/GR ratio and thus to memory impairment. In another study, maternal major life events during early pregnancy were negatively associated with attention/concentration in children aged between 5 and 8 years (Gutteling et al. 2006). These findings supplied preliminary evidence for impaired working memory and concentration due to prenatal exposure to traumas and major life events.

To sum up, experience of traumatic and negative major life events during pregnancy may be related to the risk of shortened gestational length and low birth weight. There is also some evidence that negative major life events during pregnancy may negatively affect metabolism, immune function, and cognitive functions of the child.

6.3.6 Anxiety and Depression During Pregnancy

It is well known that psychiatric disorders, such as anxiety and depression, are associated with increased circulation of stress hormones. Thus, a possible impact of the affected mother on the unborn child cannot be ruled out. There are different approaches to assess anxiety or depression during pregnancy. For example, psychiatric diagnoses, i.e., clinically significant depression or anxiety that requires treatment, can be confirmed by psychiatric interviews. The personality trait anxiety describes one person's individual dispositional anxiety proneness which is relatively stable over the time. Current depressive and anxiety symptoms which indicate the present state or cover a short time period (e.g., past week) can be obtained by questionnaires. Additionally, pregnancy-specific anxiety refers to the pregnant women's worries and anxiety regarding pregnancy and birth (e.g., medical complications, physical changes, fear of giving birth, the child's health). In the following, the question whether anxiety or depression during pregnancy may affect birth outcomes, biological outcomes, temperament and behavior, as well as mental and motor development is discussed based on a selection of the literature. To address the matter of different operationalization of anxiety or depression among the following studies, we first report findings for psychiatric disorders followed by nonclinical depression or anxiety, and finally pregnancy-specific anxiety.

Effects on birth outcomes. One longitudinal study found that mothers who were diagnosed with a depressive or anxiety disorder during mid-pregnancy were more likely to have low birth weight neonates compared to mothers without such a diagnosis (Maina et al. 2008). Studying nonclinical depression and anxiety in a cross-sectional design, maternal depressive symptoms during mid-pregnancy, but not the mothers' trait anxiety, was negatively associated with estimated fetal weight (Diego et al. 2006). Henrichs et al. (2010) have found that state anxiety in mid-pregnancy was related to lower fetal weight estimated in late pregnancy and low birth weight. A longitudinal study has shown that pregnancy-specific anxiety, but not state anxiety during mid- and late pregnancy, was associated with shortening of gestation length (Mancuso et al. 2004). In another longitudinal study, pregnancy- and birth-specific anxiety and worries during mid- and late pregnancy were found not to be associated with birth weight (Bolten et al. 2011). These findings suggest that maternal emotional state during pregnancy, such as depressive or anxiety symptoms, regardless of whether they are clinical, nonclinical, or pregnancy-specific, may contribute to the risk for negative birth outcomes. The pregnant body undergoes dramatic hormonal and physical changes. It is possible that the maternal psychological state amplifies these alterations and may thus influence the course of pregnancy and subsequent birth outcomes.

Effects on biological outcomes. There is some evidence that mothers, who reported depressive symptoms during mid-pregnancy (Field et al. 2004) or late pregnancy (Lundy et al. 1999), and their neonates have elevated cortisol levels as well as decreased dopamine levels in urine samples collected within 24 h after delivery, indicating that maternal and fetal endocrine systems are synchronized at the time of birth. This finding implies that the body functions of an unborn child respond and adjust to the current environment. One longitudinal study found that maternal prenatal anxiety during late pregnancy was positively associated with awakening cortisol levels in children of 10–11 years (O'Connor et al. 2005). Another longitudinal study reported associations between maternal anxiety during mid-pregnancy and a flattened cortisol daytime profile in adolescents of 14–15 years (Van den Bergh et al. 2008). Augmented cortisol responses to a vaccination procedure were found in 3–6-year-old children whose mothers reported high pregnancy-specific anxiety (Gutteling et al. 2004). These studies imply that depression or anxiety during pregnancy may result in an enhanced HPA axis activity and reactivity, as indicated by high basal cortisol levels and cortisol responses to a stressful challenge.

Effects on temperament and behavior. High levels of depression or anxiety symptoms during late pregnancy were associated with slower behavioral recovery after the heel prick test, indicated as enhanced levels of distress behavior following the end of this painful stress (Davis et al. 2011a). Further, positive associations between maternal depression or anxiety symptoms during late pregnancy, and behavioral reactivity to novelty, indicated as high motor activity and crying in response to standard laboratory protocol, have been reported (Davis et al. 2004a). In line with this finding, depressive symptoms during mid- and late pregnancy have been found to be related to high negative affectivity of infants, assessed by the fear reaction to a novel or surprising stimulus in daily life (Davis et al. 2007). Depressive symptoms reported by the mother during late pregnancy were found to predict externalizing problems (delinquent and aggressive behavior) of their children at 8–9 years (Luoma et al. 2001). Another longitudinal study found that anxiety symptoms in mid-pregnancy, but not during late pregnancy, predicted the child's hyperactivity/attention problems, externalizing problems, and self-reported anxiety at the age of 8–9 years (Van den Bergh and Marcoen 2004). High scores on anxiety symptoms during mid- or late pregnancy were also related to emotional problems in 4-year-old children in another study (O'Connor et al. 2002).

The aforementioned studies reported findings about the effect of nonspecific anxiety during pregnancy on the infant. Pregnant women may also experience fear that is particularly related to their ongoing pregnancy or expected birth, i.e., pregnancy- or birth-related anxiety. The average fear of bearing a handicapped child during mid- and late pregnancy was associated with heightened levels of restless/disruptive behavior (e.g., the child's general activity, impatience, distractibility) and more attention regulation problems in infants 27 months in another study (Gutteling et al. 2005b). Pregnancy-specific anxiety during mid-pregnancy has been found to be associated with deficient attention regulation but no difficult behavior (e.g., fussiness) at 3 months of age (Huizink et al. 2002). Taken together, the literature indicates that children whose mothers reported anxiety or depressive symptoms during pregnancy may show increased affective reactivity to novelty and more behavioral problems. People suffering from depression or anxiety have a characteristic pattern of circulating neurotransmitters and hormones (e.g., increased cortisol levels). Since both maternal and fetal organisms are connected in the placenta, it is conceivable that the unborn child is exposed to these neurotransmitters and hormones to a similar extent as the mother which may have possible maladaptive consequences for the child's temperament and behavior. Further, the psychological state during pregnancy may partly determine the mother's relationship to and behavior (e.g., overprotective behavior) towards the child which in turn influence the child's behavior.

Effects on mental and motor development. Maternal state anxiety in late pregnancy was found to be related to decreased mental development of infants at 2 years of age (Brouwers et al. 2001). In another study, adolescents of mothers with high prenatal state anxiety during mid-pregnancy had lower scores on a vocabulary subtest and reacted more impulsively in an encoding task, as indicated by faster responses and more errors (Van den Bergh et al. 2005). Another longitudinal study found that pregnancy-specific anxiety during early mid-pregnancy, but not state anxiety, was associated with mental development at the age of 12 months (Davis and Sandman 2010). Both mental and motor development at the age of 8 months were negatively affected when the mother reported high levels of pregnancy-specific anxiety during late pregnancy (Huizink et al. 2003). However, another longitudinal study failed to find associations between pregnancy-specific anxiety and attention/concentration and learning and memory function in 5–6-year-old children (Gutteling et al. 2006). In summary, the majority of findings suggest that the mother's psychological state during pregnancy may negatively impact the child's mental and motor development.

Most of the studies found that occurrence of anxiety or depression during pregnancy may negatively affect birth outcomes. These current findings indicate that the maternal psychological state influences endocrine diurnal patterns and responses to natural stressors, as well as the infant's affectivity. Further, maternal anxiety during pregnancy seems to be related to behavioral problems, impaired mental and motor development.

As stated above, the vast majority of human studies focus on maladaptive consequences of prenatal stress. Therefore, the previously presented findings reported mainly on negative effects. However, some of the behavioral consequences of prenatal stress appear to be also *adaptive* at a closer look. More specifically, both anxious and aggressive behavior can be adaptive in a dangerous environment by enhancing the individual's survival chances. Anxiety may prevent the individual from dangerous situations, since anxious individuals might be more prone to anticipate and avoid threatening situations. Aggressive behavior on the other hand might be adaptive in unavoidably dangerous situations by mobilizing energy resources to overcome dangerous situations. However, both anxious and aggressive behavior turn out to be inappropriate and problematic in most daily life interactions. Further, an ongoing state of anxiety or aggression may lead to additional stress.

It appears that experiencing stress during pregnancy per se does not lead to negative consequences for infants. Some women and their infants seem to be less vulnerable to negative sequelae of stress during pregnancy. Thus, there might be both person-related and environmental protective factors that contribute to this stress resistance. It is conceivable that genetic predispositions may protect the individual from maladaptive consequences following stress exposure. A most recent study investigated whether an infant's behavioral problems at the age of 7 and 11 years might be associated with maternal stress at the end of pregnancy and the infant's catecholamine-O-methyltransferase (COMT) polymorphism, which encodes an enzyme which in turn inactivates catecholamines. The study found an increased risk of behavioral problems in children with the homozygote COMT gene variant, resulting in low COMT gene activity, but only when the mother reported elevated stress levels at the end of pregnancy; no such effect of prenatal stress on the risk for behavioral problems was found for the other COMT gene variants (Thompson et al. 2012). This finding suggests that there might be genetic predispositions for developmental stress taking effect.

Besides such genetic dispositions, environmental factors might moderate the impact of prenatal stress on infants' development. Some studies found that social support during pregnancy was associated with a reduced risk for negative birth outcomes, such as preterm birth or low birth weight (Da Costa et al. 2000; Elsenbruch et al. 2007; Nkansah-Amankra et al. 2010; Turner et al. 1990), whereas others did not find such associations (Dole et al. 2003, 2004). A possible explanation for this inconsistency might be that pregnant women at high risk for experiencing stress only may benefit from the stress-buffering effect of social support. In line with that, a study found positive associations between high social support and birth weight in women who experienced more life events during pregnancy; such an association was not found in women who experienced less life events during pregnancy (Collins et al. 1993). This finding suggests that a supportive environment might buffer stress experience.

6.4 Integration and Interpretation of Findings

The findings presented in this chapter highlight the potential effect of prenatal stress on various health outcomes, depending on the type of prenatal stress. We propose a psychobiological stress model that may be used to integrate and interpret these findings (Fig. 6.1).

Possible short-term effects of prenatal stress constitute birth outcomes (e.g., gestation length or birth weight) which in turn are known to be a risk factor for developing chronic diseases, such as cardiovascular diseases, hypertension, and diabetes (e.g., see review in Osmond and Barker 2000). Reviewing the literature revealed that various types of prenatal stress (all except for perceived stress and daily hassles[7]) appear to be related to shortened gestation length or low birth weight. It is possible that this association is mediated by stress hormones. There is evidence that pCRH has regulating functions on gestation length, which is generally referred to as "placental clock" (McLean et al. 1995). Elevated pCRH during pregnancy may lead to the shortening of gestation length and may consequently increase the risk of preterm delivery, which in turn is often associated with low birth weight.

The majority of findings suggest that infants whose mothers experienced prenatal stress show augmented endocrine responses to a stress, e.g., increased cortisol levels when the HPA axis is challenged. Possible explanations for this pronounced HPA axis reactivity may be altered receptor sensitivity to glucocorticoids or impaired HPA axis feedback regulation. Some authors proposed that prenatal glucocorticoids may have a determining effect on the transcription of mineralocorticoid receptor (MR) and glucocorticoid receptor (GR) genes and thus on the receptor density in

[7]Studies have shown a lack of or only a weak association between perceived stress and stress hormone levels in response to stress (Schommer et al. 2003) or during pregnancy (Davis and Sandman 2010).

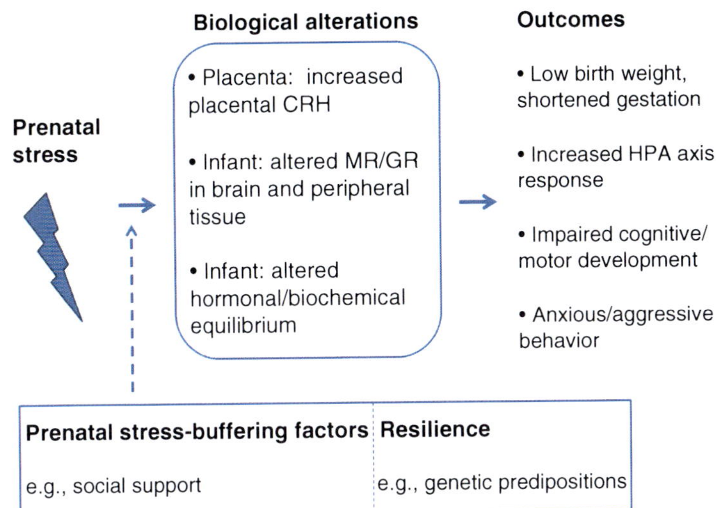

Fig. 6.1 Psychobiological stress model

both brain and peripheral tissue, resulting in permanently altered tissue sensitivity to steroids (Welberg and Seckl 2001). While MRs appear to play a main part in the control of basal HPA axis activity, GRs are involved in the coordination of negative feedback regulation after stress. Thus, reduced GRs sensitivity may explain the augmented HPA axis reactivity in prenatally stressed infants (Cottrell and Seckl 2009).

We discussed the infants' temperament and behavior as possible psychological outcomes of prenatal stress. The reviewed literature indicates that prenatal stress is associated with the infant's distress (as indicated by startle reaction, fussiness) in response to novelty as well as behavioral problems (e.g., attention problems, aggressive/delinquent behavior, or anxiety/depression). These consequences of prenatal stress may be a result of prenatal programming. From an evolutionary point of view, prenatal programming prepares the unborn child for the expected environmental demands. Specifically, increased anxiety (precaution) or aggressive behavior may increase the survival changes in a stressful or dangerous environment. In accordance to this hypothesis, the reviewed studies suggest that prenatally stressed infants are more likely to show such behavior.

Prenatal stressors, particularly with high emotional value, such as traumas/major life events or maternal anxiety/depression during pregnancy, seem to have a negative impact on the infants' mental or motor development. Specifically, children whose mothers experienced stress during pregnancy are more likely to show impairment in attention/concentration, verbal comprehension, reasoning, or memory performances, as well as difficulties in body control and gross and fine motor skills. Prenatal stress possibly alters the structure and function of brain and peripheral tissue and thus provides an explanation for these observed developmental changes. For example, impairment in memory performance due to prenatal stress may be the consequence of structural alterations of the hippocampus and frontal cortex, both of which are particularly sensitive to glucocorticoids.

There is some evidence for protective factors that may moderate the impact of prenatal stress on an infant's development. Genetic predispositions may contribute to the vulnerability or resilience to maladaptive consequences due to prenatal stress. Another moderator might be stress-buffering factors, such as social support or coping strategies. Social support may dampen perceived and physiological stress responses of the pregnant women and thus protect the unborn child from excessive exposure to maternal stress hormones.

Summarizing, it is conceivable that altered biological mechanisms (e.g., HPA axis dysregulation) may mediate the link between prenatal stress and the risk for negative physiological, psychological, behavioral, and developmental outcomes. However, an individual's resilience as well as stress-buffering factors may moderate the direction and magnitude of these consequences.

6.5 Conclusion

We set out to review the literature regarding the question whether prenatal stress of the mother have a negative influence on birth outcomes, biological functions, temperament and behavior, as well as mental and motor development of the newborn child. The existing data indicate that prenatal stress appears to increase the risk for these negative outcomes. We want to point out that prenatal stress does not inevitably lead to negative outcomes. Interindividual differences in these outcomes suggest possible genetic vulnerabilities. It is conceivable that these genetic vulnerabilities unfold their negative effects under conditions of an adverse environment (e.g., prenatal stress). However, prevention and interventions programs in risk population may minimize or even compensate negative consequences of prenatal stress. To the best of our knowledge, only very few intervention studies have been conducted. For instance, one study investigated whether an intervention during pregnancy may influence basal cortisol levels of infants whose mothers are at high risk for depression. For this purpose, a group of low-income women at high risk for depression participated in a stress management training program during pregnancy and was compared to a high-risk comparison group receiving usual care and to a low-risk control group. Infants at the age of 6 months of high-risk mothers receiving usual care showed higher average cortisol levels compared to those of high-risk mothers who received treatment and low-risk mothers (Urizar and Munoz 2011). A randomized control trial study found that pregnant women with anxiety levels, who participated in a relaxation training during pregnancy, were less likely to deliver low-birth weight neonates compared to highly anxious pregnant women receiving usual care (Bastani et al. 2006). Future studies which aim to examine the efficacy of prevention and intervention programs may contribute to improving the health outcomes of high-risk mothers and their children.

The presented studies underline the potential mediating role of the HPA axis in the association between stress and health outcomes. It is well studied that both *HPA axis* and *SAM* have regulating effects on inflammation (Bierhaus et al. 2003). However, the association between prenatal stress and SAM has been scarcely investigated. Clearly, further research is needed.

There are numerous studies in this field using a *prospective longitudinal study design*, i.e., maternal stress during pregnancy and the outcome in the infant are assessed at the time they occur. This procedure allows for accurate information in contrast to retrospective data (cross-sectional designs) that may be confounded by current psychological state or recall bias. Most of these studies investigated a relatively short time period (e.g., birth outcome up to early childhood or early adolescence). However, it would be of great interest to study developmental changes using repeated measures over a time course of several decades. This approach would enable to study short-term as well as long-term effects. Further, possible postnatal stressors (e.g., childhood maltreatment; see this chapter) which may have an additive effect on negative outcomes can be studied.

Studies using animal models strikingly show that the nature of effect depends on the timing when the stressor occurs (Cottrell and Seckl 2009). The reviewed human studies imply that particularly stressors occurring during mid- and late pregnancy (first and second trimester) are periods of high sensitivity to stress. However, the majority of prospective longitudinal studies have been carried out in women during their second or third trimester of pregnancy, possibly because the recruitment of women within 13 weeks of gestation is difficult to implement, compared to women in advanced pregnancy. The few studies that examined stress during early pregnancy often used a cross-sectional study design, i.e., prenatal stress (e.g., occurrence of a traumatic effect) was assessed retrospectively. However, retrospective data are susceptible to recall biases. Based on the literature, it seems impossible to conclude whether or what period of pregnancy is most vulnerable for prenatal programming.

References

APA (ed) (2000) Diagnostic statistical manual of mental disorders (revised 4th edition). APA, Washington, DC

Bastani F, Hidarnia A, Montgomery KS, Aguilar-Vafaei ME, Kazemnejad A (2006) Does relaxation education in anxious primigravid Iranian women influence adverse pregnancy outcomes?: A randomized controlled trial. J Perinat Neonatal Nurs 20(2):138–146

Bierhaus A, Wolf J, Andrassy M, Rohleder N, Humpert PM, Petrov D et al (2003) A mechanism converting psychosocial stress into mononuclear cell activation. Proc Natl Acad Sci U S A 100(4):1920–1925

Bolten MI, Wurmser H, Buske-Kirschbaum A, Papousek M, Pirke KM, Hellhammer D (2011) Cortisol levels in pregnancy as a psychobiological predictor for birth weight. Arch Womens Ment Health 14(1):33–41

Brouwers EPM, van Baar AL, Pop VJM (2001) Maternal anxiety during pregnancy and subsequent infant development. Infant Behav Dev 24(1):95–106

Cannon WB (1935) Stresses and strains of homeostasis. Am J Med Sci 189:1–14
Chrousos GP (2009) Stress and disorders of the stress system. Nat Rev Endocrinol 5(7):374–381
Class QA, Lichtenstein P, Langstrom N, D'Onofrio BM (2011) Timing of prenatal maternal exposure to severe life events and adverse pregnancy outcomes: a population study of 2.6 million pregnancies. Psychosom Med 73(3):234–241
Collins NL, Dunkel-Schetter C, Lobel M, Scrimshaw SC (1993) Social support in pregnancy: psychosocial correlates of birth outcomes and postpartum depression. J Pers Soc Psychol 65(6):1243–1258
Cottrell EC, Seckl JR (2009) Prenatal stress, glucocorticoids and the programming of adult disease. Front Behav Neurosci 3:19
Da Costa D, Dritsa M, Larouche J, Brender W (2000) Psychosocial predictors of labor/delivery complications and infant birth weight: a prospective multivariate study. J Psychosom Obstet Gynaecol 21(3):137–148
Dalziel SR, Liang A, Parag V, Rodgers A, Harding JE (2004) Blood pressure at 6 years of age after prenatal exposure to betamethasone: follow-up results of a randomized, controlled trial. Pediatrics 114(3):373–377
Dalziel SR, Lim VK, Lambert A, McCarthy D, Parag V, Rodgers A et al (2005a) Antenatal exposure to betamethasone: psychological functioning and health related quality of life 31 years after inclusion in randomised controlled trial. BMJ 331(7518):665
Dalziel SR, Walker NK, Parag V, Mantell C, Rea HH, Rodgers A et al (2005b) Cardiovascular risk factors after antenatal exposure to betamethasone: 30-year follow-up of a randomised controlled trial. Lancet 365(9474):1856–1862
Dancause KN, Laplante DP, Oremus C, Fraser S, Brunet A, King S (2011) Disaster-related prenatal maternal stress influences birth outcomes: project ice storm. Early Hum Dev 87(12):813–820
Davis EP, Sandman CA (2010) The timing of prenatal exposure to maternal cortisol and psychosocial stress is associated with human infant cognitive development. Child Dev 81(1):131–148
Davis EP, Snidman N, Wadhwa PD, Glynn LM, Schetter CD, Sandman C (2004a) Prenatal maternal anxiety and depression predict negative behavioral reactivity in infancy. Infancy 6(3):319–331
Davis EP, Townsend EL, Gunnar MR, Georgieff MK, Guiang SF, Ciffuentes RF et al (2004b) Effects of prenatal betamethasone exposure on regulation of stress physiology in healthy premature infants. Psychoneuroendocrinology 29(8):1028–1036
Davis EP, Glynn LM, Dunkel Schetter C, Hobel C, Chicz-Demet A, Sandman CA (2005) Corticotropin-releasing hormone during pregnancy is associated with infant temperament. Dev Neurosci 27(5):299–305
Davis EP, Townsend EL, Gunnar MR, Guiang SF, Lussky RC, Cifuentes RF et al (2006) Antenatal betamethasone treatment has a persisting influence on infant HPA axis regulation. J Perinatol 26(3):147–153
Davis EP, Glynn LM, Schetter CD, Hobel C, Chicz-Demet A, Sandman CA (2007) Prenatal exposure to maternal depression and cortisol influences infant temperament. J Am Acad Child Adolesc Psychiatry 46(6):737–746
Davis EP, Glynn LM, Waffarn F, Sandman CA (2011a) Prenatal maternal stress programs infant stress regulation. J Child Psychol Psychiatry 52(2):119–129
Davis EP, Waffarn F, Sandman CA (2011b) Prenatal treatment with glucocorticoids sensitizes the HPA axis response to stress among full-term infants. Dev Psychobiol 53(2):175–183
de Kloet ER, Oitzl MS, Joels M (1999) Stress and cognition: are corticosteroids good or bad guys? Trends Neurosci 22(10):422–426
Diego MA, Jones NA, Field T, Hernandez-Reif M, Schanberg S, Kuhn C et al (2006) Maternal psychological distress, prenatal cortisol, and fetal weight. Psychosom Med 68(5):747–753
Dole N, Savitz DA, Hertz-Picciotto I, Siega-Riz AM, McMahon MJ, Buekens P (2003) Maternal stress and preterm birth. Am J Epidemiol 157(1):14–24
Dole N, Savitz DA, Siega-Riz AM, Hertz-Picciotto I, McMahon MJ, Buekens P (2004) Psychosocial factors and preterm birth among African American and White women in central North Carolina. Am J Public Health 94(8):1358–1365

Elsenbruch S, Benson S, Rucke M, Rose M, Dudenhausen J, Pincus-Knackstedt MK et al (2007) Social support during pregnancy: effects on maternal depressive symptoms, smoking and pregnancy outcome. Hum Reprod 22(3):869–877

Entringer S, Kumsta R, Nelson EL, Hellhammer DH, Wadhwa PD, Wust S (2008a) Influence of prenatal psychosocial stress on cytokine production in adult women. Dev Psychobiol 50(6):579–587

Entringer S, Wust S, Kumsta R, Layes IM, Nelson EL, Hellhammer DH et al (2008b) Prenatal psychosocial stress exposure is associated with insulin resistance in young adults. Am J Obstet Gynecol 199(5):498.e1–498.e7

Entringer S, Buss C, Kumsta R, Hellhammer DH, Wadhwa PD, Wust S (2009a) Prenatal psychosocial stress exposure is associated with subsequent working memory performance in young women. Behav Neurosci 123(4):886–893

Entringer S, Kumsta R, Hellhammer DH, Wadhwa PD, Wust S (2009b) Prenatal exposure to maternal psychosocial stress and HPA axis regulation in young adults. Horm Behav 55(2):292–298

Entringer S, Epel ES, Kumsta R, Lin J, Hellhammer DH, Blackburn EH et al (2011) Stress exposure in intrauterine life is associated with shorter telomere length in young adulthood. Proc Natl Acad Sci U S A 108(33):E513–E518

Eskenazi B, Marks AR, Catalano R, Bruckner T, Toniolo PG (2007) Low birthweight in New York City and upstate New York following the events of September 11th. Hum Reprod 22(11):3013–3020

Field T, Diego M, Dieter J, Hernandez-Reif M, Schanberg S, Kuhn C et al (2004) Prenatal depression effects on the fetus and the newborn. Infant Behav Dev 27(2):216–229

Finken MJ, Keijzer-Veen MG, Dekker FW, Frolich M, Walther FJ, Romijn JA et al (2008) Antenatal glucocorticoid treatment is not associated with long-term metabolic risks in individuals born before 32 weeks of gestation. Arch Dis Child Fetal Neonatal Ed 93(6):442–447

Fries E, Dettenborn L, Kirschbaum C (2009) The cortisol awakening response (CAR): facts and future directions. Int J Psychophysiol 72(1):67–73

Glynn LM, Wadhwa PD, Dunkel-Schetter C, Chicz-DeMet A, Sandman CA (2001) When stress happens matters: effects of earthquake timing on stress responsivity in pregnancy. Am J Obstet Gynecol 184(4):637–642

Gutteling BM, de Weerth C, Buitelaar JK (2004) Maternal prenatal stress and 4–6 year old children's salivary cortisol concentrations pre- and post-vaccination. Stress 7(4):257–260

Gutteling BM, de Weerth C, Buitelaar JK (2005a) Prenatal stress and children's cortisol reaction to the first day of school. Psychoneuroendocrinology 30(6):541–549

Gutteling BM, de Weerth C, Willemsen-Swinkels SH, Huizink AC, Mulder EJ, Visser GH et al (2005b) The effects of prenatal stress on temperament and problem behavior of 27-month-old toddlers. Eur Child Adolesc Psychiatry 14(1):41–51

Gutteling BM, de Weerth C, Zandbelt N, Mulder EJ, Visser GH, Buitelaar JK (2006) Does maternal prenatal stress adversely affect the child's learning and memory at age six? J Abnorm Child Psychol 34(6):789–798

Henrichs J, Schenk JJ, Roza SJ, van den Berg MP, Schmidt HG, Steegers EA et al (2010) Maternal psychological distress and fetal growth trajectories: the Generation R Study. Psychol Med 40(4):633–643

Henriksen TB, Hedegaard M, Secher NJ (1994) The relation between psychosocial job strain, and preterm delivery and low birthweight for gestational age. Int J Epidemiol 23(4):764–774

Homer CJ, James SA, Siegel E (1990) Work-related psychosocial stress and risk of preterm, low birthweight delivery. Am J Public Health 80(2):173–177

Huizink AC, de Medina PG, Mulder EJ, Visser GH, Buitelaar JK (2002) Psychological measures of prenatal stress as predictors of infant temperament. J Am Acad Child Adolesc Psychiatry 41(9):1078–1085

Huizink AC, Robles de Medina PG, Mulder EJ, Visser GH, Buitelaar JK (2003) Stress during pregnancy is associated with developmental outcome in infancy. J Child Psychol Psychiatry 44(6):810–818

Karasek R, Brisson C, Kawakami N, Houtman I, Bongers P, Amick B (1998) The Job Content Questionnaire (JCQ): an instrument for internationally comparative assessments of psychosocial job characteristics. J Occup Health Psychol 3(4):322–355

Kudielka BM, Wust S (2010) Human models in acute and chronic stress: assessing determinants of individual hypothalamus-pituitary-adrenal axis activity and reactivity. Stress 13(1):1–14

Lederman SA, Rauh V, Weiss L, Stein JL, Hoepner LA, Becker M et al (2004) The effects of the World Trade Center event on birth outcomes among term deliveries at three lower Manhattan hospitals. Environ Health Perspect 112(17):1772–1778

Lee BE, Ha M, Park H, Hong YC, Kim Y, Kim YJ et al (2011) Psychosocial work stress during pregnancy and birthweight. Paediatr Perinat Epidemiol 25(3):246–254

Lundy BL, Jones NA, Field T, Nearing G, Davalos M, Pietro PA et al (1999) Prenatal depression effects on neonates. Infant Behav Dev 22(1):119–129

Luoma I, Tamminen T, Kaukonen P, Laippala P, Puura K, Salmelin R et al (2001) Longitudinal study of maternal depressive symptoms and child well-being. J Am Acad Child Adolesc Psychiatry 40(12):1367–1374

MacArthur BA, Howie RN, Dezoete JA, Elkins J (1981) Cognitive and psychosocial development of 4-year-old children whose mothers were treated antenatally with betamethasone. Pediatrics 68(5):638–643

Maina G, Saracco P, Giolito MR, Danelon D, Bogetto F, Todros T (2008) Impact of maternal psychological distress on fetal weight, prematurity and intrauterine growth retardation. J Affect Disord 111(2–3):214–220

Mancuso RA, Schetter CD, Rini CM, Roesch SC, Hobel CJ (2004) Maternal prenatal anxiety and corticotropin-releasing hormone associated with timing of delivery. Psychosom Med 66(5):762–769

McEwen BS (2003) Interacting mediators of allostasis and allostatic load: towards an understanding of resilience in aging. Metabolism 52(10 suppl 2):10–16

McEwen BS, Gianaros PJ (2011) Stress- and allostasis-induced brain plasticity. Annu Rev Med 62:431–445

McLean M, Bisits A, Davies J, Woods R, Lowry P, Smith R (1995) A placental clock controlling the length of human pregnancy. Nat Med 1(5):460–463

Meyer-Bahlburg HF, Dolezal C, Baker SW, Carlson AD, Obeid JS, New MI (2004) Cognitive and motor development of children with and without congenital adrenal hyperplasia after early-prenatal dexamethasone. J Clin Endocrinol Metab 89(2):610–614

Nkansah-Amankra S, Dhawain A, Hussey JR, Luchok KJ (2010) Maternal social support and neighborhood income inequality as predictors of low birth weight and preterm birth outcome disparities: analysis of South Carolina Pregnancy Risk Assessment and Monitoring System survey, 2000–2003. Matern Child Health J 14(5):774–785

O'Connor TG, Heron J, Golding J, Beveridge M, Glover V (2002) Maternal antenatal anxiety and children's behavioural/emotional problems at 4 years. Report from the Avon Longitudinal Study of Parents and Children. Br J Psychiatry 180:502–508

O'Connor TG, Ben-Shlomo Y, Heron J, Golding J, Adams D, Glover V (2005) Prenatal anxiety predicts individual differences in cortisol in pre-adolescent children. Biol Psychiatry 58(3):211–217

Osmond C, Barker DJ (2000) Fetal, infant, and childhood growth are predictors of coronary heart disease, diabetes, and hypertension in adult men and women. Environ Health Perspect 108(suppl 3):545–553

Oths KS, Dunn LL, Palmer NS (2001) A prospective study of psychosocial job strain and birth outcomes. Epidemiology 12(6):744–746

Paarlberg KM, Vingerhoets AJ, Passchier J, Dekker GA, Heinen AG, van Geijn HP (1999) Psychosocial predictors of low birthweight: a prospective study. Br J Obstet Gynaecol 106(8):834–841

Peltoniemi OM, Kari MA, Lano A, Yliherva A, Puosi R, Lehtonen L et al (2009) Two-year follow-up of a randomised trial with repeated antenatal betamethasone. Arch Dis Child Fetal Neonatal Ed 94(6):402–406

Rothenberger SE, Resch F, Doszpod N, Moehler E (2011) Prenatal stress and infant affective reactivity at five months of age. Early Hum Dev 87(2):129–136

Schlotz W, Kumsta R, Layes I, Entringer S, Jones A, Wust S (2008) Covariance between psychological and endocrine responses to pharmacological challenge and psychosocial stress: a question of timing. Psychosom Med 70(7):787–796

Schommer NC, Hellhammer DH, Kirschbaum C (2003) Dissociation between reactivity of the hypothalamus-pituitary-adrenal axis and the sympathetic-adrenal-medullary system to repeated psychosocial stress. Psychosom Med 65(3):450–460

Seckl JR, Cleasby M, Nyirenda MJ (2000) Glucocorticoids, 11beta-hydroxysteroid dehydrogenase, and fetal programming. Kidney Int 57(4):1412–1417

Selye H, Fortier C (1950) Adaptive reaction to stress. Psychosom Med 12(3):149–157

Spinillo A, Viazzo F, Colleoni R, Chiara A, Maria Cerbo R, Fazzi E (2004) Two-year infant neurodevelopmental outcome after single or multiple antenatal courses of corticosteroids to prevent complications of prematurity. Am J Obstet Gynecol 191(1):217–224

Thompson JM, Sonuga-Barke EJ, Morgan AR, Cornforth CM, Turic D, Ferguson LR et al (2012) The catechol-O-methyltransferase (COMT) Val158Met polymorphism moderates the effect of antenatal stress on childhood behavioural problems: longitudinal evidence across multiple ages. Dev Med Child Neurol 54(2):148–154

Trautman PD, Meyer-Bahlburg HF, Postelnek J, New MI (1995) Effects of early prenatal dexamethasone on the cognitive and behavioral development of young children: results of a pilot study. Psychoneuroendocrinology 20(4):439–449

Turner RJ, Grindstaff CF, Phillips N (1990) Social support and outcome in teenage pregnancy. J Health Soc Behav 31(1):43–57

Urizar GG Jr, Munoz RF (2011) Impact of a prenatal cognitive-behavioral stress management intervention on salivary cortisol levels in low-income mothers and their infants. Psychoneuroendocrinology 36(10):1480–1494

Van den Bergh BR, Marcoen A (2004) High antenatal maternal anxiety is related to ADHD symptoms, externalizing problems, and anxiety in 8- and 9-year-olds. Child Dev 75(4):1085–1097

Van den Bergh BR, Mennes M, Oosterlaan J, Stevens V, Stiers P, Marcoen A et al (2005) High antenatal maternal anxiety is related to impulsivity during performance on cognitive tasks in 14- and 15-year-olds. Neurosci Biobehav Rev 29(2):259–269

Van den Bergh BR, Van Calster B, Smits T, Van Huffel S, Lagae L (2008) Antenatal maternal anxiety is related to HPA-axis dysregulation and self-reported depressive symptoms in adolescence: a prospective study on the fetal origins of depressed mood. Neuropsychopharmacology 33(3):536–545

Vrijkotte TG, van der Wal MF, van Eijsden M, Bonsel GJ (2009) First-trimester working conditions and birthweight: a prospective cohort study. Am J Public Health 99(8):1409–1416

Welberg LAM, Seckl JR (2001) Prenatal stress, glucocorticoids and the programming of the brain. J Neuroendocrinol 13(2):113–128

Xiong X, Harville EW, Mattison DR, Elkind-Hirsch K, Pridjian G, Buekens P (2008) Exposure to Hurricane Katrina, post-traumatic stress disorder and birth outcomes. Am J Med Sci 336(2):111–115

Ziemssen T, Kern S (2007) Psychoneuroimmunology—cross-talk between the immune and nervous systems. J Neurol 254(2):II8–II11

Chapter 7
Consequences of Developmental Stress in Humans: Adversity Experienced During Childhood and Adolescence

Urs M. Nater and Nadine Skoluda

Abstract Early life stress is a term used to describe adversities which may occur during childhood and adolescence. The majority of human studies focus on negative outcomes resulting from early life stress. There is an emerging research field which aims to investigate possible positive outcomes of early life stress. In this chapter, human studies are reviewed with regard to both maladaptive and adaptive consequences of early life stress. Recent findings suggest that early life stress is associated with central nervous alterations, altered physiological stress responses, impaired cognitive functioning as well as an increased risk of developing behavior problems, and somatic and psychiatric illnesses. Early life stress may also result in posttraumatic growth (PTG) which describes changes in self-perception, interpersonal relationships, and worldview. A psychobiological stress model is proposed to integrate the findings.

7.1 Introduction

There is a growing body of literature reporting that early life stress during childhood and adolescence appear to have a negative impact on development and numerous health outcomes later in life. Periods of heightened brain plasticity mark developmental phases during which the brain is most vulnerable to stress. It is well known that exposure to adverse conditions during these critical sensitive phases in development may lead to long-term alterations in brain structures, various bodily functions (e.g., immunological or endocrinological), and behavior, which may facilitate the manifestation of disease.

U.M. Nater (✉) • N. Skoluda
Department of Psychology, Philipps University of Marburg, Gutenbergstrasse 18, Marburg 35032, Germany
e-mail: nater@uni-marburg.de; skoluda@uni-marburg.de

G. Laviola and S. Macrì (eds.), *Adaptive and Maladaptive Aspects of Developmental Stress*, Current Topics in Neurotoxicity 3, DOI 10.1007/978-1-4614-5605-6_7,

There is sufficient evidence that stress-responsive systems[1] play a mediating role in the association between stress exposure and various developmental or health consequences. Stress processing originates in specific brain regions that are responsible for the early perception and evaluation of potentially threatening or stressful features of the current situation. The outcome of such an evaluation may lead to the activation of the stress axes, the sympatho-adrenal-medullary (SAM) system and the hypothalamus–pituitary–adrenal (HPA) axis (for details, see Sect. 2 of Chap. 6). Briefly, the SAM system elicits the release of norepinephrine and epinephrine, both of which are involved in the "flight-or-fight" response. The HPA axis includes the paraventricular nucleus (PVN) in the hypothalamus which releases the corticotrophin-releasing hormone or factor (CRH or CRF, respectively) which in turn induces the secretion of the adrenocorticotropic hormone (ACTH) from the anterior lobe of the pituitary gland into the blood stream. Finally, cortisol is secreted from the adrenal cortex, exerting its effects throughout the body (e.g., on metabolism, immune system, behavior, or cognition). Cortisol has counter-regulating effects on the HPA axis by inhibiting the further release of CRH and ACTH and thus acts as a negative feedback factor. The pulsatile secretion of cortisol follows a diurnal rhythm characterized by high morning levels and a decrease of cortisol levels during the course of the day. The cortisol awakening response (CAR), an important marker for cortisol regulation, describes the morning peak of cortisol levels measured 30 min after awakening and has been related to a number of negative health outcomes (Kudielka and Wust 2010). Thus, studying cortisol release during the morning and/or during the day constitutes a well-researched approach to study HPA axis *activity*. In human research, the most prominent and effective stressors to study HPA axis *reactivity* are (a) performing a cognitive task (e.g., giving a speech and/or solving a difficult arithmetic task) in the presence of an evaluating audience or (b) administration of pharmacological substances (pharmacological challenge test). The latter is used to study HPA axis regulation, particularly via activating the system's inherent feedback mechanisms (Ditzen et al. 2013). Pharmacological challenge tests are used to investigate how HPA axis activity is affected by temporary suppression due to administration of synthetic glucocorticoids (e.g., dexamethasone suppression test, DST), by stimulation due to administration of CRH or ACTH (CRH stimulation test, $ACTH_{1-24}$ stimulation test), or by a time-lagged combination of suppressing and stimulating substances (e.g., dexamethasone/corticotrophin-releasing hormone test, Dex/CRH test).

---

[1] *Stress-responsive systems* refer to various physiological systems that are activated in response to stress. The most prominent systems are the hypothalamus-pituitary-adrenal (HPA) axis, autonomic nervous system (ANS), particularly the sympatho-adrenal-medullary (SAM) system, the central nervous system, and the immune system. We introduce the term *stress axes*, which refer to the HPA axis and the ANS, since these well-studied stress-responsive systems are known to interact and partly regulate the other aforementioned stress-responsive systems.

Considering the amount of literature and the limited space of a book chapter, a selective overview of the literature on stress experienced during childhood and adolescence will be presented.[2] The overall goal of this chapter is to review the literature and give an overview of the current state of knowledge about possible short- and long-term consequences of stress occurring in childhood and adolescence (early life stress).

7.2 Stress During Childhood and Adolescence: Early Life Stress

We introduce the term *early life stress* which refers to stress that occurs during childhood and adolescence. Early life stress describes here a relatively heterogeneous concept which encompasses various adverse events that most people perceive as stressful and thus possibly lead to the activation of the above-mentioned stress-responsive systems. One source of early life stress constitutes childhood maltreatment which in turn can be divided into emotional abuse, physical abuse, sexual abuse, emotional neglect, or physical neglect. Growing up in an orphanage or institutional care provides a setting with possibly increased risk for neglect. One of the most extreme cases of neglect became public after the fall of the Ceausescu regime in Romania. Living conditions in institutional care during that time were characterized by deprivation of basic needs as well as emotional and physical neglect (most of the cited studies in this chapter have been conducted in Romanian orphans). These adverse living circumstances in institutions seem to be more likely under conditions of social poverty or under strict political regimes. Another source of early life stress may be the separation or loss of a parent or someone close. Finally, traumatic events such as natural disasters or war experiences are considered here as early life stress.

Children and adolescents might be particularly prone to stress since certain developmental stages (e.g., puberty) are characterized by various biological and hormonal changes. Early life stress during such sensitive periods may have consequences for biological outcomes, temperament and behavior, mental development, or general health later in life.

---

[2] A systematic literature search was conducted on databases PubMed and Google Scholar using a combination of the following keywords: stress brain, child(hood) maltreatment (outcome), maltreatment infant, early stress, child adversity health outcome(s), parental loss, and posttraumatic growth. The list of abstracts was reviewed with regard to the type of stress and outcome (neuroendocrine, behavioral, psychological, or health-related consequences). Only English-language, peer-reviewed original research papers were included. Additionally, the references of the papers and those of published review articles have been checked for further articles meeting our search criteria. Although we have tried to consider as many publications as possible, we are aware that the presented literature is not comprehensive.

In this chapter we present findings organized along the following outcomes:

- *Effects on the central nervous system (CNS).* This includes alterations in brain structure or brain functions (indicated by volume or activity of certain brain regions) using neuroimaging methods. Since recent research has revealed important central nervous effects of the neuropeptide oxytocin in humans (e.g., interpersonal bonding, affiliation behavior, and stress reduction), we present preliminary findings for this hormone here.
- *Effects on the HPA axis.* Findings are separately presented for basal HPA axis activity (e.g., cortisol diurnal profile or CAR) and HPA axis reactivity in response to a psychological stressor or a pharmacological challenge test.
- *Effects on temperament and behavior.* Measurement of personality or temperament is often achieved through assessing the infant's attachment style, which describes how secure the bonding between infant and primary caregiver is. Negative behavior is usually operationalized as internalizing problems (withdrawal, somatic complaints, anxiety/depression) or externalizing problems (aggressive, delinquent behavior) which are thought to be disadvantageous manifestations of behavior.
- *Effects on cognitive functioning and academic performance.* Developmental tests and intelligent tests are used to measure executive functions (e.g., decision making, response control), attention, memory, verbal skills, reasoning, and general intellectual/academic performance.
- *Effects on somatic health.* Somatic health is measured either as perceived general health or as the occurrence of specific somatic illnesses (e.g., infectious disease).
- *Effects on psychological health.* Psychological health may be measured via assessing symptoms such as depression, anxiety, or suicidal thoughts, and via diagnosing clinical disorders, such as personality disorders, major depressive disorder, anxiety disorders, and substance abuse.

In general, the traditional research focus in this field lies on negative outcomes of stress. Therefore, the majority of findings present maladaptive consequences in this chapter. In the next paragraphs we will present a selection of findings related to these effects, with maladaptive outcomes measured during childhood and adolescence (2.1.), maladaptive outcomes measured during adulthood (2.2.), and adaptive outcomes in children, adolescents, and adults (2.3.).

7.2.1 Maladaptive Outcomes Measured During Childhood and Adolescence

Studying outcomes of early life stress in children and adolescents may provide valuable information about short-term effects, since adverse events may have taken place only recently. In the following, possible effects on the CNS, the HPA axis, temperament and behavior, cognitive functioning and academic performance, somatic health, and psychological health are discussed.

Effects on the CNS. Several neuroimaging studies suggest that maltreated children and adolescents have small volumes of selected temporal regions (De Bellis et al. 2002) and the orbitofrontal cortex (Hanson et al. 2010), both of which are involved in executive and cognitive functions (e.g., reasoning, memory) and emotion regulation, as well as a small corpus callosum (Teicher et al. 2004), which connects both hemispheres. In another study, adopted orphans showed decreased activity in some of the aforementioned brain regions (specifically in the orbitofrontal cortex, temporal cortex, and brain stem) compared to same-aged children (7-11 years) with medically intractable partial epilepsy or to healthy adults (Chugani et al. 2001). De Bellis and colleagues, using both a cross-sectional (De Bellis et al. 2002) and a longitudinal approach (De Bellis et al. 2001), failed to show that childhood maltreatment was related to altered hippocampus or amygdala volumes, i.e., brain regions which are thought to be associated with emotion regulation and memory. In summary, there are some findings suggesting that early life stress may be associated with structural and functional alterations of brain regions that are related to cognitive and emotional functioning.

Effects on the HPA axis. There are numerous studies that investigated HPA axis activity (using basal cortisol levels) with regard to early life stress. One study showed that adoptees who spent more than 9 months in an orphanage displayed augmented daily cortisol levels, suggesting a generally enhanced HPA axis activity, compared to early adopted children and non-adopted children (Gunnar et al. 2001). Other studies focused on cortisol levels at a specific time of the day, mostly morning cortisol levels. One study found that neglected children (aged 3–31 months), still living with their birth parents, showed lower waking cortisol levels and significantly higher cortisol levels at bedtime compared to maltreated children who had been placed in foster care and compared to non-maltreated children (Bernard et al. 2010). In line with this finding, another study has found that sexually abused 5–7-year-old girls with posttraumatic stress disorder (PTSD) showed lower morning cortisol levels than control girls (King et al. 2001). In contrast to the aforementioned findings, sexually and physically abused children (mean age, 9.25 years) displayed higher morning cortisol levels than non-maltreated, emotionally maltreated, neglected, or physically abused subgroups who were studied during a day camp research program (Cicchetti and Rogosch 2001). A possible explanation for the inconsistent results in morning cortisol levels may be differences in age (range, 3 months–9 years) and time of morning sampling (range, 5–12 am) among these studies.

In addition to basal measurement, studying HPA axis reactivity specifically contributes to the understanding of underlying mechanisms of HPA axis (dys-)regulation. Healthy non-abused adolescents, aged 12–16 years, showed increased cortisol levels and a gradual flattening over time following an acute stress test, while nondepressed maltreated adolescents did not show such a pattern in cortisol levels (MacMillan et al. 2009). Another study found that the presence of any anxiety disorder, early life stress, or chronic stress during adolescence predicted peak cortisol response to a psychological stress test, whereby the interaction of early life stress and chronic stress during adolescence was the best predictor of this stress response (Rao et al. 2008). Another study reported that intravenous CRH administration led to greater ACTH peaks in depressed abused children (age range, 7–13 years) than in depressed, non-abused children or children with no history of either depression or abuse, whereas no group differences were apparent for cortisol responses (Kaufman et al. 1997). These studies hint at an altered HPA axis response under challenge; however, the direction of this response appears to be unclear. It is conceivable that anxious or depressive symptoms partly determine the direction and extent of the stress response, leading to either hyper- or hyporeactivity of the HPA axis.

Effects on temperament and behavior. In institutions such as orphanages, caregivers have to provide care and supervise several children at the same time, which may make it difficult to meet the child's individual needs and to form selective interpersonal bonding between a primary caregiver and the child (Drury et al. 2011). As forming a child-caregiver bond may thus be limited in such an environment, several studies have examined attachment styles in children with a history of institutional care. One study in adoptees found insecure attachment styles and high occurrence of indiscriminately friendly behavior, as determined by parents' reports (Chisholm 1998). In another study, adoptees showed barely any attachment behavior in a strange situation procedure, which often serves as a tool to assess an infant's attachment style to a primary caregiver (Zeanah et al. 2005). These findings suggest that experiences of institutional care may be a risk factor for maladaptive temperament, as measured via attachment styles.

There is also some evidence that early life stress may be related to behavioral problems. For example, one study found that adolescents, who had been physically abused in early childhood, were more likely to have been arrested for violent and nonviolent offenses than non-abused adolescents (Lansford et al. 2007). Another study found gender differences among sexually abused adolescents, with boys showing more sexual risk behavior and delinquent behavior than girls (Chandy et al. 1996). A summer camp study, using a longitudinal design, found that maltreated children exhibited significantly greater externalizing behavior problems and more impaired affect and behavior regulation (lower levels of ego-resiliency and higher levels of ego-undercontrol[3]) than non-maltreated children (Kim et al. 2009). In another study, children with a history of institutional care were more likely to show externalized or internalized disorders than never-institutionalized children (Zeanah et al. 2009). Similarly, sexually abused children and adolescents had more externalizing and internalizing symptoms than those without a history of abuse in another study (Maikovich-Fong and Jaffee 2010). In line with these findings,

---

[3]Ego-control describes to the degree to which individuals express or inhibit their impulses. Individuals with ego-undercontrol are characterized by spontaneous, emotionally expressive behavior, and pursuit of immediate gratification of desires. Ego-resiliency refers to the individuals' ability to modulate these impulses adaptively in response to situational demands and affordances (Huey and Weisz 1997; Letzring et al. 2005).

maltreatment-related PTSD was associated with internalizing and externalizing symptoms in children and adolescents (De Bellis et al. 2002). Finally, parent–child aggression was found to be associated with conduct problems, attention problems, anxiety-withdrawal, and motor excess in boys, whereas girls only showed increased anxiety-withdrawal (Jouriles et al. 1987). Considering the available literature, early life stress seems to raise the risk for attachment disturbance and behavioral problems.

Effects on cognitive functioning and academic performance. Evidence for possible cognitive alterations in affective children or adolescents comes from studies examining academic performance or performance in standardized tests. One study found that physically abused adolescents were less likely to have graduated from high school than non-abused peers (Lansford et al. 2007). Studying gender differences in maltreated adolescents revealed that male adolescents with a history of sexual abuse tended to report performing below average and had a higher dropout risk than sexually abused female adolescents (Chandy et al. 1996). In another study, maltreated adolescents were found to score lower on tests assessing reading performance and reasoning compared to non-maltreated adolescents (14 years) (Mills et al. 2011). A longitudinal study found that children living in an orphanage (measured at the age of 42 and 54 months) showed more cognitive impairment (e.g., verbal abilities) than orphans who were placed in a foster family before the age of 31 months (Nelson et al. 2007). This same study suggested an age-associated beneficial effect of foster care, i.e., the younger the child at the time of entry in foster care, the more cognitive improvement due to foster care was found. In another study, 8-year-old children with any orphanage experiences in early life were found to perform worse on tests requiring visual memory and executive function (e.g., decision making or response control) compared to peers without a history of institutional care (Bos et al. 2009). Another study partly confirmed these findings, i.e., adoptees at the age of 8 years showed deficits in visual memory, attention, and associative learning measures but not in executive functions (Pollak et al. 2010). Mild impairment in verbal skills, impulsivity, and attention were reported for adoptees compared to never-institutionalized children with medically intractable partial epilepsy (Chugani et al. 2001). In a longitudinal study, non-maltreated children showed a pronounced decline in attention problems over the course of time, while in maltreated children attention problems slightly increased between age 4 and age 6 and then remained consistently on a high level until age 10 (Thompson and Tabone 2010). These findings clearly indicate that early life stress has negative effects on a child's cognitive development, as indicated by early school leaving and poor academic performance, as well as poor results in tests assessing memory, verbal skills, or attention.

Effects on somatic health. Several findings suggest that early life stress may negatively affect a child's health. One study showed that parents of maltreated 4–6-year-old children rated their child's health as poor more often and reported more serious illness when the child suffered from repeated adverse events than children without such adversities (Flaherty et al. 2006). In another study, childhood maltreatment occurring before the age of 11 years was found to increase the risk of asthma by 73% and even double the risk of non-asthma cardiorespiratory and nonsexually transmitted infectious diseases in adolescents (Lanier et al. 2010). These studies indicate a heightened risk for diseases later in life due to early life stress.

Effects on psychological health. Apart from detrimental effects of early life stress on somatic health, the psyche may be affected as well. One study found that single and multiple negative early life events (e.g., loss of someone close, parental separation, serious conflicts) were associated with depression disorder in adolescents (Patton et al. 2003). In another study, the report of repeated childhood sexual abuse was associated with personality disorder in a dose–response manner, going together with higher scores on neuroticism and lower agreeableness (Moran et al. 2011). In a longitudinal study of children at the age between 4 and 10 years, maltreated children were found to show a steeper increase of anxiety/depression over time (Thompson and Tabone 2010). Other studies have confirmed this finding for high levels of depressive symptoms (Schilling et al. 2007), whereby girls seemed to be more affected (Chandy et al. 1996). High depression scores have been found in female adolescents who experienced a relatively mild life event in the past year (i.e., new stepbrother or stepsister; (step)brother or (step)sister leaving home and father losing his job) and who had at least one parent having a history of depression or anxiety (Silberg et al. 2001). These studies suggest that early life stress may enhance the risk for showing high levels of depression and personality disorders.

To summarize these findings, early life stress appears to be associated with decreased volumes of certain brain regions which are known to regulate memory and emotion (e.g., corpus callosum, orbitofrontal cortex, and temporal regions). Inconsistent results were reported for HPA axis activity and reactivity, a finding that may be partly explained by different study designs. Externalizing and internalizing problems, as well as antisocial behavior were predominately present in children and adolescents with early life stress. Also, early life stress was associated with early school leaving and impaired cognitive performance (e.g., memory, verbal skills, or attention). Children or adolescents, who experienced early life stress, seem to be at high risk for somatic and psychiatric conditions.

7.2.2 Maladaptive Outcomes Measured During Adulthood

Numerous studies investigated adults who retrospectively reported having experienced early life stress. These studies provide valuable information whether early life stress may have long-lasting effects on health later in life. In the following paragraphs, a selection of findings for the effect on the CNS, the HPA axis, temperament and behavior, cognitive functioning and academic performance, somatic health, and psychological health are presented.

Effects on CNS. Several neurocognitive studies have found small left hippocampus volumes, a brain structure which is associated with memory, in abused females with depression (Vythilingam et al. 2002) or anxiety disorder (Stein et al. 1997). Adults with a history of early life stress were found to have small anterior cingulated cortex and caudate nuclei (located within basal ganglia), brain regions that are associated with attention and motor control, respectively (Cohen et al. 2006). In another study, PTSD women with childhood abuse showed increased affective responses to a trauma script as well as decreased activity in the anterior cingulated cortex and right hippocampus, both of which are related to memory (Bremner et al. 1999). These studies suggest that childhood maltreatment seems to be associated with decreased volume and activation of memory-related brain regions. Oxytocin is suggested to play an important role in bonding and attachment and thus may influence interpersonal relationships and interactions. One study found that early life stress was related to reduced plasma oxytocin levels (Opacka-Juffry and Mohiyeddini 2011). Similarly, another study found that reduced cerebrospinal fluid oxytocin levels were associated with childhood maltreatment, especially in females with severe childhood maltreatment (Heim et al. 2009). These two studies indicate that early life stress is related to decreased oxytocin levels later in life, which may be associated with maladaptive bonding and attachment in personal relationships.

Effects on the HPA axis. Measuring basal cortisol levels enables to study alterations in HPA axis activity in adults who have experienced early life stress. One study found that adults with a current anxiety disorder, who also had been severely maltreated before being adopted, demonstrated lower morning cortisol levels and a flattened diurnal cortisol secretion (van der Vegt et al. 2010). Another study found positive associations between severe childhood maltreatment and elevated cortisol levels throughout the day in women with either fibromyalgia or osteoarthritis (Nicolson et al. 2010). However, physically and psychologically healthy women, who experienced childhood maltreatment, did not differ in their diurnal cortisol pattern from those without such adverse experiences (Klaassens et al. 2009). The inconsistency in these findings for diurnal cortisol levels might be partly ascribed to the different samples (i.e., patient sample with an anxiety disorder, pain patients, and healthy subjects without a current somatic or psychiatric disorder). In contrast, a relatively consistent picture appears for the effect of early life stress on the CAR, which describes a rapid increase of cortisol concentrations within 30 min after awakening and is usually considered a valid index of an intact HPA axis regulation. A decreased CAR has been found in adults with chronic fatigue who have been maltreated during childhood (Heim et al. 2009), in adults who were severely maltreated before adoption (van der Vegt et al. 2009), or in adults who lost a parent before the age of 14 (Meinlschmidt and Heim 2005). Based on these findings, a preliminary conclusion can be drawn in that adults with childhood maltreatment experiences may show attenuated basal morning cortisol levels.

In addition, there is a variety of studies which investigated the impact of early life stress on HPA axis reactivity using various approaches. Studying the HPA axis response to psychological stress, females with a history of childhood abuse and current PTSD showed higher cortisol levels before, during, and immediately after exposure to a trauma script than abused women without current PTSD (Elzinga et al. 2003). Giving a speech about a controversial topic was used as a stressor in another study, with the finding that adults with parental loss experiences before the age 16 showed increases of cortisol concentrations in response to a stress test in contrast to a control group (Luecken 1998). In another study, both abused women with and without current major depression showed increased ACTH responses to a psychological stress test compared to non-abused women (Heim et al. 2000). In the same study, only depressed women with a history of childhood abuse showed higher cortisol responses to the stressor than depressed women without childhood abuse and nondepressed women with or without childhood abuse experiences. In the same sample both the maximum ACTH and cortisol responses to a psychological stress test were predicted by the history of childhood abuse and the number of distinct types of abuse events (Heim et al. 2002). Another study found that adults, who retrospectively reported multiple adverse events, exhibited a smaller cortisol response to a psychological stress test than control subject; interestingly, this effect was more pronounced in male subjects (Elzinga et al. 2008). Lovallo et al. (2012) investigated whether early adverse life events (e.g., being mugged, threatened with a weapon, experiencing a robbery, sexual assault, or parental separation before age 15) have an impact on endocrine stress reactivity in adults without a psychiatric disorder. Adults who retrospectively reported more such adversities showed smaller cortisol responses to a psychological stressor. These findings provide some evidence that early life stress may be associated with altered HPA axis responses to psychological stressors.

Another approach to study HPA axis reactivity is using pharmacological challenge tests. Using the low-dose dexamethasone suppression test (DST), a test which is used to examine HPA axis negative feedback sensitivity, depressed women with childhood trauma were found to have lower cortisol and ACTH afternoon levels (super-suppression) compared to abused nondepressed women or non-abused healthy controls, whereas no difference was found when the test was conducted using the standard dose (Newport et al. 2004). The administration of CRH or ACTH stimulates the HPA axis with results in secretion of subsequent hormones of the HPA axis and thus is used to study the integrity of top-down HPA axis regulation. In one study, abused nondepressed women showed higher mean stimulated ACTH concentrations 5 and 30 min after the CRH stimulation test compared to healthy non-abused controls; in contrast, depressed women (regardless of the presence of history of childhood) showed lower CRH-stimulated ACTH levels than controls. In the same study both abused women with and without major depression demonstrated lower baseline and stimulated cortisol levels than control subjects (Heim et al. 2001). Further, the same sample underwent an $ACTH_{1-24}$ stimulation test, with the finding that abused females without depressive symptoms demonstrated decreased basal and stimulated cortisol concentrations, whereas abused depressed females had only lower baseline cortisol values compared to controls (Heim et al. 2001). The combined dexamethasone/corticotrophin-releasing hormone (Dex/CRH) test is a useful tool to study HPA axis sensitivity. One study found that abused men with current major depression demonstrated augmented ACTH and cortisol responses to the Dex/CRH test (Heim et al. 2008). Another study corroborated these increased cortisol levels in response to the Dex/CRH test in adults who had lost a parent before the age of 18 years. However, no effect of parental loss was present for ACTH responses (Tyrka et al. 2008). In another study, adults with a history of childhood maltreatment exhibited nonsignificant decreased ACTH and cortisol responses to a Dex/CRH test after adjusting for potential confounders (age, menopausal stage, body mass index, use of contraceptives) (Klaassens et al. 2009). In contrast to the aforementioned findings for cortisol levels, Carpenter et al. (2009) found attenuated cortisol responses to the Dex/CRH test in adults who had been emotionally abused as a child. These contradicting results might be partly explained by the type or severity of early life stress, as well as by specific genetic factors. As for the latter, a protective CRHR1 receptor polymorphism in males has been found, i.e., maltreated men with this polymorphism showed low cortisol responses to the Dex/CRH test (Heim et al. 2009). These preliminary findings indicate generally low basal HPA axis activity, accompanied by altered HPA axis reactivity. Whether hyper- or hyporeactivity is predominant appears to be partly determined by the type of HPA axis stimulation, genetic factors, or comorbidity with a psychological disorder.

Effects on temperament and behavior. Using a longitudinal design, childhood maltreatment was found to be associated with internal and external problems in early adulthood (Silverman et al. 1996). One cross-sectional study found that adults, who were maltreated as child, were at high risk of showing antisocial behavior and conduct disorders (Nelson et al. 2002). This was corroborated in two longitudinal studies which found that young adults reporting early life stress, such as parental separation or childhood maltreatment, had an increased rate of antisocial behavior (Schilling et al. 2007; Thornberry et al. 2010). There is some evidence for genetic contributions that seem to moderate the association between childhood maltreatment and antisocial behavior. In particular, several studies found a moderating role of the polymorphism in the monoamine oxidase (MAO)-A gene, which encodes the activity of the enzyme MAO-A, which in turn metabolizes monoamines, i.e., neurotransmitters such as norepinephrine, serotonin, and dopamine. It was found that men who had a history of childhood maltreatment and a polymorphism associated with low MAO-A activity showed an increased risk for antisocial behavior compared to maltreated men with high MAO-A activity (Beach et al. 2010a, b; Caspi et al. 2002). These studies suggest that early life stress may increase the risk for behavioral problems in adulthood.

Effects on cognitive functioning and academic performance. One study showed that adults with a history of childhood abuse had greater verbal short-term memory deficits than controls but no difference was present in intelligence scores (Bremner et al. 1995). In accordance to this finding, two other studies found impaired memory, particularly affecting the verbal domain, in adults who reported childhood maltreatment, whereas no negative effect on attention or executive function (e.g., reasoning or planning abilities) was observed (Majer et al. 2010; Navalta et al. 2006). These studies indicate memory impairment in adults who experienced early life stress.

Effects on somatic health. Emotional abuse was found to be related to obesity in men but not in women (Gunstad et al. 2006). A representative survey found that adults, who retrospectively reported childhood physical abuse, showed a higher risk for having a heart disease than non-abused adults (Fuller-Thomson et al. 2010). In another study, adults with a history of multiple adverse childhood experiences (such as abuse, domestic violence, criminal familial environment) were more likely to be hospitalized due to autoimmune disease compared to adults without such a history (Dube et al. 2009). One study found that women with a history of childhood sexual abuse were more likely to experience chronic fatigue, asthma, or cardiovascular problems than female controls, whereas physically abused women suffered more from chronic pain compared to controls (Romans et al. 2002). In accordance with the aforementioned finding, physically abused women were found to suffer more from chronic pelvic pain and chronic low-back pain, and women with severe sexual abuse during childhood reported chronic pelvic pain more often compared to controls (Lampe et al. 2003). Two studies provide some evidence for positive associations between childhood maltreatment experiences and heightened risk of chronic fatigue syndrome (Heim et al. 2009; Heim et al. 2006). Another study reported that emotional abuse or physical neglect raised the risk for chronic fatigue syndrome, irritable bowel syndrome, fibromyalgia, and arthritis (Tietjen et al. 2010). In summary, early life stress appears to increase the risk for negative health outcomes later in life.

Effects on psychological health. There are several studies which investigated the association between early life stress and the occurrence of psychological disorders based on clinical diagnoses according to international classification systems, such as the ICD-10 (WHO 2004) or the DSM-IV (APA 2000). For example, one epidemiological survey found that early life stress increased the risk for psychological disorders in adulthood, including affective, anxiety, or substance disorders (Fujiwara et al. 2011). Another study found that both childhood sexual or physical abuse was related to an increased risk for depression, PTSD, alcohol, or drug dependence (Silverman et al. 1996). Another study reported similar findings, with self-reported childhood sexual abuse being associated with depression, general anxiety disorder, panic disorder, alcohol, or drug dependence (Kendler et al. 2000). Children and adolescents (age range, 6–18 years) who had lost their father during the Kosovo War were more likely to show major depression, posttraumatic disorder, or panic disorder 10 years later compared to peers with similar war experiences but no father loss (Morina et al. 2011). Two other studies found more symptoms of personality disorders in adults who retrospectively reported childhood maltreatment compared to non-maltreated adults (Johnson et al. 1999; Tyrka et al. 2009). One study found that adults, who retrospectively reported childhood sexual abuse, were at high risk to develop depression (Kendler et al. 2004). Another study showed that depressive patients reported the loss of the mother before the age of 17 years more often than controls; this negative effect was not found for the loss of the father (Kunugi et al. 1995). Another study found that men with alcohol dependence did not differ from controls with regard to parental loss before the age of 16 (Furukawa et al. 1998).

Numerous studies operationalized psychological health by assessing psychological symptoms (e.g., depression or anxiety levels measured by questionnaires) or high-risk behavior (e.g., regular use of drugs), which do not meet fully the criteria of psychological disorders. However, these psychological symptoms and risk behavior may give indications for psychological well-being or risk to develop a psychological disorder in the future. One cross-sectional study (Nelson et al. 2002) and two longitudinal studies (Schilling et al. 2007; Thornberry et al. 2010) found an increased rate of drug use in young adults when they had experienced early life stress. Thoughts and attempts of suicide can be seen as indicators for depression. Several studies found that the risk for suicide thoughts or attempts was dramatically increased in adults with a history of early life stress (Dube et al. 2001; Thornberry et al. 2010). A great number of studies give evidence that early life stress, in particular childhood maltreatment or parental loss, may be associated with increased levels of depression (Kiecolt-Glaser et al. 2011; Lampe et al. 2003; Nelson et al. 2002; Wingo et al. 2010), or both anxiety and depression (Handa et al. 2008; Heim et al. 2009; Heim et al. 2006; Nicolson et al. 2010; Tyrka et al. 2008). Summarizing these findings, early life stress seems to be associated with an increased risk for psychiatric disorders in adulthood.

To summarize the findings, early life stress may be associated with reduced volumes of memory-related brain regions (e.g., hippocampus). Adults, who retrospectively reported early life stress, appear to have decreased levels of oxytocin, which is known to be related to bonding. Low morning cortisol levels seem to be characteristic for adults with a history of early life stress. Early life stress may be associated with altered HPA axis responses to psychological stressors and to pharmacological challenge tests. Further, early life stress may be related to behavioral problems, such as antisocial behavior. The findings suggest that early life stress may influence negatively memory, particularly of verbal material. Adults with a history of early life stress may be more likely to suffer from negative somatic health outcomes (e.g., pain disorder, chronic fatigue) or negative psychological health outcomes (e.g., depression, anxiety, drug abuse, personality disorders) later in life.

7.2.3 Adaptive Outcomes Measured in Children, Adolescents, and Adults

The previous sections have delineated maladaptive outcomes of early life stress, which is the traditional focus in human research. Adaptive consequences of early life stress are often disregarded in human research. In the last years, however, researchers have started to investigate adaptive outcomes of developmental stress, which will be briefly summarized in the next paragraphs.

There are three concepts that may provide approaches to study adaptive consequences of early life stress: (a) resilience, (b) compensating factors to reduce negative effects of stress, and (c) posttraumatic growth (PTG); we present findings particularly for the latter concept, since PTG comes most closely to the idea of adaptive consequences.

Resilience in this context describes the individual's ability to develop in a normal range despite experienced early life stress (see also review, Feder et al. 2009). A resilient individual may have genetic dispositions (protective polymorphism; see also conclusion) which protect the individual from negative consequences of adverse conditions. Also adaptive coping styles (e.g., cognitive reappraisal) and resilient personality factors (e.g., dispositional optimism) may characterize resilience in an individual.

Individuals who are vulnerable to stress and consequently are at high risk for maladaptive consequences due to early life stress may benefit from compensating factors. These include various types of social support (e.g., instrumental, emotional support) and a supportive social network which may buffer the stress impact (Ozbay et al. 2007). Interventions such as educational support, assistance in academic performance (tutoring, coaching), and public health programs may diminish developmental deficits and enable a "normal development" of the infant.

Finally, the concept of PTG proposes possible positive effects resulting from experiencing and overcoming an adverse event (see review, Meyerson et al. 2011). According to the theory of PTG, an adverse event (trauma) may change the individuals' concept of themselves, relationship to others, and life in general. The Posttraumatic Growth Inventory (PTGI) has been developed to assess PTG; there are separate versions for both children and adults (Kilmer et al. 2009; Tedeschi and Calhoun 1996). This scale comprises 5 domains, which may be changed due to trauma experience: (1) perception of self/personal strength (e.g., being stronger than individual thought), (2) relationship to others (e.g., closer relationship to family and friends), (3) appreciation of life (e.g., changing priorities), (4) new possibilities (e.g., new interests and goals), and (5) spiritual change (e.g., strengthening faith). There is some evidence that children who have experienced a traumatic event (e.g., injury or death of a loved one, natural disaster, or accident) report higher PTG scores than children without such trauma experiences (Alisic et al. 2008). The PTGI has been successfully administered in children and adults of various types of early life stress: cancer during childhood or adolescence (Barakat et al. 2006; Kamibeppu et al. 2010; Turner-Sack et al. 2012), serious or chronic illness in childhood (Devine et al. 2010), death of a close one (Ho et al. 2008; Wolchik et al. 2008), traffic accidents (Salter and Stallard 2004), childhood sexual abuse (Shakespeare-Finch and de Dassel 2009), terror exposure (Laufer et al. 2010; Levine et al. 2008; Levine et al. 2009), or exposure to natural disasters (Cryder et al. 2006; Hafstad et al. 2010; Kilmer and Gil-Rivas 2010; Yu et al. 2010).

This research trend contributes to the understanding of a comprehensive picture of early life stress and its consequences on the individual and may further provide applicable interventions and treatments for individuals who suffered from early life stress.

7.3 Integration and Interpretation of the Findings

The findings presented in this book chapter suggest that stress experienced during childhood or adolescence may have long-lasting consequences long after exposure to adversity has stopped. We propose a psychobiological stress model that is used to integrate and interpret some of these findings (Fig. 7.1).

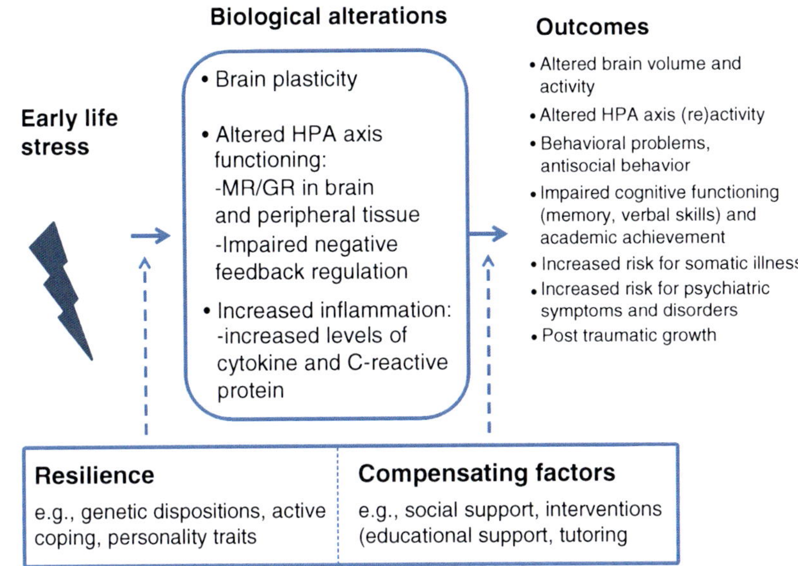

Fig. 7.1 Psychobiological model

It is established that the human brain is particularly sensitive to stress in early life due to the heightened brain plasticity during this vulnerable period. The HPA axis appears to be involved in such functional and structural brain alterations. Specifically, there are several brain regions with a particularly high density of glucocorticoid receptors (GR), e.g., hippocampus and prefrontal cortex, both of which regulate HPA axis activity by negative feedback mechanisms. Consequently, excessive cortisol secretion due to early life stress may result in persistent altered GR sensitivity in these brain regions. This in turn may lead to functional impairments, particularly when the individual is exposed to stress and thus to glucocorticoids. Impaired memory, which was found in both children and adults with a history of early life stress, may be related to such altered GR sensitivity of the memory-associated brain regions (e.g., hippocampus and prefrontal cortex). Animal models showed that glucocorticoid overexposure may have neurotoxic effects, particularly in the CA3 region of the hippocampus, which may explain decreased hippocampal volumes as the result of excessive stress.

Specifically, the findings summarized above suggest that early life stress may negatively affect HPA axis activity and reactivity. Heim and colleagues (Heim et al. 2008) hypothesized that early life stress may lead to the development of highly sensitive stress-responsive systems which is indicated by increased psychological and biological responses to adulthood stress, accompanied by alterations in HPA axis regulation. In particular, the authors suggest that increased ACTH secretion, as it has been observed in the literature, may be mediated by downregulation of pituitary CRH receptors and impaired negative feedback regulation of the HPA axis under stimulation.

Early life stress was found to negatively affect both somatic and psychological health which has been also largely demonstrated in studies using animal model (Cirulli et al. 2009; O'Mahony et al. 2009; Schmidt et al. 2011). There is sufficient evidence that diseases or psychological disorders are related to hypo- or hyperactivity of the HPA axis as a result of stress (Chrousos 2009). Since the stress axes and the immune system interact (e.g., cortisol has suppressive or stimulating effects on immune processes), dysregulation of the HPA axis may contribute to the development and maintenance of conditions in which the immune system plays an important pathophysiological role. There are a handful of studies which investigated the association between early life stress and inflammation, which is an approach to study immune activity. These studies found increased levels of inflammatory markers. For example, adults who had been maltreated as child showed higher basal interleukin-6 levels and higher tumor necrosis factor (TNF)-alpha levels than controls (Kiecolt-Glaser et al. 2011) as well as increased IL-6 responses to a psychological stressor (Carpenter et al. 2010), indicating increased inflammation. A longitudinal study found that childhood maltreatment was associated with increased C-reactive protein (CRP) levels in adults, indicating systemic inflammation (Danese et al. 2009). A persistent inflammation may increase the risk for developing diseases later in life; for example, high CRP levels appear to be related to cardiovascular diseases and high TNF-alpha levels to rheumatoid arthritis.

Besides the aforementioned negative outcomes of early life stress, there is also some evidence for adaptive consequences of stress. Some individuals report that the evaluation of themselves, relationships to others, and life in general has changed in a positive direction after the exposure of a traumatic event, which in essence describes the idea of PTG. Early life stress may thus force individuals to reappraise and reorganize general concept of themselves, their world, and the future. There is some evidence that expected maladaptive consequences of early life stress may be alleviated or even inhibited by an individual's protective factors (resilience) or compensating factors (e.g., social support), which have biological correlates and pathways (Feder et al. 2009). These protective and compensating factors may act as a moderating role in the association between early life stress and (mal)adaptive outcomes.

This psychobiological stress model proposes a biological link between stress and (mal)adaptive consequences. The brain plays a central role in this model. Both maladaptive and adaptive consequences may result from brain plasticity and active cognitive reappraisals.

7.4 Conclusion

This book chapter summarized the main findings of the last few decades of research on the impact of early life stress in humans. However, despite the relative abundance of studies, some questions which should be addressed in future studies remain. In the following, we want to give an outlook on current research and future directions by addressing still insufficiently studied research questions.

7.4.1 Underlying Biological Mechanisms for Associations Between Early Life Stress and Negative Outcomes

There is great evidence that early life stress has extensive negative effects on mind and body. These negative effects are probably partly mediated by alterations of the HPA axis, one of the most prominent and most studied stress-responsive systems. However, this mediation cannot explain all findings; thus, other stress-responsive systems and factors should be taken into account in order to understand the underlying mechanisms of how early life stress may lead to negative outcomes. To the best of our knowledge, the association between early life stress and the SAM system has been scarcely investigated. However, it is known that alteration of both the HPA axis and the SAM system may increase the risk for diseases (McEwen and Seeman 1999). Thus, further research is needed incorporating assessment of both the HPA axis and the SAM system. Another stress-responsive system, which is scarcely investigated with regard to early life stress, is the above-mentioned immune system. First studies, however, suggest that early life stress may be associated with enhanced inflammation, indicating an enhanced activity of the immune system. Clearly, the intricate relationships between stress axis and the immune system should be considered in future studies in order to further elucidate the detrimental effects of early life stress on both psychological and somatic health.

7.4.2 Risk and Protective Factors (Vulnerability and Resilience)

In general, the presented studies in this chapter suggest that early life stress may increase the likelihood or *risk* for negative outcomes, but not that early life stress per se inevitably leads to negative outcomes. It is conceivable that stress may play a triggering role, and that stress and other potential risk factors (vulnerability), but also protective factors (resilience) may be relevant in whether negative health outcomes manifest or not. There is some evidence from the literature for polymorphisms that might act as risk or protective factors. However, these alone cannot explain the variance of stress-related morbidity. We should not neglect the role of environmental risk and protective factors, e.g., social support may have a buffering effect and can be used as basis for intervention programs. Future studies are required to investigate interindividual differences by considering possible risk and protective factors. These vulnerability factors may help to identify high-risk individuals, who may benefit from prevention and intervention programs.

7.4.3 Epigenetics

There is some evidence that early life stress may interact with genetic predispositions for psychological conditions. For example, one study found that certain glucocorticoid receptor gene polymorphisms in interaction with occurrence of early life stress may increase the risk of (recurrent) depression symptoms (Bet et al. 2009). The severity of depression was found to be influenced by an interaction of childhood maltreatment with a CRHR1 polymorphism (Bradley et al. 2008; Heim et al. 2009) or with a polymorphism associated with high MAO-A activity (Beach et al. 2010). A similar gene x environment interaction was found for PTSD symptoms, with the severity of PTSD symptoms being predicted by the FKBP5 polymorphism and childhood maltreatment (Binder et al. 2008). These polymorphisms are related to monoamine metabolism (e.g., catecholamine, serotonin) or glucocorticoid receptors. There is some evidence that these neurotransmitters (e.g., serotonin) and hormones (e.g., cortisol) may be at least partly involved in the development and manifestation of psychological conditions.

It is well established that the environment can alter gene activity; this can be seen by the extent of DNA methylation. The extent of DNA methylation in the promoter regions of genes regulates gene expression and thus the transcription activity of genes, i.e., the synthesis of gene products, mostly proteins. Protein synthesis is fundamental and essential for body functioning. Studies have shown that childhood abuse was associated with overall methylation at the SLC6A4 region (Beach et al. 2010) and methylation at the 5HTT promoter region (Beach et al. 2011); in these studies, the extent of methylation was suggested to influence the synthesis of the serotonin transporter, which in turn plays an important role in the pathophysiology of depression.

Another genetic marker, which appears to be influenced by environmental conditions, is the shortening of telomere length. As a marker of cellular aging, shortening of telomere length is suggested to be indicative for biological aging, and to be a risk factor for early onset of a variety of age-related diseases. Telomeres, i.e., repetitive DNA sequences at the end of chromosomes, shorten with age due to repetitive cell divisions. The shortening of telomeres appears to be promoted by unhealthy lifestyle (e.g., smoking) or stress and thus may constitute a possible stress-sensitive marker (Tyrka et al. 2010). One study found that the time reared in institutional care before adoption was associated with shorter telomere length in children aged 4–10 years, indicating telomere length shortening due to early life stress (Drury et al. 2011). Further evidence for telomere shortening due to early life stress was provided by several cross-sectional studies which were conducted in adult samples (Kananen et al. 2010; Kiecolt-Glaser et al. 2011; O'Donovan et al. 2011; Tyrka et al. 2010).

Clearly, these studies, still small in number, hold great promise of elucidating mechanisms that translate early life stress into negative outcomes later in life by studying biological systems on a molecular level.

7.4.4 The Type of Early Life Stress

In this book chapter we introduced early life stress as a concept that encompasses various stressful events, such as childhood maltreatment (i.e., forms of abuse and neglect), parental loss, and traumatic events (e.g., war). Thus, we cannot give any differentiated conclusion about the impact of a single type of early life stress. It seems plausible that the type of early life stress may be a determining factor. To our knowledge, there is a lack of comparison studies, e.g., comparison among the type of childhood maltreatment (emotional neglect, physical neglect, emotional abuse, physical abuse, and sexual abuse). One reason for this lack of such studies might be that individuals often experienced several types (physical and sexual abuse); thus, a more precise analysis may not be possible. Other factors, such as the number, severity, and duration of such adverse events, possibly have additional effects and thus should be assessed in future studies.

In summary, stress throughout the entire life span seems to be associated with negative consequences for health and well-being. Future research is needed to explore possible risk factors, protective factors, and the underlying biological mechanisms between the association of stress and health. This research will hopefully advance the development of appropriate interventions.

References

Alisic E, van der Schoot TA, van Ginkel JR, Kleber RJ (2008) Looking beyond posttraumatic stress disorder in children: posttraumatic stress reactions, posttraumatic growth, and quality of life in a general population sample. J Clin Psychiatry 69(9):1455–1461

APA (2000) Diagnostic statistical manual of mental disorders (Revised 4th edition). Author, Washington, DC

Barakat LP, Alderfer MA, Kazak AE (2006) Posttraumatic growth in adolescent survivors of cancer and their mothers and fathers. J Pediatr Psychol 31(4):413–419

Beach SR, Brody GH, Gunter TD, Packer H, Wernett P, Philibert RA (2010a) Child maltreatment moderates the association of MAOA with symptoms of depression and antisocial personality disorder. J Fam Psychol 24(1):12–20

Beach SR, Brody GH, Todorov AA, Gunter TD, Philibert RA (2010b) Methylation at SLC6A4 is linked to family history of child abuse: an examination of the Iowa adoptee sample. Am J Med Genet B Neuropsychiatr Genet 153B(2):710–713

Beach SR, Brody GH, Todorov AA, Gunter TD, Philibert RA (2011) Methylation at 5HTT mediates the impact of child sex abuse on women's antisocial behavior: an examination of the Iowa adoptee sample. Psychosom Med 73(1):83–87

Bernard K, Butzin-Dozier Z, Rittenhouse J, Dozier M (2010) Cortisol production patterns in young children living with birth parents vs children placed in foster care following involvement of Child Protective Services. Arch Pediatr Adolesc Med 164(5):438–443

Bet PM, Penninx BWJH, Bochdanovits Z, Uitterlinden AG, Beekman ATF, van Schoor NM et al (2009) Glucocorticoid receptor gene polymorphisms and childhood adversity are associated with depression: new evidence for a gene-environment interaction. Am J Med Genet B Neuropsychiatr Genet 150B(5):660–669

Binder EB, Bradley RG, Liu W, Epstein MP, Deveau TC, Mercer KB et al (2008) Association of FKBP5 polymorphisms and childhood abuse with risk of posttraumatic stress disorder symptoms in adults. JAMA 299(11):1291–1305

Bos KJ, Fox N, Zeanah CH, Nelson CA (2009) Effects of early psychosocial deprivation on the development of memory and executive function. Front Behav Neurosci 3:16

Bradley RG, Binder EB, Epstein MP, Tang Y, Nair HP, Liu W et al (2008) Influence of child abuse on adult depression: moderation by the corticotropin-releasing hormone receptor gene. Arch Gen Psychiatry 65(2):190–200

Bremner JD, Randall P, Scott TM, Capelli S, Delaney R, McCarthy G et al (1995) Deficits in short-term memory in adult survivors of childhood abuse. Psychiatry Res 59(1–2):97–107

Bremner JD, Narayan M, Staib LH, Southwick SM, McGlashan T, Charney DS (1999) Neural correlates of memories of childhood sexual abuse in women with and without posttraumatic stress disorder. Am J Psychiatry 156(11):1787–1795

Carpenter LL, Tyrka AR, Ross NS, Khoury L, Anderson GM, Price LH (2009) Effect of childhood emotional abuse and age on cortisol responsivity in adulthood. Biol Psychiatry 66(1):69–75

Carpenter LL, Gawuga CE, Tyrka AR, Lee JK, Anderson GM, Price LH (2010) Association between plasma IL-6 response to acute stress and early-life adversity in healthy adults. Neuropsychopharmacology 35(13):2617–2623

Caspi A, McClay J, Moffitt TE, Mill J, Martin J, Craig IW et al (2002) Role of genotype in the cycle of violence in maltreated children. Science 297(5582):851–854

Chandy JM, Blum RW, Resnick MD (1996) Gender-specific outcomes for sexually abused adolescents. Child Abuse Negl 20(12):1219–1231

Chisholm K (1998) A three year follow-up of attachment and indiscriminate friendliness in children adopted from Romanian orphanages. Child Dev 69(4):1092–1106

Chrousos GP (2009) Stress and disorders of the stress system. Nat Rev Endocrinol 5(7):374–381

Chugani HT, Behen ME, Muzik O, Juhasz C, Nagy F, Chugani DC (2001) Local brain functional activity following early deprivation: a study of postinstitutionalized Romanian orphans. Neuroimage 14(6):1290–1301

Cicchetti D, Rogosch FA (2001) Diverse patterns of neuroendocrine activity in maltreated children. Dev Psychopathol 13(3):677–693

Cirulli F, Francia N, Berry A, Aloe L, Alleva E, Suomi SJ (2009) Early life stress as a risk factor for mental health: role of neurotrophins from rodents to non-human primates. Neurosci Biobehav Rev 33(4):573–585

Cohen RA, Grieve S, Hoth KF, Paul RH, Sweet L, Tate D et al (2006) Early life stress and morphometry of the adult anterior cingulate cortex and caudate nuclei. Biol Psychiatry 59(10):975–982

Cryder CH, Kilmer RP, Tedeschi RG, Calhoun LG (2006) An exploratory study of posttraumatic growth in children following a natural disaster. Am J Orthopsychiatry 76(1):65–69

Danese A, Moffitt TE, Harrington H, Milne BJ, Polanczyk G, Pariante CM et al (2009) Adverse childhood experiences and adult risk factors for age-related disease: depression, inflammation, and clustering of metabolic risk markers. Arch Pediatr Adolesc Med 163(12):1135–1143

De Bellis MD, Hall J, Boring AM, Frustaci K, Moritz G (2001) A pilot longitudinal study of hippocampal volumes in pediatric maltreatment-related posttraumatic stress disorder. Biol Psychiatry 50(4):305–309

De Bellis MD, Keshavan MS, Shifflett H, Iyengar S, Beers SR, Hall J et al (2002) Brain structures in pediatric maltreatment-related posttraumatic stress disorder: a sociodemographically matched study. Biol Psychiatry 52(11):1066–1078

Devine KA, Reed-Knight B, Loiselle KA, Fenton N, Blount RL (2010) Posttraumatic growth in young adults who experienced serious childhood illness: a mixed-methods approach. J Clin Psychol Med Settings 17(4):340–348

Ditzen B, Nater UM, Heim C (2013) HPA axis challenge tests. In: Gellman MD, Turner JR (eds) Encyclopedia of behavioral medicine, Springer, Berlin pp. 1468–1471

Drury SS, Theall K, Gleason MM, Smyke AT, De Vivo I, Wong JY et al (2011) Telomere length and early severe social deprivation: linking early adversity and cellular aging. Mol Psychiatry 17(7):719–27

Dube SR, Anda RF, Felitti VJ, Chapman DP, Williamson DF, Giles WH (2001) Childhood abuse, household dysfunction, and the risk of attempted suicide throughout the life span: findings from the Adverse Childhood Experiences Study. JAMA 286(24):3089–3096

Dube SR, Fairweather D, Pearson WS, Felitti VJ, Anda RF, Croft JB (2009) Cumulative childhood stress and autoimmune diseases in adults. Psychosom Med 71(2):243–250

Elzinga BM, Schmahl CG, Vermetten E, van Dyck R, Bremner JD (2003) Higher cortisol levels following exposure to traumatic reminders in abuse-related PTSD. Neuropsychopharmacology 28(9):1656–1665

Elzinga BM, Roelofs K, Tollenaar MS, Bakvis P, van Pelt J, Spinhoven P (2008) Diminished cortisol responses to psychosocial stress associated with lifetime adverse events a study among healthy young subjects. Psychoneuroendocrinology 33(2):227–237

Feder A, Nestler EJ, Charney DS (2009) Psychobiology and molecular genetics of resilience. Nat Rev Neurosci 10(6):446–457

Flaherty EG, Thompson R, Litrownik AJ, Theodore A, English DJ, Black MM et al (2006) Effect of early childhood adversity on child health. Arch Pediatr Adolesc Med 160(12):1232–1238

Fujiwara T, Kawakami N, World Mental Health Japan Survey Group (2011) Association of childhood adversities with the first onset of mental disorders in Japan: results from the World Mental Health Japan, 2002–2004. J Psychiatr Res 45(4):481–487

Fuller-Thomson E, Brennenstuhl S, Frank J (2010) The association between childhood physical abuse and heart disease in adulthood: findings from a representative community sample. Child Abuse Negl 34(9):689–698

Furukawa T, Harai H, Hirai T, Fujihara S, Kitamura T, Takahashi K (1998) Childhood parental loss and alcohol dependence among Japanese men: a case–control study. Group for Longitudinal Affective Disorders Study (GLADS). Acta Psychiatr Scand 97(6):403–407

Gunnar MR, Morison SJ, Chisholm K, Schuder M (2001) Salivary cortisol levels in children adopted from Romanian orphanages. Dev Psychopathol 13(3):611–628

Gunstad J, Paul RH, Spitznagel MB, Cohen RA, Williams LM, Kohn M et al (2006) Exposure to early life trauma is associated with adult obesity. Psychiatry Res 142(1):31–37

Hafstad GS, Gil-Rivas V, Kilmer RP, Raeder S (2010) Parental adjustment, family functioning, and posttraumatic growth among Norwegian children and adolescents following a natural disaster. Am J Orthopsychiatry 80(2):248–257

Handa M, Nukina H, Hosoi M, Kubo C (2008) Childhood physical abuse in outpatients with psychosomatic symptoms. Biopsychosoc Med 2:8

Hanson JL, Chung MK, Avants BB, Shirtcliff EA, Gee JC, Davidson RJ et al (2010) Early stress is associated with alterations in the orbitofrontal cortex: a tensor-based morphometry investigation of brain structure and behavioral risk. J Neurosci 30(22):7466–7472

Heim C, Newport DJ, Heit S, Graham YP, Wilcox M, Bonsall R et al (2000) Pituitary-adrenal and autonomic responses to stress in women after sexual and physical abuse in childhood. JAMA 284(5):592–597

Heim C, Newport DJ, Bonsall R, Miller AH, Nemeroff CB (2001) Altered pituitary-adrenal axis responses to provocative challenge tests in adult survivors of childhood abuse. Am J Psychiatry 158(4):575–581

Heim C, Newport DJ, Wagner D, Wilcox MM, Miller AH, Nemeroff CB (2002) The role of early adverse experience and adulthood stress in the prediction of neuroendocrine stress reactivity in women: a multiple regression analysis. Depress Anxiety 15(3):117–125

Heim C, Wagner D, Maloney E, Papanicolaou DA, Solomon L, Jones JF et al (2006) Early adverse experience and risk for chronic fatigue syndrome: results from a population-based study. Arch Gen Psychiatry 63(11):1258–1266

Heim C, Mletzko T, Purselle D, Musselman DL, Nemeroff CB (2008a) The dexamethasone/corticotropin-releasing factor test in men with major depression: role of childhood trauma. Biol Psychiatry 63(4):398–405

Heim C, Newport DJ, Mletzko T, Miller AH, Nemeroff CB (2008b) The link between childhood trauma and depression: insights from HPA axis studies in humans. Psychoneuroendocrinology 33(6):693–710

Heim C, Bradley B, Mletzko TC, Deveau TC, Musselman DL, Nemeroff CB et al (2009a) Effect of childhood trauma on adult depression and neuroendocrine function: sex-specific moderation by CRH receptor 1 gene. Front Behav Neurosci 3:41

Heim C, Nater UM, Maloney E, Boneva R, Jones JF, Reeves WC (2009b) Childhood trauma and risk for chronic fatigue syndrome: association with neuroendocrine dysfunction. Arch Gen Psychiatry 66(1):72–80

Heim C, Young LJ, Newport DJ, Mletzko T, Miller AH, Nemeroff CB (2009c) Lower CSF oxytocin concentrations in women with a history of childhood abuse. Mol Psychiatry 14(10):954–958

Ho SM, Chu KW, Yiu J (2008) The relationship between explanatory style and posttraumatic growth after bereavement in a non-clinical sample. Death Stud 32(5):461–478

Huey SJ Jr, Weisz JR (1997) Ego control, ego resiliency, and the five-factor model as predictors of behavioral and emotional problems in clinic-referred children and adolescents. J Abnorm Psychol 106(3):404–415

Johnson JG, Cohen P, Brown J, Smailes EM, Bernstein DP (1999) Childhood maltreatment increases risk for personality disorders during early adulthood. Arch Gen Psychiatry 56(7):600–606

Jouriles EN, Barling J, Oleary KD (1987) Predicting child-behavior problems in maritally violent families. J Abnorm Child Psychol 15(2):165–173

Kamibeppu K, Sato I, Honda M, Ozono S, Sakamoto N, Iwai T et al (2010) Mental health among young adult survivors of childhood cancer and their siblings including posttraumatic growth. J Cancer Surviv 4(4):303–312

Kananen L, Surakka I, Pirkola S, Suvisaari J, Lonnqvist J, Peltonen L et al (2010) Childhood adversities are associated with shorter telomere length at adult age both in individuals with an anxiety disorder and controls. PLoS One 5(5):e10826

Kaufman J, Birmaher B, Perel J, Dahl RE, Moreci P, Nelson B et al (1997) The corticotropin-releasing hormone challenge in depressed abused, depressed nonabused, and normal control children. Biol Psychiatry 42(8):669–679

Kendler KS, Bulik CM, Silberg J, Hettema JM, Myers J, Prescott CA (2000) Childhood sexual abuse and adult psychiatric and substance use disorders in women: an epidemiological and cotwin control analysis. Arch Gen Psychiatry 57(10):953–959

Kendler KS, Kuhn JW, Prescott CA (2004) Childhood sexual abuse, stressful life events and risk for major depression in women. Psychol Med 34(8):1475–1482

Kiecolt-Glaser JK, Gouin JP, Weng NP, Malarkey WB, Beversdorf DQ, Glaser R (2011) Childhood adversity heightens the impact of later-life caregiving stress on telomere length and inflammation. Psychosom Med 73(1):16–22

Kilmer RP, Gil-Rivas V (2010) Exploring posttraumatic growth in children impacted by Hurricane Katrina: correlates of the phenomenon and developmental considerations. Child Dev 81(4):1211–1227

Kilmer RP, Gil-Rivas V, Tedeschi RG, Cann A, Calhoun LG, Buchanan T et al (2009) Use of the revised posttraumatic growth inventory for children. J Trauma Stress 22(3):248–253

Kim J, Cicchetti D, Rogosch FA, Manly JT (2009) Child maltreatment and trajectories of personality and behavioral functioning: implications for the development of personality disorder. Dev Psychopathol 21(3):889–912

King JA, Mandansky D, King S, Fletcher KE, Brewer J (2001) Early sexual abuse and low cortisol. Psychiatry Clin Neurosci 55(1):71–74

Klaassens ER, van Noorden MS, Giltay EJ, van Pelt J, van Veen T, Zitman FG (2009) Effects of childhood trauma on HPA-axis reactivity in women free of lifetime psychopathology. Prog Neuropsychopharmacol Biol Psychiatry 33(5):889–894

Kudielka BM, Wust S (2010) Human models in acute and chronic stress: assessing determinants of individual hypothalamus-pituitary-adrenal axis activity and reactivity. Stress 13(1):1–14

Kunugi H, Sugawara N, Aoki H, Nanko S, Hirose T, Kazamatsuri H (1995) Early parental loss and depressive disorder in Japan. Eur Arch Psychiatry Clin Neurosci 245(2):109–113

Lampe A, Doering S, Rumpold G, Solder E, Krismer M, Kantner-Rumplmair W et al (2003) Chronic pain syndromes and their relation to childhood abuse and stressful life events. J Psychosom Res 54(4):361–367

Lanier P, Jonson-Reid M, Stahlschmidt MJ, Drake B, Constantino J (2010) Child maltreatment and pediatric health outcomes: a longitudinal study of low-income children. J Pediatr Psychol 35(5):511–522

Lansford JE, Miller-Johnson S, Berlin LJ, Dodge KA, Bates JE, Pettit GS (2007) Early physical abuse and later violent delinquency: a prospective longitudinal study. Child Maltreat 12(3):233–245

Laufer A, Solomon Z, Levine SZ (2010) Elaboration on posttraumatic growth in youth exposed to terror: the role of religiosity and political ideology. Soc Psychiatry Psychiatr Epidemiol 45(6):647–653

Letzring TD, Block J, Funder DC (2005) Ego-control and ego-resiliency: generalization of self-report scales based on personality descriptions from acquaintances, clinicians, and the self. J Res Pers 39(4):395–422

Levine SZ, Laufer A, Hamama-Raz Y, Stein E, Solomon Z (2008) Posttraumatic growth in adolescence: examining its components and relationship with PTSD. J Trauma Stress 21(5):492–496

Levine SZ, Laufer A, Stein E, Hamama-Raz Y, Solomon Z (2009) Examining the relationship between resilience and posttraumatic growth. J Trauma Stress 22(4):282–286

Lovallo WR, Farag NH, Sorocco KH, Cohoon AJ, Vincent AS (2012) Lifetime adversity leads to blunted stress axis reactivity: studies from the oklahoma family health patterns project. Biol Psychiatry 71(4):344–349

Luecken LJ (1998) Childhood attachment and loss experiences affect adult cardiovascular and cortisol function. Psychosom Med 60(6):765–772

MacMillan HL, Georgiades K, Duku EK, Shea A, Steiner M, Niec A et al (2009) Cortisol response to stress in female youths exposed to childhood maltreatment: results of the youth mood project. Biol Psychiatry 66(1):62–68

Maikovich-Fong AK, Jaffee SR (2010) Sex differences in childhood sexual abuse characteristics and victims' emotional and behavioral problems: findings from a national sample of youth. Child Abuse Negl 34(6):429–437

Majer M, Nater UM, Lin JM, Capuron L, Reeves WC (2010) Association of childhood trauma with cognitive function in healthy adults: a pilot study. BMC Neurol 10:61

McEwen BS, Seeman T (1999) Protective and damaging effects of mediators of stress. Elaborating and testing the concepts of allostasis and allostatic load. Ann N Y Acad Sci 896:30–47

Meinlschmidt G, Heim C (2005) Decreased cortisol awakening response after early loss experience. Psychoneuroendocrinology 30(6):568–576

Meyerson DA, Grant KE, Carter JS, Kilmer RP (2011) Posttraumatic growth among children and adolescents: a systematic review. Clin Psychol Rev 31(6):949–964

Mills R, Alati R, O'Callaghan M, Najman JM, Williams GM, Bor W et al (2011) Child abuse and neglect and cognitive function at 14 years of age: findings from a birth cohort. Pediatrics 127(1):4–10

Moran P, Coffey C, Chanen A, Mann A, Carlin JB, Patton GC (2011) Childhood sexual abuse and abnormal personality: a population-based study. Psychol Med 41(06):1311–1318

Morina N, von Lersner U, Prigerson HG (2011) War and bereavement: consequences for mental and physical distress. PLoS One 6(7):e22140

Navalta CP, Polcari A, Webster DM, Boghossian A, Teicher MH (2006) Effects of childhood sexual abuse on neuropsychological and cognitive function in college women. J Neuropsychiatry Clin Neurosci 18(1):45–53

Nelson EC, Heath AC, Madden PA, Cooper ML, Dinwiddie SH, Bucholz KK et al (2002) Association between self-reported childhood sexual abuse and adverse psychosocial outcomes: results from a twin study. Arch Gen Psychiatry 59(2):139–145

Nelson CA, Zeanah CH, Fox NA, Marshall PJ, Smyke AT, Guthrie D (2007) Cognitive recovery in socially deprived young children: the Bucharest Early Intervention Project. Science 318(5858):1937–1940

Newport DJ, Heim C, Bonsall R, Miller AH, Nemeroff CB (2004) Pituitary-adrenal responses to standard and low-dose dexamethasone suppression tests in adult survivors of child abuse. Biol Psychiatry 55(1):10–20

Nicolson NA, Davis MC, Kruszewski D, Zautra AJ (2010) Childhood maltreatment and diurnal cortisol patterns in women with chronic pain. Psychosom Med 72(5):471–480

O'Donovan A, Epel E, Lin J, Wolkowitz O, Cohen B, Maguen S et al (2011) Childhood trauma associated with short leukocyte telomere length in posttraumatic stress disorder. Biol Psychiatry 70(5):465–71

O'Mahony SM, Marchesi JR, Scully P, Codling C, Ceolho AM, Quigley EM et al (2009) Early life stress alters behavior, immunity, and microbiota in rats: implications for irritable bowel syndrome and psychiatric illnesses. Biol Psychiatry 65(3):263–267

Opacka-Juffry J, Mohiyeddini C (2011) Experience of stress in childhood negatively correlates with plasma oxytocin concentration in adult men. Stress 15(1):1–10

Ozbay F, Johnson DC, Dimoulas E, Morgan CA, Charney D, Southwick S (2007) Social support and resilience to stress: from neurobiology to clinical practice. Psychiatry (Edgmont) 4(5):35–40

Patton GC, Coffey C, Posterino M, Carlin JB, Bowes G (2003) Life events and early onset depression: cause or consequence? Psychol Med 33(7):1203–1210

Pollak SD, Nelson CA, Schlaak MF, Roeber BJ, Wewerka SS, Wiik KL et al (2010) Neurodevelopmental effects of early deprivation in postinstitutionalized children. Child Dev 81(1):224–236

Rao U, Hammen C, Ortiz LR, Chen LA, Poland RE (2008) Effects of early and recent adverse experiences on adrenal response to psychosocial stress in depressed adolescents. Biol Psychiatry 64(6):521–526

Romans S, Belaise C, Martin J, Morris E, Raffi A (2002) Childhood abuse and later medical disorders in women. An epidemiological study. Psychother Psychosom 71(3):141–150

Salter E, Stallard P (2004) Posttraumatic growth in child survivors of a road traffic accident. J Trauma Stress 17(4):335–340

Schilling EA, Aseltine RH Jr, Gore S (2007) Adverse childhood experiences and mental health in young adults: a longitudinal survey. BMC Public Health 7:30

Schmidt MV, Wang XD, Meijer OC (2011) Early life stress paradigms in rodents: potential animal models of depression? Psychopharmacology (Berl) 214(1):131–140

Shakespeare-Finch J, de Dassel T (2009) Exploring posttraumatic outcomes as a function of childhood sexual abuse. J Child Sex Abus 18(6):623–640

Silberg J, Rutter M, Neale M, Eaves L (2001) Genetic moderation of environmental risk for depression and anxiety in adolescent girls. Br J Psychiatry 179:116–121

Silverman AB, Reinherz HZ, Giaconia RM (1996) The long-term sequelae of child and adolescent abuse: a longitudinal community study. Child Abuse Negl 20(8):709–723

Stein MB, Koverola C, Hanna C, Torchia MG, McClarty B (1997) Hippocampal volume in women victimized by childhood sexual abuse. Psychol Med 27(4):951–959

Tedeschi RG, Calhoun LG (1996) The Posttraumatic Growth Inventory: measuring the positive legacy of trauma. J Trauma Stress 9(3):455–471

Teicher MH, Dumont NL, Ito Y, Vaituzis C, Giedd JN, Andersen SL (2004) Childhood neglect is associated with reduced corpus callosum area. Biol Psychiatry 56(2):80–85

Thompson R, Tabone JK (2010) The impact of early alleged maltreatment on behavioral trajectories. Child Abuse Negl 34(12):907–916

Thornberry TP, Henry KL, Ireland TO, Smith CA (2010) The causal impact of childhood-limited maltreatment and adolescent maltreatment on early adult adjustment. J Adolesc Health 46(4):359–365

Tietjen GE, Brandes JL, Peterlin BL, Eloff A, Dafer RM, Stein MR et al (2010) Childhood maltreatment and migraine (part III). Association with comorbid pain conditions. Headache 50(1):42–51

Turner-Sack AM, Menna R, Setchell SR (2012) Posttraumatic growth, coping strategies, and psychological distress in adolescent survivors of cancer. J Pediatr Oncol Nurs 29(2):70–79

Tyrka AR, Wier L, Price LH, Ross N, Anderson GM, Wilkinson CW et al (2008) Childhood parental loss and adult hypothalamic-pituitary-adrenal function. Biol Psychiatry 63(12):1147–1154

Tyrka AR, Wyche MC, Kelly MM, Price LH, Carpenter LL (2009) Childhood maltreatment and adult personality disorder symptoms: influence of maltreatment type. Psychiatry Res 165(3):281–287

Tyrka AR, Price LH, Kao HT, Porton B, Marsella SA, Carpenter LL (2010) Childhood maltreatment and telomere shortening: preliminary support for an effect of early stress on cellular aging. Biol Psychiatry 67(6):531–534

van der Vegt EJ, van der Ende J, Kirschbaum C, Verhulst FC, Tiemeier H (2009) Early neglect and abuse predict diurnal cortisol patterns in adults A study of international adoptees. Psychoneuroendocrinology 34(5):660–669 van der Vegt EJ, van der Ende J, Huizink AC, Verhulst FC, Tiemeier H (2010) Childhood adversity modifies the relationship between anxiety disorders and cortisol secretion. Biol Psychiatry 68(11):1048–1054

Vythilingam M, Heim C, Newport J, Miller AH, Anderson E, Bronen R et al (2002) Childhood trauma associated with smaller hippocampal volume in women with major depression. Am J Psychiatry 159(12):2072–2080

WHO (2004) International Statistical Classification of Diseases and Health Related Problems. World Health Organization, Geneva

Wingo AP, Wrenn G, Pelletier T, Gutman AR, Bradley B, Ressler KJ (2010) Moderating effects of resilience on depression in individuals with a history of childhood abuse or trauma exposure. J Affect Disord 126(3):411–414

Wolchik SA, Coxe S, Tein JY, Sandler IN, Ayers TS (2008) Six-year longitudinal predictors of posttraumatic growth in parentally bereaved adolescents and young adults. Omega (Westport) 58(2):107–128

Yu XN, Lau JT, Zhang J, Mak WW, Choi KC, Lui WW et al (2010) Posttraumatic growth and reduced suicidal ideation among adolescents at month 1 after the Sichuan earthquake. J Affect Disord 123(1–3):327–331

Zeanah CH, Smyke AT, Koga SF, Carlson E (2005) Attachment in institutionalized and community children in Romania. Child Dev 76(5):1015–1028

Zeanah CH, Egger HL, Smyke AT, Nelson CA, Fox NA, Marshall PJ et al (2009) Institutional rearing and psychiatric disorders in Romanian preschool children. Am J Psychiatry 166(7):777–785

Part III
Adaptive and Maladaptive Consequences of Developmental Stress in Animal Models

Chapter 8
Behavioural and Neuroendocrine Consequences of Prenatal Stress in Rat

Sara Morley-Fletcher, Jérôme Mairesse, and Stefania Maccari

Abstract Chronic hyperactivation of the hypothalamus–pituitary axis is associated with the suppression of reproductive, growth, thyroid and immune functions that may lead to various pathological states. Although many individuals experiencing stressful events do not develop pathologies, stress seems to be a provoking factor in those individuals with particular vulnerability, determined by genetic factors or earlier experience. Exposure of the developing brain to severe and/or prolonged stress may result in hyperactivity of the stress system, defective glucocorticoid negative feedback, altered cognition, novelty seeking, increased vulnerability to addictive behaviours and mood-related disorders. Therefore, stress-related events that occur in the perinatal period can permanently change brain and behaviour of the developing individual. Prenatal restraint stress (PRS) in rats is a well-documented model of early stress known to induce long-lasting neurobiological and behavioural alterations including impaired feedback mechanisms of the HPA axis, disruption of circadian rhythms and altered neuroplasticity. Together with the HPA axis the glutamate system is particularly impaired, and such impairment appears to be involved in the anxious profile of PRS rats.

Chronic treatments with antidepressants at adulthood have proven high predictive validity of the PRS rat as animal model of depression/anxiety and reinforce the idea of the usefulness of the PRS rat as an interesting animal model for the design and testing of new pharmacologic strategies in the treatment of stress-related disorders.

S. Morley-Fletcher (✉) • J. Mairesse • S. Maccari
Neural Plasticity Team-UMR CNRS/USTL n° 8576 Structural and Functional Glycobiology Unit, North University of Lille, Lille, France
e-mail: sara.morley-fletcher@univ-lille1.fr

G. Laviola and S. Macrì (eds.), *Adaptive and Maladaptive Aspects of Developmental Stress*, Current Topics in Neurotoxicity 3, DOI 10.1007/978-1-4614-5605-6_8,

8.1 Early Environmental Influences on HPA Axis Development

Although many individuals experiencing stressful events do not develop pathologies, stress seems to be a provoking factor in those individuals with particular vulnerability, determined by genetic factors or earlier experience (McEwen 2011). The chronic hyperactivation of HPA axis can be determined by multiple factors including genetic and environmental factors. The perinatal life, infancy, childhood and adolescence are periods of increased plasticity for the stress system and are, therefore, particularly sensitive to stressors. Adverse stressors during these critical periods of life may affect behaviours and physiologic functions, such as growth, metabolism, reproduction and inflammatory/immune response (Seckl 2008; Seckl and Holmes 2007). These environmental triggers or stressors may have not a transient but rather a permanent effect on the organism. Barker has emphasised how adult vulnerability to disease may be programmed during the fetal period (Barker 1995; Barker et al. 2006). Indeed, nongenetic factors that could act early in life to organise or imprint permanently physiological systems are known as perinatal *programming*. It can be speculated that prenatal plasticity of physiological systems allows environmental factors, acting on the mother and/or the fetus, to alter the set point or "hard-wire" the differentiated functions of an organ or tissue system to prepare the unborn animal optimally for the environmental conditions ex utero (see Chap. 1).

Intrauterine growth restriction and low birth weight are considered indices of prenatal stress in humans. Glucocorticoids may underlie the association between low birth weight and adult stress-related cardiovascular, metabolic and neuroendocrine disorders such as hypertension, type 2 diabetes, ischaemic heart disease and affective disorders. These intriguing findings have spawned the *fetal origins hypothesis* of adult disease. The brain is very sensitive to prenatal programming, and glucocorticoids in particular have powerful brain-programming properties. In rats, substantial evidence suggests that prenatal stress programmes the HPA axis as well as behaviour and that plasticity of developing brain monoamine systems underlies, in part, these changes. Because an important feature of the stress response is the secretion of high levels of glucocorticoids, these steroids have become an obvious candidate for the role of *programming factor* in the prenatal stress paradigm. A large number of animal studies have described the effects of prenatal exposure to the synthetic glucocorticoid dexamethasone, which relatively readily passes the placenta (see Chap. 9). Moreover, prenatal dexamethasone exposure has recently been implicated in the development of adult hyperglycaemia and hypertension as well as behavioural changes and HPA activation (Nyirenda et al. 2001).

8.1.1 Programming of a Phenotype by Stress: The Prenatal Restraint Stress Model in the Rat

Numerous animal models of early stress are currently being developed because early stress results in long-term disruptions of neuronal functions and the development of

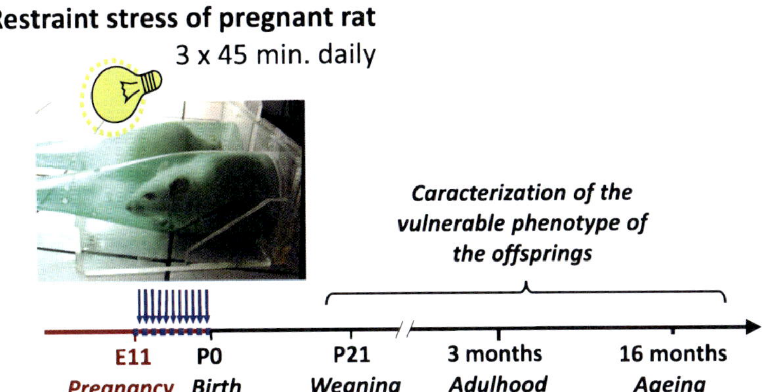

Fig. 8.1 The prenatal restraint stress (PRS) model in the rat

long-term behavioural disorders. During the last 20 years we have studied the influences of a prenatal restraint stress (PRS) in a rat animal model (Fig. 8.1).

The prenatal stress procedure we have used consisted in restraining the pregnant rat—in a transparent Plexiglas cylinder, three times/day for 45 min under bright light—at the day 11 of pregnancy until delivery at 21–22 days (Maccari et al. 1995, 2003; Morley-Fletcher et al. 2003a). The HPA axis functioning of the PRS offspring is long-term impaired (Fig. 8.2) with a prolonged corticosterone stress response (Maccari et al. 1995, 2003; Koehl et al. 1999) and reduced levels of both mineralocorticoid and glucocorticoid receptors in the hippocampus at the adolescent and adult stage (Henry et al. 1994; Maccari et al. 1995; Van Waes et al. 2006). During neonatal development, the HPA axis of rats undergoes a period of hyporesponsiveness (SHRP, PD2–PD14), when most stress stimuli fail to induce a stress response. Beginning on about PD 14, glucocorticoid levels rise sharply, marking the end of the SHRP. Remarkably, the age-related HPA axis dysfunctions are enhanced by PRS, since SHRP is abolished in newborn PRS rats (Henry et al. 1994) and circulating glucocorticoid levels of PRS middle-aged animals are similar to those found in old non-stressed animals (Vallée et al. 1999). These findings suggest an exacerbation in the maturation process of HPA induced by PRS.

The impact of PRS is already detectable at the fetal stage, giving further support to prenatal stress programming in adult pathophysiology (Mairesse et al. 2007a). In the placenta of PRS rats, the expression of glucose transporters type 1 (GLUT1) was decreased, whereas GLUT3 and GLUT4 were slightly increased. Moreover, placental expression and activity of the glucocorticoid barrier enzyme 11beta-hydroxysteroid dehydrogenase type 2 was strongly reduced. At E21, PRS fetuses exhibited reduced body weight and decreased weight of the adrenals, pancreas and testis. These alterations were associated in the offspring with reduced pancreatic beta-cells mass, plasma levels of glucose, growth hormone and ACTH, whereas corticosterone, insulin, IGF-1 and CBG levels were unaffected (Mairesse et al. 2007a). Moreover, PRS increases the levels of 5-HT_2 receptors (Peters 1988),

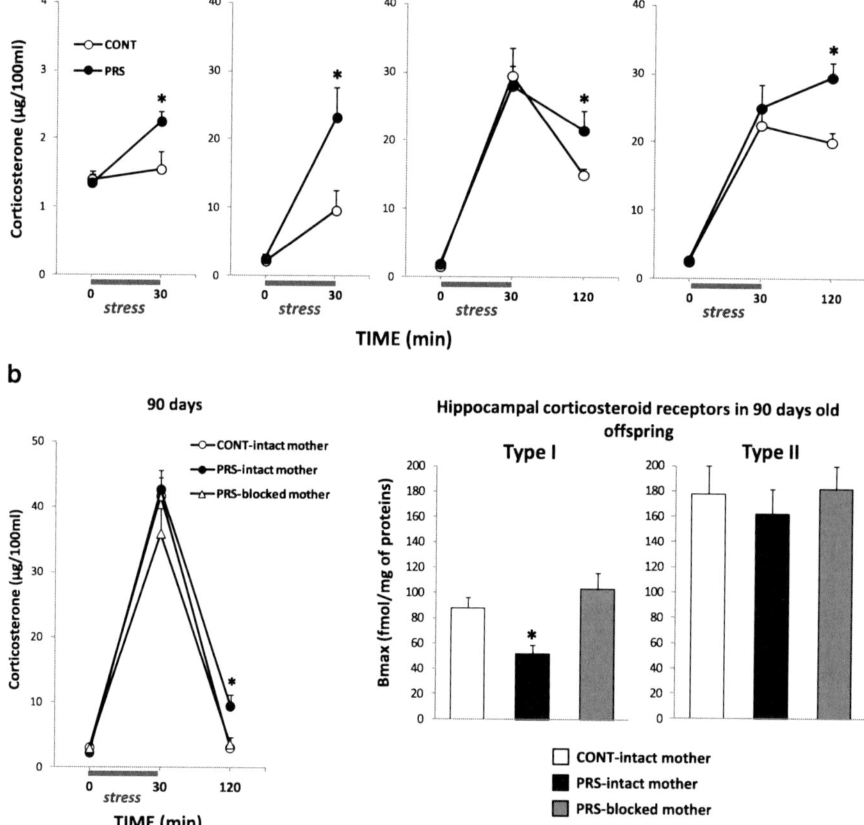

Fig. 8.2 HPA axis alterations in PRS rats. (**a**) PRS enhances corticosterone response to stress during the life span. (**b**) Blockade of the mother's stress-induced glucocorticoid secretion suppresses the prolonged stress-induced corticosteroid response and the decrease in type I hippocampal corticosteroid receptors usually observed in PRS adults. $*P<0.05$ and $**P<0.01$ vs. controls. Original data are reported in Henry et al. (1995) and Vallée et al. (1999) (**a**) Barbazanges et al. (1996) (**b**)

induces a higher expression of 5-HT_{1A} mRNA in the prefrontal cortex (Morley-Fletcher et al. 2004a) and increases acetylcholine release in the hippocampus after mild stress (Day et al. 1998). High maternal corticosterone levels may contribute to the long-term effect (Salomon et al. 2011) described in the offspring after PRS (in addition to possible internal vassal constrictions that would affect blood supply to the placenta). Indeed, the reduction or the increase of maternal glucocorticoids by adrenalectomy of stressed mothers results in suppressed or reinstated PRS effects on HPA axis offspring (Barbazanges et al. 1996). The immediate postnatal environment plays also an important role on PRS-related outcome on HPA axis that can be reversed by an early postnatal manipulation such as early adoption

(Maccari et al. 1995). Adoption modifies maternal behaviour, increasing pup-directed behaviour in foster mothers and decreases the stress-induced corticosterone secretion in the adult PRS offspring.

8.1.2 Behavioural Abnormalities Induced by PRS

The hyperactivity of the HPA axis observed in PRS rats is accompanied by increased anxious/depression-like behaviour during the life span. This is evidenced by increased ultrasonic vocalisations in infancy (Laloux et al. 2012), reduced social play during adolescence (Morley-Fletcher et al. 2003b), reduced exploration in the elevated-plus maze test and open-field test (Vallée et al. 1997, 1999) or increased immobility in the forced swim test during adulthood (Morley-Fletcher et al. 2003a, 2004a, 2011). Rats subjected to PRS emitted significantly more ultrasonic vocalisations in response to isolation at postnatal day 10 as compared to controls. In addition, PRS pups do not show the phenomenon of "maternal potentiation", i.e. the increase in USVs in response to a brief maternal reunion after isolation, normally seen at postnatal day 10 and postnatal day 14 (Laloux et al. 2012). Taken together, these data offer a clear-cut demonstration that PRS causes anxiety-like behaviour in the early developmental phase. They also suggest the attractive possibility that early interventions interfering with the epigenetic programming may correct the pathological phenotype of adult PRS rats. Remarkably, a number of behavioural alterations observed in PRS are positively correlated to enhanced corticosterone response to stress (see Fig. 8.2). Strategies aimed at restoring HPA axis functioning, i.e. environmental enrichment during adolescence (Morley-Fletcher et al. 2003a) or pharmacological manipulations during adulthood (Morley-Fletcher et al. 2004a), restore the associated behavioural patterns.

Also, PRS induces persistent behavioural and neurobiological alterations leading to a greater consumption of psychostimulants during adulthood. This is expressed by enhanced cocaine self-administration (Deminière et al. 1992; Henry et al. 1995) and increased sensitivity to locomotor-activating effect of nicotine (Koehl et al. 2000), MDMA (Morley-Fletcher et al. 2004b) or alcohol (Van Waes et al. 2009). Although PRS has no effect on alcohol consumption, it potentiates alcohol-induced ΔFosB levels in the nucleus accumbens (Van Waes et al. 2009). This suggests that negative events occurring in utero do not modulate alcohol preference in male rats but potentiate chronic alcohol-induced molecular neuroadaptation in the brain reward circuitry. Further studies are needed to determine whether the exacerbated ΔFosB upregulation in PRS rats could be extended to other reinforcing stimuli.

PRS induces also learning impairments in aged animals as evidenced in the Morris water maze and Y-maze (Vallée et al. 1999; Darnaudery et al. 2006), and learning deficit can be reversed by a chronic treatment with the neurotrophin IGF-1 (Darnaudery et al. 2006).

PRS alters not only reactive adaptation but also predictive adaptation by changing circadian rhythms. Significant phase advances are observed in the circadian rhythms of locomotor activity relative to the entraining light–dark cycle in both male and female PRS rats. Interestingly, when subjected to an abrupt shift in the light–dark cycle, PRS rats resynchronised their activity rhythm to the new light–dark cycle slower than control rats (Maccari and Van Reeth 2007). Also, PRS induced higher levels of corticosterone secretion at the end of the light period in both males and females, and hypercorticism over the entire diurnal cycle only in females (Koehl et al. 1997). The sleep–wake cycle is dramatically modified by PRS (Dugovic et al. 1999; Mairesse et al. 2012a), with a significant increase in the amount of REM sleep over the 24-h recording session, positively correlated to plasma corticosterone levels (see Fig. 8.3a, b). Other changes include increased sleep fragmentation, total light slow-wave sleep time and a slight decrease in the percentage of deep slow-wave sleep relative to total sleep time, thus providing a polygraphic demonstration of long-term effects of PRS on the sleep–wake cycle when the animals reach adulthood.

8.1.3 Neuroplastic Programming Induced by PRS

We have identified some of anatomical substrates and neural mechanisms sustaining the HPA axis hyperactivity classically described in PRS rats after stress exposure. Fos protein expression after exposure to a mild stressor (open arm of the elevated-plus maze) was evaluated in hippocampus and locus coeruleus, brain areas involved in the feedback control of the HPA axis and in the PVN, that reflect the magnitude of the hormonal response to stress (Viltart et al. 2006). At basal level, PRS rats exhibited higher number of Fos-immunoreactive neurons than controls in the hippocampus and locus coeruleus, whereas they presented a higher basal expression of hypothalamic vGAT, a marker of GABAergic synapses. After exposure to the open arm, number of Fos-immunoreactive neurons increased in the PVN, whereas no changes were observed in the hippocampus and locus coeruleus of PRS rats compared to basal condition. Moreover, only PRS rats presented an elevation of the number of activated catecholaminergic neurons in the locus coeruleus. Mairesse et al. (2007b) also examined whether behavioural reactivity was correlated to neuronal activation, by assessing Fos expression in limbic regions of rats submitted to forced exposure to a low or high anxiogenic environment (the closed or open arms of the elevated-plus maze, respectively). A negative correlation was found between behavioural and neuronal activation, with a lower behavioural reactivity and a higher neuronal response observed in rats exposed to the more anxiogenic environment (the open arm) with respect to the less anxiogenic environment (the closed arm). Interestingly, the variation in the neurobehavioural response between the two arms of the maze was less pronounced in rats that had been subjected to PRS. These studies provide evidence of long-lasting changes in brain plasticity induced by PRS that affect the ability of limbic neurons to cope with anxiogenic stimuli of different strength.

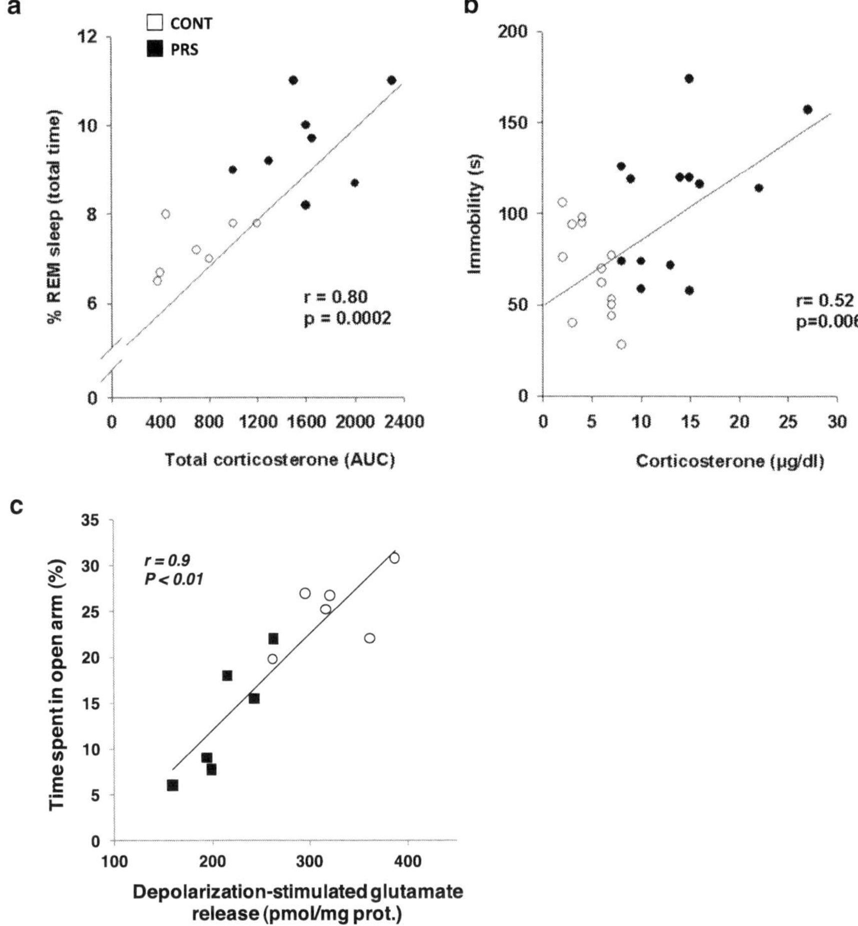

Fig. 8.3 HPA axis- and glutamate-related neurobehavioural alterations in PRS rats. (**a**) Positive correlations between individual stress-induced plasma corticosterone and amounts of REM sleep expressed as percentage of total recording time in controls and PRS rats. (**b**) Positive correlation between individual stress-induced (60 min after stress extinction) time spent in immobility in the forced swim test in controls and PRS rats. (**c**) Negative correlation between anxiety-like behaviour and evoked release of glutamate in the ventral hippocampus of adult PRS rats. $^{*}P<0.05$ and $^{**}P<0.01$ vs. controls. Original data are reported in Dugovic et al. (1999) (**a**), Morley-Fletcher et al. (2003a) (**b**) Marrocco et al. (2012) (**c**)

PRS induces a long-lasting reduction in hippocampal neurogenesis in male (Lemaire et al. 2000; Zuena et al. 2008; Morley-Fletcher et al. 2011) but not female rats (Darnaudery et al. 2006) together with an increase in BDNF in males and not in females (Zuena et al. 2008). Also, as revealed by a recent mass spectrometry analysis in the hippocampus of adult male PRS rats (Mairesse et al. 2012b), we found that PRS rats display changes in the expression profile of a number of proteins involved in the regulation of signal transduction, synaptic vesicles, protein synthesis, cytoskeleton dynamics and energetic metabolism. In particular, PRS causes a dramatic reduction of the proteins of the SNARE complex which regulate neurotransmitter release (VAMP, syntaxin), and other proteins regulating the trafficking of synaptic vesicles, such as synaptophysin, synapsins, munc-18 and Rab3A. Synaptophysin acts as a regulator of the SNARE complex (Hinz et al. 2001), and is also considered as a marker protein of presynaptic nerve endings (Thome et al. 2001; Grillo et al. 2005). Synapsins are involved in the clustering of synaptic vesicles to the reserve pool near the release sites in presynaptic terminals (Valtorta et al. 1992; Greengard et al. 1993). Munc-18 is a molecular chaperone of syntaxin-1, which is involved in mechanisms of SNARE-mediated membrane fusion and docking of large dense-core vesicles to the plasma membrane (Han et al. 2010). Rab3A, a member of a large family of monomeric GTP-binding proteins, regulates the trafficking of synaptic vesicles and cooperates with synapsin II in promoting the latest steps of neurotransmitter release (Sakane et al. 2006; Coleman and Bykhovskaia 2010).

Remarkably, most of the aforementioned neuroplastic alterations induced by PRS in males occur prominently in the ventral portion of the hippocampus. A recent review (Fanselow and Dong 2010) highlights the distinct functions performed by the ventral (or temporal) and dorsal (or septal) portions of the hippocampus. The ventral hippocampus (corresponding to the anterior hippocampus in primates) is mainly related to stress, emotions and affect, whereas the dorsal hippocampus (the posterior hippocampus in primates) performs primarily cognitive functions (Henke 1990; Moser et al. 1995). Lesions of the most ventral quarters of the hippocampus affect anxiety-like behaviour at the EPM, increasing the entry in the open (unprotected) arms of the maze (Kjelstrup et al. 2002). Thus, reduction of neuroplastic markers found selectively in the ventral hippocampus of PRS rats would appear consistent with the anxious phenotype of these animals.

8.1.4 PRS Effect on the Glutamate System

The effects of stress on the brain have long been associated with the onset and exacerbation of several neuropsychiatric disorders such as depression, anxiety, drug addiction and epilepsy (McEwen 2011). Alterations in glutamate neurotransmission are believed to play a role in the pathophysiology of such disorders (Ongür et al. 2008; Chen et al. 2010). As recently reviewed by Popoli et al. (2012), acute exposure to stress or treatment with glucocorticoids enhances glutamate release in the hippocampus, amygdala and prefrontal cortex, three brain regions that are critically involved in the pathophysiology of psychiatric disorders. The study of glutamate release in response to chronic stress is still at its infancy (Moghaddam 2002; Yamamoto and Reagan 2006), and there are no data on how early life stress affects glutamate release in the adult life. The latter issue is particularly relevant because early life stress causes long-lasting changes in neuroplasticity that result into an increased vulnerability to stress-related disorders in the adult life (Meaney et al. 2007; Darnaudéry and Maccari 2008; Lupien et al. 2009).

Our group has proven increasing evidence of an involvement of the glutamate transmission and machinery in particular at the level of metabotropic glutamate receptors in the neuroplastic programming and anxious phenotype induced by PRS.

PRS rats show a reduced expression and function of group-I and group-II metabotropic glutamate receptors in the hippocampus (Zuena et al. 2008; Van Waes et al. 2009; Morley-Fletcher et al. 2011; Laloux et al. 2012). Hippocampal levels of mGlu1 and mGlu5 receptors are already reduced in infant PRS rats at postnatal day 10, whereas expression of mGlu2/3 receptors declined only after weaning (Laloux et al. 2012). We have recently shown that PRS causes a selective impairment of glutamate release in the ventral hippocampus, a brain region which specifically encodes memories related to stress and emotions (Fanselow and Dong 2010). Remarkably, we found a high negative correlation between the extent of depolarisation-evoked glutamate release in the ventral hippocampus and anxiety-like behaviour in both control and PRS rats (Fig. 8.3c). This is extremely relevant since it provides evidence that glutamate release in the ventral hippocampus is associated with trait anxiety, a stress-related disorder (Marrocco et al. 2012).

We measured glutamate release from superfused isolated synaptosomes using a method that specifically allows the detection of exocytotic, Ca^{2+}-dependent release and eliminates any interference by endogenous ligands that activate presynaptic receptors (Raiteri et al. 1974; Raiteri and Raiteri 2000). The reduction in the evoked release of glutamate found in the ventral hippocampus of PRS rats was not due to an impaired glutamate synthesis in presynaptic terminals because it was also seen in synaptosomes preloaded with D-[^{3}H]-aspartate, a nonmetabolisable analogue of glutamate, but very probably it was associated to the marked reduction of synaptic vesicle proteins found in the ventral hippocampus (Marrocco et al. 2012). Thus, PRS would lead to an impairment of the intrinsic machinery of exocytotic glutamate release in the ventral hippocampus. This profile of expression of vesicle-associated proteins fits nicely with the finding that glutamate release was reduced in the ventral hippocampus, but not in the dorsal hippocampus of PRS rats. Remarkably, at least two proteins that were found to be reduced in the ventral hippocampus of PRS rats, i.e. munc-18 and Rab3A, have been specifically associated to the regulation of glutamate release. Accordingly, munc-18 regulates the size of the readily releasable vesicle pool in glutamatergic but not GABAergic terminals (Augustin et al. 1999), and Rab3A is preferentially, albeit not exclusively, expressed at glutamatergic terminals (Geppert et al. 1994). Therefore, one of the most striking findings of our study is that PRS had profound effects on glutamate release, but it failed to affect GABA release.

Our data suggest that PRS causes an imbalance between excitatory and inhibitory neurotransmission in the ventral hippocampus, an effect that might perturb cognitive functions related to stress and emotions (reviewed by Bannerman et al. 2004; Fanselow and Dong 2010). Presynaptic alterations in the glutamate/GABA balance have been associated with anxiety, depressive-like behaviour and memory impairment (Tordera et al. 2007; Garcia-Garcia et al. 2009; Chen et al. 2010). Thus, the imbalance between excitatory and inhibitory neurotransmission in the ventral hippocampus might contribute to explain the anxious/depressive phenotype of PRS rats (Vallée et al. 1997; Zuena et al. 2008; Morley-Fletcher et al. 2011).

The lack of changes in glutamate release in the dorsal hippocampus is in agreement with the finding that PRS rats do not show abnormalities in spatial memory unless they have >10 months of age (Vallée et al. 1999; Van Waes et al. 2009). At this old age, PRS rats show changes in the hippocampal expression and activity of group-I metabotropic glutamate receptors (Van Waes et al. 2009), which are localised post-synaptically and are involved in the regulation of learning and memory (reviewed by Nicoletti et al. 2011). We cannot exclude that changes in postsynaptic glutamate receptors in the hippocampus contribute to the behavioural phenotype of PRS rats particularly during ageing. The involvement of postsynaptic glutamate machinery in the dorsal hippocampus in the alterations of spatial memory has been reported in other models of stress during prenatal of juvenile/adult stage (Yaka et al. 2007; Schmidt et al. 2010).

8.1.5 Predictive Validity of PRS Rat as Model of Depression

A scientific consensus is emerging that the origins of adult disease are often found among developmental and biological disruptions occurring during the early years of life. These early experiences can affect adult health in two ways: either by cumulative damage over time or by the biological embedding of adversities during sensitive developmental periods. According to the developmental hypothesis of mood disorders, stressful events occurring during critical periods of brain development trigger a maladaptive programme that alters mechanisms of resilience to stress across the entire life span. Along this line, major depression can be seen as a latent outcome of stressful early life events that become more influential in genetically predisposed individuals (Darnaudéry and Maccari 2008; Weinstock 2008; Lupien et al. 2009; Shonkoff et al. 2009; Krishnan and Nestler 2008). The rat PRS model is particularly valuable for the study of the pathological consequences of early life stress and for the identification of novel therapeutic strategies in the treatment of depression and anxiety (Darnaudéry and Maccari 2008; Maccari et al. 2003; Maccari and Morley-Fletcher 2007). The value of PRS rats as a model of depression is strengthened by the finding that these animals show a persistent deficit in hippocampal neurogenesis and abnormalities in transcription factors and surface receptors that have been related to the pathophysiology of depression and anxiety (Darnaudery et al. 2006; Lemaire et al. 2000; Schmitz et al. 2002; Zuena et al. 2008).

PRS rats display biobehavioural alterations that can parallel to some extent indices in human depression research, thus becoming a useful tool for the design and testing of new pharmacologic strategies in mood and sleep disorders. The criteria proposed by Willner and Mitchell (2002) require that animal models of depression exhibit face, predictive and construct validity. Face validity refers to the phenomenological similarity, whereas predictive validity refers to the accuracy of a model in forecasting the course and outcome of a human syndrome. Finally, construct validity represents the degree to which both the human syndrome and the animal model are unambiguously defined such that a rational theory can be constructed to explain the pathophysiology of disorder. However, because mental disorder is a human pathology, the perfect homology of an animal model to a human psychiatric condition cannot be absolutely demonstrated. In contrast, it is possible to use animal models to highlight some similar symptoms and develop new pharmacological strategies. Various clinical observations in humans suggest a possible pathophysiological link between depression and disturbances in hypothalamus–pituitary axis, circadian rhythmicity, body temperature fluctuations, various peripheral hormone concentrations and urinary levels of neurotransmitter metabolites (Holsboer 2001). Added to our previous findings in PRS rats of high anxiety and emotionality, dysfunction of the HPA axis and circadian timing abnormalities, the observation of long-term changes in their sleep structure (Dugovic et al. 1999; Mairesse et al. 2012a) supports the validity of the PRS model as a valid animal model of anxiety/depression.

Our group has provided increasing evidence for the predictive validity of PRS model by mean of chronic treatment with different classes of antidepressants in adult rats (Morley-Fletcher et al. 2003a, 2004a, 2011). Indeed, imipramine (tricyclic) and tianeptine (a selective serotonin reuptake enhancer, structurally similar to the tricyclic antidepressants) reverse several PRS-induced alterations at the behavioural, neurochemical and neuroanatomical level (Morley-Fletcher et al. 2003a, 2004a). Thus, following antidepressant treatment, PRS rats displayed reduced immobility behaviour in the forced swim test, increased exploration of the open arm in the elevated-plus maze, enhanced mineralocorticoid and glucocorticoid receptor densities in the hippocampus and modified 5-HT_{1A} mRNA expression.

We have also tested the therapeutic efficacy of agomelatine, a novel antidepressant that behaves as a mixed MT1/MT2 melatonin receptor agonist/5-HT_{2c} serotonin receptor antagonist (Morley-Fletcher et al. 2011). Agomelatine treatment corrected *all* abnormalities displayed by PRS rats, suggesting that the drug impacts mechanisms that lie at the core of the maladaptive programming induced by PRS (see Fig. 8.4). We wish to highlight that agomelatine had no effect on any of the parameters we have tested in control rats. This suggests that agomelatine, at least in PRS rats, acts as an etiopathogenetic drug and its action is specific to the pathological state (i.e. agomelatine behaves as a "disease-dependent" drug).

PRS rats showed a deficit in adult neurogenesis, which was more prominent in the ventral portion of the hippocampus. Agomelatine treatment preferentially enhanced neurogenesis in the ventral hippocampus of PRS rats, as it does in transgenic mice with defective glucocorticoid receptors (Païzanis et al. 2010). Other antidepressant drugs also affect neurogenesis in the ventral hippocampus in different animal models of depression (Bisgaard et al. 2007; Jayatissa et al. 2006; Marais et al. 2009; Surget et al. 2008). To what extent the increase in hippocampal neurogenesis contributes to the antidepressant-like effect of agomelatine in PRS rats is unclear. In general, the role of a defective hippocampal neurogenesis in the pathophysiology of depression is debated (Sapolsky 2004). A decreased number of progenitor cells in the hippocampal dentate gyrus are seen in animal models of depression—including chronic stress and chronic high-corticosterone injections (Brummelte and Galea 2010; Czéh et al. 2001; Warner-Schmidt and Duman 2006;

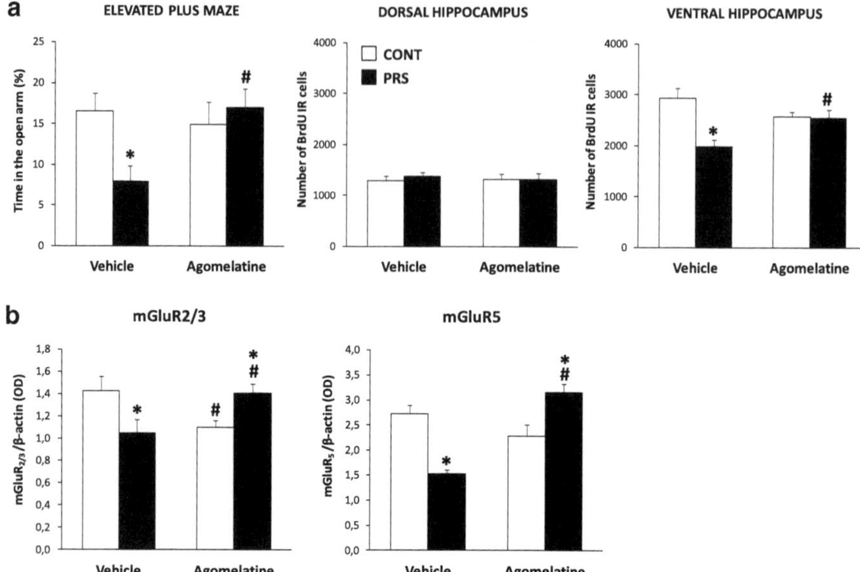

Fig. 8.4 Response to antidepressants in PRS rats. (**a**) A chronic treatment with agomelatine reduces anxiety-like behaviour of PRS rats in the EPM and reverses PRS-induced deficit in hippocampal neurogenesis specifically in the ventral hippocampus, a key region involved in anxiety regulation. (**b**) A chronic treatment with the antidepressant agomelatine increases mGluR5 and mGluR2/3 in adult PRS rats. $*P<0.05$ and $**P<0.01$ vs. controls. Original data are reported in Morley-Fletcher et al. (2011)

Zhang et al. 2010)—as well as in the hippocampus of elderly depressed patients (Lucassen et al. 2010a). However, there is also evidence that a reduced neurogenesis in the dentate gyrus does not associate with a depressive phenotype in response to chronic stress (Vollmayr et al. 2003; Jayatissa et al. 2009). It is a general belief that hippocampal neurogenesis is enhanced by antidepressant drugs (Santarelli et al. 2003; Malberg and Schechter 2005). However, there are exceptions (Marlatt et al. 2010), and, in addition, the action of antidepressants on neurogenesis might be critically influenced by age, gender and stress context (Malberg et al. 2000; Oitzl et al. 2000; Navailles et al. 2008; Lucassen et al. 2009, 2010b; David et al. 2009).

Chronic agomelatine treatment corrected all aspects of the pathological programming triggered by PRS, which includes a defective neurogenesis in the ventral hippocampus. Newly formed neurons in the hippocampal dentate gyrus are involved in the temporal encoding of new memories acting as "pattern integrators" (reviewed by Aimone et al. 2010). In response to agomelatine, the ventral hippocampus of PRS rats may reacquire the ability to link stress-related events that occur simultaneously or close in time and to separate recent and remote stress-related memories, a mechanism that may critically affect resilience to stress.

Agomelatine also reversed the reduction in the levels of p-CREB, mGlu2/3 receptors and mGlu5 receptors found in the hippocampus of PRS rats. CREB phosphorylation, a process that enhances the transcriptional activity of CREB, is induced by chronic antidepressant treatment in heterologous expression systems (Abdel-Razaq et al. 2007), as well as in the rat hippocampus and cerebral cortex (Tiraboschi et al. 2004). In addition, p-CREB levels in peripheral blood T lymphocytes positively correlate with the response of depressed patients to antidepressant medication (Koch et al. 2002). Reductions in hippocampal levels of mGlu2/3 receptors have been found not only in PRS rats (Zuena et al. 2008; Morley-Fletcher et al. 2011) but also in Flinders Sensitive Line (FSL) rats (Matrisciano et al. 2008), which are considered as a genetic model of depression (Overstreet et al. 2005). Chronic treatment with the antidepressant imipramine upregulates the expression and function of mGlu2/3 receptors in the rat hippocampus (Matrisciano et al. 2002), and pharmacological activation of mGlu2/3 receptors shortens the latency of antidepressant medication in FSL rats (Matrisciano et al. 2007, 2008). mGlu5 receptors are critically involved in the regulation of synaptic plasticity in the hippocampus (reviewed by Nicoletti et al. 2011), and a reduced expression of these receptors in the hippocampus might contribute to the pathophysiology of cognitive dysfunction and to the low resilience to stress in PRS rats (Darnaudéry and Maccari 2008). Interestingly, CREB, mGlu3 receptors and mGlu5 receptors are potentially linked to changes in neurogenesis in PRS rats. CREB represents a convergence point of multiple intracellular signalling pathways that regulate hippocampal neurogenesis, although its precise role is unclear (reviewed by Gass and Riva 2007; Dworkin and Mantamadiotis 2010). Activation of mGlu3 and mGlu5 receptors enhances proliferation of neuroprogenitors in brain niches of adult neurogenesis, including the subgranular zone of the hippocampal dentate gyrus (Di Giorgi-Gerevini et al. 2005; Melchiorri et al. 2007). Thus, cellular and biochemical abnormalities corrected by agomelatine in the hippocampus of PRS rats might be interconnected in the context of an epigenetic programming that alters the resilience to stress and leads to an anxious/depressive phenotype.

8.2 Conclusions

In an animal model of early stress, it has been shown that stress-related events that occur during the fetal and early postnatal period may have lifelong programming effects on HPA axis functioning and different body functions with a considerable impact on disease susceptibility. Stress-related disease can be interpreted broadly, including cardiovascular disease, components of the metabolic syndrome and emotional alterations, for which the evidence of fetal origins is most abundant. Preclinical and clinical studies have shown that lifelong programming of the HPA axis function by fetal life conditions is likely to be a key factor in mediating associations with these disorders. It is therefore highly plausible that susceptibility to different stress-related disorders originates in a similar manner during early life, although direct evidence is still lacking. The development of animal models involving early-life environmental manipulations should allow to study the concept of vulnerability applied to stress-related disorders and help to improve the prevention by developing new therapeutic strategies. In this framework, it appears that an impairment of glutamate transmission in the hippocampus may lie at the core of the neuroplastic programme induced by prenatal stress, which is translated into an anxious/depressive phenotype in the adult life. Therapeutic strategies aimed at restoring the balance between excitatory and inhibitory neurotransmission in the hippocampus could be effective in correcting the pathological phenotype caused by early life stress.

References

Abdel-Razaq W, Bates TE, Kendall DA (2007) The effects of antidepressants on cyclic AMP-response element-driven gene transcription in a model cell system. Biochem Pharmacol 73:1995–2003

Aimone JB, Deng W, Gage FH (2010) Adult neurogenesis: integrating theories and separating functions. Trends Cogn Sci 14:325–337

Augustin I, Rosenmund C, Südhof TC, Brose N (1999) Munc13-1 is essential for fusion competence of glutamatergic synaptic vesicles. Nature 400(6743):457–461

Bannerman DM, Rawlins JNP, McHugh SB, Deacon RMJ, Yee BK, Bast T, Zhang WN, Pothuizen HH, Feldon J (2004) Regional dissociations within the hippocampus—memory and anxiety. Neurosci Biobehav Rev 28(3):273–283

Barbazanges A, Piazza PV, Le Moal M, Maccari S (1996) Maternal glucocorticoid secretion mediates long-term effects of prenatal stress. J Neurosci 16(12):3943–3949

Barker DJ (1995) Intrauterine programming of adult disease. Mol Med Today 1(9):418–423

Barker DJ, Bagby SP, Hanson MA (2006) Mechanisms of disease: in utero programming in the pathogenesis of hypertension. Nat Clin Pract Nephrol 2(12):700–707

Bisgaard CF, Jayatissa MN, Enghild JJ, Sanchéz C et al (2007) Proteomic investigation of the ventral rat hippocampus links DRP-2 to escitalopram treatment resistance and SNAP to stress resilience in the chronic mild stress model of depression. J Mol Neurosci 32:132–144

Brummelte S, Galea LA (2010) Chronic corticosterone during pregnancy and postpartum affects maternal care, cell proliferation and depressive-like behavior in the dam. Horm Behav 58:769–779

Chen G, Henter ID, Manji HK (2010) Presynaptic glutamatergic dysfunction in bipolar disorder. Biol Psychiatry 67(11):1007–1009

Coleman WL, Bykhovskaia M (2010) Cooperative regulation of neurotransmitter release by Rab3a and synapsin II. Mol Cell Neurosci 44(2):190–200

Czéh B, Michaelis T, Watanabe T, Frahm J, de Biurrun G, van Kampen M, Bartolomucci A, Fuchs E (2001) Stress-induced changes in cerebral metabolites, hippocampal volume, and cell proliferation are prevented by antidepressant treatment with tianeptine. Proc Natl Acad Sci U S A 98:12796–12801

Darnaudéry M, Maccari S (2008) Epigenetic programming of the stress response in male and female rats by prenatal restraint stress. Brain Res Rev 57:571–585

Darnaudery M, Perez-Martin M, Belizaire G, Maccari S, Garcia-Segura LM (2006) Insulin-like growth factor 1 reduces age-related disorders induced by prenatal stress in female rats. Neurobiol Aging 27:119–127

David DJ, Samuels BA, Rainer Q, Wang JW et al (2009) Neurogenesis-dependent and -independent effects of fluoxetine in an animal model of anxiety/depression. Neuron 62:479–493

Day JC, Koehl M, Deroche V, Le Moal M, Maccari S (1998) Prenatal stress enhances stress- and corticotropin-releasing factor-induced stimulation of hippocampal acetylcholine release in adult rats. J Neurosci 18(5):1886–1892

Deminière JM, Piazza PV, Guegan G, Abrous N, Maccari S, Le Moal M, Simon H (1992) Increased locomotor response to novelty and propensity to intravenous amphetamine self-administration in adult offspring of stressed mothers. Brain Res 586(1):135–139

Di Giorgi-Gerevini V, Melchiorri D, Battaglia G, Ricci-Vitiani L, Ciceroni C, Busceti CL, Biagioni F, Iacovelli L, Canudas AM, Parati E, De Maria R, Nicoletti F (2005) Endogenous activation of metabotropic glutamate receptors supports the proliferation and survival of neural progenitor cells. Cell Death Differ 12:1124–1133

Dugovic C, Maccari S, Weibel L, Turek FW et al (1999) High corticosterone levels in prenatally stressed rats predict persistent paradoxical sleep alterations. J Neurosci 19:8656–8664

Dworkin S, Mantamadiotis T (2010) Targeting CREB signalling in neurogenesis. Expert Opin Ther Targets 14:869–879

Fanselow MS, Dong HW (2010) Are the dorsal and ventral hippocampus functionally distinct structures? Neuron 65:7–19

Garcia-Garcia AL, Elizalde N, Matrov D, Harro J, Wojcik SM, Venzala E, Ramírez MJ, Del Rio J, Tordera RM (2009) Increased vulnerability to depressive-like behavior of mice with decreased expression of VGLUT1. Biol Psychiatry 66(3):275–282

Gass P, Riva MA (2007) CREB, neurogenesis and depression. Bioessays 29:957–961

Geppert M, Bolshakov VY, Siegelbaum SA, Takei K, De Camilli P, Hammer RE, Südhof TC (1994) The role of Rab3A in neurotransmitter release. Nature 369(6480):493–497

Greengard P, Valtorta F, Czernik AJ, Benfenati F (1993) Synaptic vesicle phosphoproteins and regulation of synaptic function. Science 259:780–785

Grillo CA, Piroli GG, Wood GE, Reznikov LR, McEwen BS, Reagan LP (2005) Immunocytochemical analysis of synaptic proteins provides new insights into diabetes-mediated plasticity in the rat hippocampus. Neuroscience 136(2):477–486

Han GA, Malintan NT, Collins BM, Meunier FA, Sugita S (2010) Munc18-1 as a key regulator of neurosecretion. J Neurochem 115(1):1–10

Henke PG (1990) Granule cell potentials in the dentate gyrus of the hippocampus: coping behavior and stress ulcers in rats. Behav Brain Res 36:97–103

Henry C, Kabbaj M, Simon H, Le Moal M, Maccari S (1994) Prenatal stress increases the hypothalamo-pituitary-adrenal axis response in young and adult rats. J Neuroendocrinol 6(3):341–345

Henry C, Guegant G, Cador M, Arnauld E, Arsaut J, Moal ML, Demotes-Mainard J (1995) Prenatal stress in rats facilitates amphetamine-induced sensitization and induces long-lasting changes in dopamine receptors in the nucleus accumbens. Brain Res 685(1–2):179–186

Hinz B, Becher A, Mitter D, Schulze K, Heinemann U, Draguhn A, Ahnert-Hilger G (2001) Activity-dependent changes of the presynaptic synaptophysin-synaptobrevin complex in adult rat brain. Eur J Cell Biol 80(10):615–619

Holsboer F (2001) Antidepressant drug discovery in the postgenomic era. World J Biol Psychiatry 2(4):165–177

Jayatissa MN, Bisgaard C, Tingström A, Papp M, Wiborg O (2006) Hippocampal cytogenesis correlates to escitalopram-mediated recovery in a chronic mild stress rat model of depression. Neuropsychopharmacology 31:2395–2404

Jayatissa MN, Henningsen K, West MJ, Wiborg O (2009) Decreased cell proliferation in the dentate gyrus does not associate with development of anhedonic-like symptoms in rats. Brain Res 1290:133–141

Kjelstrup KG, Tuvnes FA, Steffenach HA, Murison R et al (2002) Reduced fear expression after lesions of the ventral hippocampus. Proc Natl Acad Sci U S A 99:10825–10830

Koch JM, Kell S, Hinze-Selch D, Aldenhoff JB (2002) Changes in CREB-phosphorylation during recovery from major depression. J Psychiatr Res 36:369–375

Koehl M, Barbazanges A, Le Moal M, Maccari S (1997) Prenatal stress induces a phase advance of circadian corticosterone rhythm in adult rats which is prevented by postnatal stress. Brain Res 759(2):317–320

Koehl M, Darnaudéry M, Dulluc J, Van Reeth O, Le Moal M, Maccari S (1999) Prenatal stress alters circadian activity of hypothalamo-pituitary-adrenal axis and hippocampal corticosteroid receptors in adult rats of both gender. J Neurobiol 40(3):302–315

Koehl M, Bjijou Y, Le Moal M, Cador M (2000) Nicotine-induced locomotor activity is increased by preexposure of rats to prenatal stress. Brain Res 882(1–2):196–200

Krishnan V, Nestler EJ (2008) The molecular neurobiology of depression. Nature 455:894–902

Laloux C, Mairesse J, Van Camp G, Giovine A, Branchi I, Bouret S, Morley-Fletcher S, Bergonzelli G, Malagodi M, Gradini R, Nicoletti F, Darnaudéry M, Maccari S (2012) Anxiety-like behaviour and associated neurochemical and endocrinological alterations in male pups exposed to prenatal stress. Psychoneuroendocrinology 37(10):1646–1658

Lemaire V, Koehl M, Le Moal M, Abrous DN (2000) Prenatal stress produces learning deficits associated with an inhibition of neurogenesis in the hippocampus. Proc Natl Acad Sci U S A 97:11032–11037

Lucassen PJ, Bosch OJ, Jousma E, Krömer SA, Andrew R, Seckl JR, Neumann ID (2009) Prenatal stress reduces postnatal neurogenesis in rats selectively bred for high, but not low, anxiety: possible key role of placental 11beta-hydroxysteroid dehydrogenase type 2. Eur J Neurosci 29:97–103

Lucassen PJ, Stumpel MW, Wang Q, Aronica E (2010a) Decreased numbers of progenitor cells but no response to antidepressant drugs in the hippocampus of elderly depressed patients. Neuropharmacology 58:940–949

Lucassen PJ, Meerlo P, Naylor AS, van Dam AM, Dayer AG, Fuchs E, Oomen CA, Czéh B (2010b) Regulation of adult neurogenesis by stress, sleep disruption, exercise and inflammation: implications for depression and antidepressant action. Eur Neuropsychopharmacol 20:1–17

Lupien SJ, McEwen BS, Gunnar MR, Heim C (2009) Effects of stress throughout the lifespan on the brain, behavior and cognition. Nat Rev Neurosci 10:434–445

Maccari S, Morley-Fletcher S (2007) Effects of prenatal restraint stress on the hypothalamus-pituitary-adrenal axis and related behavioural and neurobiological alterations. Psychoneuroendocrinology 32:S10–S15

Maccari S, Van Reeth O (2007) Circadian rhythms, effects of prenatal stress in rodents: an animal model for human depression. In: Fink G (ed) Encyclopedia of stress. Elsevier, Amsterdam

Maccari S, Piazza PV, Kabbaj M, Barbazanges A, Simon H, Le Moal M (1995) Adoption reverses the long-term impairment in glucocorticoid feedback induced by prenatal stress. J Neurosci 15:110–116

Maccari S, Darnaudery M, Morley-Fletcher S, Zuena AR, Cinque C, Van Reeth O (2003) Prenatal stress and long-term consequences: implications of glucocorticoid hormones. Neurosci Biobehav Rev 27:119–127

Mairesse J, Viltart O, Salomé N, Giuliani A, Catalani A, CasoliniP M-FS, Nicoletti F, Maccari S (2007a) Prenatal stress alters the negative correlation between neuronal activation in limbic regions and behavioral responses in rats exposed to high and low anxiogenic environments. Psychoneuroendocrinology 32(7):765–776

Mairesse J, Lesage J, Breton C, Bréant B, Hahn T, Darnaudéry M, Dickson SL, Seckl J, Blondeau B, Vieau D, Maccari S, Viltart O (2007b) Maternal stress alters endocrine function of the feto-placental unit in rats. Am J Physiol Endocrinol Metab 292(6):E1526–E1533

Mairesse J, Silletti V, Laloux C, Zuena AR, Giovine A, Consolazione M, van Camp G, Malagodi M, Gaetani S, Cianci S, Catalani A, Mennuni G, Mazzetta A, van Reeth O, Gabriel C, Mocaër E, Nicoletti F, Morley-Fletcher S, Maccari S (2012a) Chronic agomelatine treatment corrects the abnormalities in the circadian rhythm of motor activity and sleep/wake cycle induced by prenatal restraint stress in adult rats. Int J Neuropsychopharmacol 6:1–16

Mairesse J, Vercoutter-Edouart AS, Marrocco J, Zuena AR, Giovine A, Nicoletti F, Michalski JC, Maccari S, Morley-Fletcher S (2012b) Proteomic characterization in the hippocampus of prenatally stressed rats. J Proteomics 75(6):1764–1770

Malberg JE, Schechter LE (2005) Increasing hippocampal neurogenesis: a novel mechanism for antidepressant drugs. Curr Pharm Des 11:145–155

Malberg JE, Eisch AJ, Nestler EJ, Duman RS (2000) Chronic antidepressant treatment increases neurogenesis in adult rat hippocampus. J Neurosci 20:9104–9110

Marais L, Hattingh SM, Stein DJ, Daniels WM (2009) A proteomic analysis of the ventral hippocampus of rats subjected to maternal separation and escitalopram treatment. Metab Brain Dis 24:569–586

Marlatt MW, Lucassen PJ, van Praag H (2010) Comparison of neurogenic effects of fluoxetine, duloxetine and running in mice. Brain Res 231341:93–99

Marrocco J, Mairesse J, Ngomba RT, Silletti V, Van Camp G, Bouwalerh H, Summa M, Pittaluga A, Nicoletti F, Maccari S, Morley-Fletcher S (2012) Anxiety-like behavior in the prenatally stressed rat model of depression is associated with a selective reduction of glutamate release in the ventral hippocampus. J Neurosci. DOI:10.1523/JNEUROSCI.1040-12.2012

Matrisciano F, Storto M, Ngomba RT, Cappuccio I, Caricasole A, Scaccianoce S, Riozzi B, Melchiorri D, Nicoletti F (2002) Imipramine treatment up-regulates the expression and function of mGlu2/3 metabotropic glutamate receptors in the rat hippocampus. Neuropharmacology 42:1008–1015

Matrisciano F, Panaccione I, Zusso M, Giusti P, Tatarelli R, Iacovelli L, Mathé AA, Gruber SH, Nicoletti F, Girardi P (2007) Group-II metabotropic glutamate receptor ligands as adjunctive drugs in the treatment of depression: a new strategy to shorten the latency of antidepressant medication? Mol Psychiatry 12(8):704–706

Matrisciano F, Caruso A, Orlando R, Marchiafava M, Bruno V, Battaglia G, Gruber SH, Melchiorri D, Tatarelli R, Girardi P, Mathè AA, Nicoletti F (2008) Defective group-II metaboropic glutamate receptors in the hippocampus of spontaneously depressed rats. Neuropharmacology 55:525–531

McEwen BS (2011) The ever-changing brain: cellular and molecular mechanisms for the effects of stressful experiences. Dev Neurobiol 72(6):878–890. doi:10.1002/dneu.20968

Meaney MJ, Szyf M, Seckl JR (2007) Epigenetic mechanisms of perinatal programming of hypothalamic-pituitary-adrenal function and health. Trends Mol Med 13(7):269–277

Melchiorri D, Cappuccio I, Ciceroni C, Spinsanti P, Mosillo P, Sarichelou I, Sale P, Nicoletti F (2007) Metabotropic glutamate receptors in stem/progenitor cells. Neuropharmacology 53:473–480

Moghaddam B (2002) Stress activation of glutamate neurotransmission in the prefrontal cortex: implications for dopamine-associated psychiatric disorders. Biol Psychiatry 51(10):775–787

Morley-Fletcher S, Darnaudery M, Koehl M, Casolini P, Van Reeth O, Maccari S (2003a) Prenatal stress in rats predicts immobility behavior in the forced swim test. Effects of a chronic treatment with tianeptine. Brain Res 989:246–251

Morley-Fletcher S, Rea M, Maccari S, Laviola G (2003b) Environmental enrichment during adolescence reverses the effects of prenatal stress on play behaviour and HPA axis reactivity in rats. Eur J Neurosci 18:3367–3374

Morley-Fletcher S, Darnaudéry M, Mocaer E, Froger N, Lanfumey L, Laviola G, Casolini P, Zuena AR, Marzano L, Hamon M, Maccari S (2004a) Chronic treatment with imipramine reverses immobility behaviour, hippocampal corticosteroid receptors and cortical 5-HT(1A) receptor mRNA in prenatally stressed rats. Neuropharmacology 47:841–847

Morley-Fletcher S, Puopolo M, Gentili S, Gerra G, Macchia T, Laviola G (2004b) Prenatal stress affects 3,4-methylenedioxymethamphetamine pharmacokinetics and drug-induced motor alterations in adolescent female rats. Eur J Pharmacol 489(1–2):89–92

Morley-Fletcher S, Mairesse J, Soumier A, Banasr M, Fagioli F, Gabriel C, Mocaer E, Daszuta A, McEwen B, Nicoletti F, Maccari S (2011) Chronic agomelatine treatment corrects behavioral, cellular, and biochemical abnormalities induced by prenatal stress in rats. Psychopharmacology 217(3):301–313

Moser MB, Moser EI, Forrest E, Andersen P, Morris RG (1995) Spatial learning with a minislab in the dorsal hippocampus. Proc Natl Acad Sci U S A 92:9697–9701

Navailles S, Hof PR, Schmauss C (2008) Antidepressant drug-induced stimulation of mouse hippocampal neurogenesis is age-dependent and altered by early life stress. J Comp Neurol 509:372–381

Nicoletti F, Bockaert J, Collingridge GL, Conn PJ, Ferraguti F, Schoepp DD, Wroblewski JT, Pin JP (2011) Metabotropic glutamate receptors: from the workbench to the bedside. Neuropharmacology 60(7–8):1017–1041

Nyirenda MJ, Welberg LA, Seckl JR (2001) Programming hyperglycaemia in the rat through prenatal exposure to glucocorticoids-fetal effect or maternal influence? J Endocrinol 170(3):653–660

Oitzl MS, Workel JO, Fluttert M, Frösch F, De Kloet ER (2000) Maternal deprivation affects behaviour from youth to senescence: amplification of individual differences in spatial learning and memory in senescent Brown Norway rats. Eur J Neurosci 12:3771–3780

Ongür D, Jensen JE, Prescot AP, Stork C, Lundy M, Cohen BM, Renshaw PF (2008) Abnormal glutamatergic neurotransmission and neuronal-glial interactions in acute mania. Biol Psychiatry 64(8):718–726

Overstreet DH, Friedman E, Mathé AA, Yadid G (2005) The Flinders sensitive line rat: a selectively bred putative animal model of depression. Neurosci Biobehav Rev 29:739–759

Païzanis E, Renoir T, Lelievre V, Saurini F, Melfort M, Gabriel C, Barden N, Mocaër E, Hamon M, Lanfumey L (2010) Behavioural and neuroplastic effects of the new-generation antidepressant. Int J Neuropsychopharmacol 13:759–774

Peters DA (1988) Effects of maternal stress during different gestational periods on the serotonergic system in adult rat offspring. Pharmacol Biochem Behav 31(4):839–843

Popoli M, Yan Z, McEwen BS, Sanacora G (2012) The stressed synapse: the impact of stress and glucocorticoids on glutamate transmission. Nat Rev Neurosci 13(1):22–37

Raiteri L, Raiteri M (2000) Synaptosomes still viable after 25 years of superfusion. Neurochem Res 25(9–10):1265–1274

Raiteri M, Angelini F, Levi G (1974) A simple apparatus for studying the release of neurotransmitters from synaptosomes. Eur J Pharmacol 25(3):411–414

Sakane A, Manabe S, Ishizaki H, Tanaka-Okamoto M, Kiyokage E, Toida K, Yoshida T, Miyoshi J, Kamiya H, Takai Y, Sasaki T (2006) Rab3 GTPase-activating protein regulates synaptic transmission and plasticity through the inactivation of Rab3. Proc Natl Acad Sci U S A 103(26):10029–10034

Salomon S, Bejar C, Schorer-Apelbaum D, Weinstock M (2011) Corticosterone mediates some but not other behavioural changes induced by prenatal stress in rats. J Neuroendocrinol 23(2):118–128

Santarelli L, Saxe M, Gross C, Surget A, Battaglia F, Dulawa S, WeisstaubN LJ, Duman R, Arancio O, Belzung C, Hen R (2003) Requirement of hippocampal neurogenesis for the behavioural effects of antidepressants. Science 301:805–809

Sapolsky RM (2004) Is impaired neurogenesis relevant to the affective symptoms of depression? Biol Psychiatry 56:137–139

Schmidt MV, Trümbach D, Weber P, Wagner K, Scharf SH, Liebl C, Datson N, Namendorf C, Gerlach T, Kühne C, Uhr M, Deussing JM, Wurst W, Binder EB, Holsboer F, Müller MB (2010) Individual stress vulnerability is predicted by short-term memory and AMPA receptor subunit ratio in the hippocampus. J Neurosci 30(50):16949–16958

Schmitz C, Rhodes ME, Bludau M, Kaplan S et al (2002) Depression: reduced number of granule cells in the hippocampus of female, but not male, rats due to prenatal restraint stress. Mol Psychiatry 7:810–813

Seckl JR (2008) Glucocorticoids, developmental 'programming' and the risk of affective dysfunction. Prog Brain Res 167:17–34

Seckl JR, Holmes MC (2007) Mechanisms of disease: glucocorticoids their placental metabolism and fetal 'programming' of adult pathophysiology. Nat Clin Pract Endocrinol Metab 3(6):479–488, Review. PubMed PMID: 17515892

Shonkoff JP, Boyce WT, McEwen BS (2009) Neuroscience, molecular biology, and the childhood roots of health disparities: building a new framework for health promotion and disease prevention. JAMA 301:2252–2259

Surget A, Saxe M, Leman S, Ibarguen-Vargas Y et al (2008) Drug-dependent requirement of hippocampal neurogenesis in a model of depression and of antidepressant reversal. Biol Psychiatry 64:293–301

Thome J, Pesold B, Baader M, Hu M, Gewirtz JC, Duman RS, Henn FA (2001) Stress differentially regulates synaptophysin and synaptotagmin expression in hippocampus. Biol Psychiatry 50(10):809–812

Tiraboschi E, Tardito D, Kasahara J, Moraschi S, Pruneri P, Gennarelli M, Racagni G, Popoli M (2004) Selective phosphorylation of nuclear CREB by fluoxetine is linked to activation of CaM kinase IV and MAP kinase cascades. Neuropsychopharmacology 29:1831–1840

Tordera RM, Totterdell S, Wojcik SM, Brose N, Elizalde N, Lasheras B, Del Rio J (2007) Enhanced anxiety, depressive-like behavior and impaired recognition memory in mice with reduced expression of the vesicular glutamate transporter 1 (VGLUT1). Eur J Neurosci 25(1):281–290

Vallée M, Mayo W, Dellu F, Le Moal M, Simon H, Maccari S (1997) Prenatal stress induces high anxiety and postnatal handling induces low anxiety in adult offspring: correlation with stress-induced corticosterone secretion. J Neurosci 17:2626–2636

Vallée M, Maccari S, Dellu F, Simon H, Le Moal M, Mayo W (1999) Long-term effects of prenatal stress and postnatal handling on age-related glucocorticoid secretion and cognitive performance: a longitudinal study in the rat. Eur J Neurosci 11(8):2906–2916

Valtorta F, Benfenati F, Greengard P (1992) Structure and function of the synapsins. J Biol Chem 267(11):7195–7198

Van Waes V, Enache M, Dutriez I, Lesage J, Morley-Fletcher S, Vinner E, Lhermitte M, Vieau D, Maccari S, Darnaudéry M (2006) Hypo-response of the hypothalamic-pituitary-adrenocortical axis after an ethanol challenge in prenatally stressed adolescent male rats. Eur J Neurosci 24(4):1193–1200

Van Waes V, Enache M, Zuena A, Mairesse J, Nicoletti F, Vinner E, Lhermitte M, Maccari S, Darnaudéry M (2009) Ethanol attenuates spatial memory deficits and increases mGlu1a receptor expression in the hippocampus of rats exposed to prenatal stress. Alcohol Clin Exp Res 33(8):1346–1354

Viltart O, Mairesse J, Darnaudéry M, Louvart H, Vanbesien-Mailliot C, Catalani A, Maccari S (2006) Prenatal stress alters Fos protein expression in hippocampus and locus coeruleus stress-related brain structures. Psychoneuroendocrinology 31(6):769–780

Vollmayr B, Simonis C, Weber S, Gass P, Henn F (2003) Reduced cell proliferation in the dentate gyrus is not correlated with the development of learned helplessness. Biol Psychiatry 54:1035–1040

Warner-Schmidt JL, Duman RS (2006) Hippocampal neurogenesis: opposing effects of stress and antidepressant treatment. Hippocampus 16:239–249

Weinstock M (2008) The long-term behavioural consequences of prenatal stress. Neurosci Biobehav Rev 32:1073–1086

Willner P, Mitchell PJ (2002) The validity of animal models of predisposition to depression. Behav Pharmacol 13(3):169–188

Yaka R, Salomon S, Matzner H, Weinstock M (2007) Effect of varied gestational stress on acquisition of spatial memory, hippocampal LTP and synaptic proteins in juvenile male rats. Behav Brain Res 179(1):126–132

Yamamoto BK, Reagan LP (2006) The glutamatergic system in neuronal plasticity and vulnerability in mood disorders. Neuropsychiatr Dis Treat 2:7–14

Zhang JM, Tonelli L, Regenold WT, McCarthy MM (2010) Effects of neonatal flutamide treatment on hippocampal neurogenesis and synaptogenesis correlate with depression-like behaviors in preadolescent male rats. Neuroscience 169:544–554

Zuena AR, Mairesse J, Casolini P, Cinque C, Alemà GS, Morley-Fletcher S, Chiodi V, Spagnoli LG, Gradini R, Catalani A, Nicoletti F, Maccari S (2008) Prenatal restraint stress generates two distinct behavioural and neurochemical profiles in male and female rats. PLoS One 3(5):e2170

Chapter 9
Developmental Consequences of Prenatal Administration of Glucocorticoids in Rodents and Primates

Jonas Hauser

Abstract Since their first use in 1972 by Liggins and Howie, prenatal exposure to synthetic glucocorticoids (GCs) is commonplace in antenatal medicine to impede the preterm birth-associated morbid symptoms. Synthetic GCs are ligands of the receptor of endogenous GC, the glucocorticoid receptor. Although prenatal GC is warranted for its increased survival rate of preterm infants, the repeated exposure to synthetic GC long-term effects has been questioned, and investigation of potentially harmful long-term effects in animal studies is required. I will first summarise the existing findings in animal studies, which include two robust phenotypes: a transient reduction of body weight and alteration of the hypothalamo–pituitary–adrenal gland axis activity. Several studies assessed the neurotransmitters' concentrations in animals exposed to prenatal GC and reported an overall increased activity of serotoninergic and dopaminergic systems. Prenatal GC administration has also been shown to increase anxiety and reduce cognitive abilities in the long term. All these effects have been proposed to be mediated via epigenetics programming, which is the change of gene expression caused by mechanisms other than the DNA sequence (e.g. promoter methylation). Interestingly, the same mechanism has been proposed to mediate the long-term effects of altered maternal behaviour, suggesting that the developing individual, from conception until weaning, is undergoing epigenetics programming based on its environment.

J. Hauser (✉)
Institute of Behavioural Neuroscience, Department of Cognitive, Perceptual and Brain Sciences, University College London, Bedford Way 26, London WC1H 0AP, UK
e-mail: jonas.hauser@gmail.com

G. Laviola and S. Macrì (eds.), *Adaptive and Maladaptive Aspects of Developmental Stress*, Current Topics in Neurotoxicity 3, DOI 10.1007/978-1-4614-5605-6_9,

9.1 Why Study Prenatal Synthetic Glucocorticoids?

In the USA, 7–10% of births are premature (before 37th gestational week) each year (N.I.H. Consensus 1994), and preterm birth is the leading cause of neonatal death (Mathews and MacDorman 2011). Prenatal synthetic glucocorticoids (GC) are commonly prescribed in diagnosed preterm delivery to prophylactically impede the associated morbid symptoms (e.g. respiratory distress syndrome and intraventricular haemorrhage, Liggins and Howie 1972). This treatment received support from the American National Institutes of Health (NIH) that stated in the 1994 Consensus Developmental Conference on the Effects of Corticosteroids for Fetal Maturation on Perinatal Outcomes: "All fetuses between 24 and 34 weeks' gestation at risk of preterm delivery should be considered candidates for antenatal treatment with corticosteroids" (N.I.H. Consensus 1994). The positive impact of prenatal GC treatment on preterm baby survival has been confirmed in a recent meta-analysis (Crowther et al. 2011). The efficacy of prenatal GC has been observed for birth occurring within 7 days post treatment (Roberts et al. 2006), and consequently it is common practice in clinics to repeat the treatment weekly from diagnosis until birth (Crowther et al. 2011). The positive effects associated with prenatal GC exposure have however been mitigated by reports of transient reduction in some growth indices (mean weight, mean length and mean *Z*-scores of head circumference) (Crowther et al. 2011) and other unwanted side effects (e.g. increased behavioural and hormonal responses to painful stimuli, Davis et al. 2011). In a recent review focusing on the effects of prenatal GC on the hypothalamo–pituitary–adrenal gland axis (HPA), Tegethoff et al. (2009) concluded that antenatal GC exposure resulted in reduced basal and challenged HPA activity visible from foetal developmental stage up to 2 weeks of age. These effects were more marked with increasing total amount of antenatal GC, suggesting a dose response type of effects. Because this prenatal GC treatment has only been recently adopted in clinical practice and because of the difficulty to follow-up treated infants beyond hospital discharge, there are very few long-term studies of subjects exposed to prenatal GC. These studies reported no differences in neurosensory functions at 6 months (Mazumder et al. 2008) or, in the Kaufman assessment battery for children—which provides an IQ equivalent—at 5 years (Foix-L'Helias et al. 2008). To account for the evolving practice of prenatal GC treatment, the NIH revisited the Consensus of 1994 in 2001 and one of their conclusion was "animal studies should evaluate the pathophysiologic and metabolic mechanisms of potential benefits and risks, including the effects of repeat corticosteroids on central nervous system myelination and brain development" (N.I.H. Consensus 2001).

9.2 Stress, the HPA Axis and Their Relation to Synthetic GC

A stressor has been defined by Cannon as a situation of threat to an organism, to which it will react with physiological responses, aiming at energy mobilisation and increase in arousal, that increase survival through either confrontation or avoidance, the so-called fight-or-flight response (Cannon 1939). This definition has been refined by several authors and most recently by McEwen (McEwen and Wingfield 2003) as follows: [stressful stimuli are] "... events that are threatening to an individual and which elicit physiological and behavioural responses as part of allostasis (i.e. maintenance of an organism homeostasis) in addition to that imposed by normal life cycle". The stress response is characterised by a fast activation of the sympathetic branch of the autonomous nervous system (ANS) and by a delayed slower increase of HPA axis activity. ANS activation stimulates the release of catecholamines (epinephrine and norepinephrine) from the adrenal medulla. Following termination of the stressor, the parasympathetic branch of the ANS starts a compensatory response. The slower activation of the HPA axis results in release of the 41 amino acid peptide corticotropin-releasing hormone (CRH, also known as corticotropin-releasing factor and corticoliberin) by the paraventricular nucleus of the hypothalamus. CRH induces in pituitary target cells release of adrenocorticotropic hormone (ACTH). ACTH stimulates the adrenal gland, which releases endogenous GC (cortisol in primates, including humans, and corticosterone in rats) in blood flow where it is quickly bound by corticosteroid-binding globulin (CBG, Westphal 1983). Only free GCs are active. The CBG binding of GC serves as a tissue buffer against potential deleterious effects of elevated GC and can regulate the availability of free hormone to target tissue (about 95% of GC is protein bound, Gayrard et al. 1996). Recent findings have proposed a more active role of CBG in terms of mediation of the availability of GC to specific target tissue, as opposed to just reservoir of GC, as well as intracellular transportation of bound GC in specific cell types, allowing increased free GC levels above what could be achieved through simple diffusion (reviewed in Breuner and Orchinik 2002). The HPA axis activity is returned to baseline via a negative feedback mechanism that acts at all of its three levels: the hypothalamus, the pituitary and the adrenal gland (see Fig. 9.1 for a general schematic of the HPA axis function).

GC binds to both glucocorticoid receptors (GR) and mineralocorticoid receptors (MR) (Rosenfeld et al. 1993; de Kloet et al. 1990). Ontogeny of GR and MR in the rodent foetal brain undergoes spatial, temporal and sex-specific regulations (Owen and Matthews 2003; Pryce 2008), but both receptors are visible from respectively embryonic days 15.5 and 12.5 (Diaz et al. 1998). In humans, MR and GR are expressed in hippocampus from gestational week 24 (Noorlander et al. 2006). MR and GR are ligand-activated intracytoplasmatic transcription factor composed of three domains: the N-terminal domain (GR reviewed in Wright et al. 1993, MR reviewed in Pascual-Le Tallec and Lombes 2005), responsible for transcriptional activity, the DNA-binding domain; and the C-terminal domain or ligand-binding domain (LBD). Upon binding of their ligand, GR and MR translocate to the nucleus, dimerise and recognise specific semi-palindromic DNA promoter segments; this enables direct and indirect interactions with the transcription initiation complex and thereby the upregulation of target gene expression (reviewed in Beato and Sanchez-Pacheco 1996). Additional mechanisms mediated through ligand-bound receptors can repress transcription of certain genes: binding to a negative response elements (Malkoski and Dorin 1999), heterodimerisation with other nuclear receptors

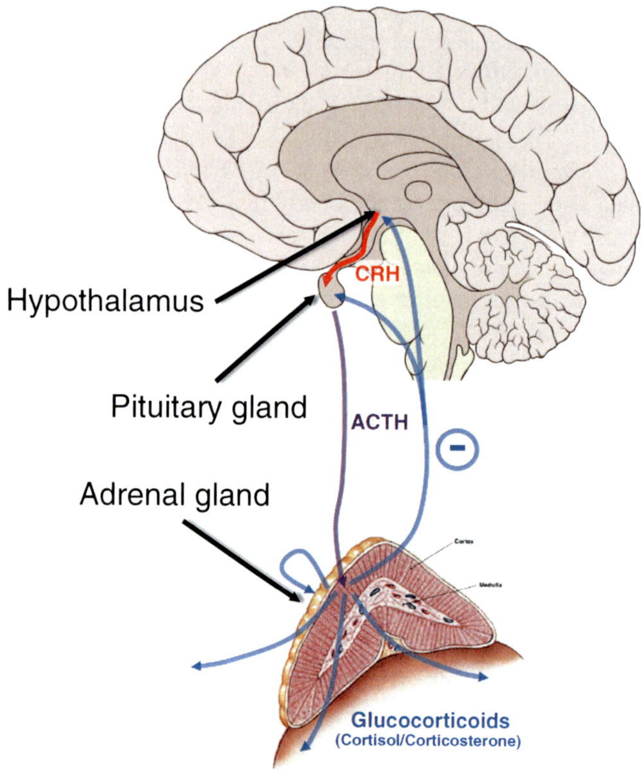

Fig. 9.1 Schematic of the HPA axis. In response to a stressor, the hypothalamus secretes CRH (*red*), which stimulates secretion of ACTH (*violet*) by the pituitary gland. ACTH in turn activates the adrenal gland release of GC (*blue*). GC exerts a negative feedback at all three levels

(Ou et al. 2001) or cross-talk with other nuclear receptors through protein–protein interactions (McKay and Cidlowski 1998, 1999, 2000). In addition to these well-studied transcriptional effects of GC, recent publications have reported the following non-genomic effects (Song and Buttgereit 2006): (1) physiochemical interaction with cellular membrane (Buttgereit and Scheffold 2002), (2) membrane-bound GR-mediated non-genomic effects (Groeneweg et al. 2011) and (3) cytosolic GR-mediated non-genomic effects (Bartholome et al. 2004).

The two main synthetic GCs are dexamethasone (DEX) and betamethasone (BETA), which differ from the endogenous GC by a fluorine atom on the ninth carbon. Due to the structural differences between endogenous and synthetic GCs, the latter exhibit increased binding to GR (25-fold endogenous GC affinity) and strongly reduced binding to MR (Grossmann et al. 2004). There are other differences between synthetic and endogenous GC: (1) The 11β-HSD2 enzyme oxidises natural GC to their inactive metabolite; it has been reported in numerous tissues, especially in the placental syncytiotrophoblast, where it provides a good protection of the foetus from maternal GC (Speirs et al. 2004). Synthetic GCs are poor substrates for

11β-HSD2; therefore they readily cross placenta and access the foetus (Gitau et al. 2001; Murphy et al. 2007). After 1 h incubation, 95% of endogenous GC but only 17% of synthetic GC is metabolised by 11β-HSD2 (Brown et al. 1996). (2) Contrarily to the natural GC, which activity can be modulated by their binding to CBG, synthetic GCs do not bind by CBG and only partially to albumin; 65–70% of synthetic GC is in bound state compared to the 95% of endogenous GC (Peets et al. 1969; Schwab and Klotz 2001). Synthetic GC exposure reproduces some elements of the GR activation following stress response; however, note the following differences between this treatment and the stress response: absence ANS stimulation, metabolisation by 11β-HSD2, low to no MR activation and different bioavailability of synthetic GC.

9.3 Summary of the Known Effects of Prenatal GC in Animal Models

Most studies assessing prenatal synthetic GC exposure used rodents (rats, guinea pigs and mice); there were also a few studies that used non-human primates or ovine, which focused on the HPA axis activity and lung functions. Studies in rodents typically used a treatment of 0.1 mg/kg during the last week of pregnancy (gestational days 14–21). In primates, treatment was usually performed in the last trimester of pregnancy; however, there were no general rules for the treatment duration or dose, which varied respectively from 2 to 42 days and from 0.1 to 15 mg/kg (Coe and Lubach 2005). The foetal treatment used in all these studies varied from subclinical exposure, in dose and duration, to overexposure. It is very difficult to select the onset, duration and dose of treatment, as there are differences in response to GC between species. Furthermore, organ- or system-specific developmental stages might be different at the same relative gestational age between precocial (primates) and altricial (rat) species. The only common element visible across studies and across species is that the onset of the treatment was systematically targeting the last trimester of pregnancy.

An extensive number of animal studies examined the impact of prenatal GC exposure on physical growth and HPA axis function. Prenatal synthetic GC has been associated with a reduction of body weight throughout life in rats (Brabham et al. 2000; Emgard et al. 2007; Hauser et al. 2006, 2009; Oliveira et al. 2006; Welberg et al. 2001), although some studies using lower synthetic GC doses have reported catch up growth (Kreider et al. 2005; McArthur et al. 2005). This effect was not observed in primates (Hauser et al. 2007, 2008; Uno et al. 1994), except in two studies using longer treatment (Johnson et al. 1981; Novy and Walsh 1983). In rodents, the general findings were that HPA basal activity was unaffected whereas its response to a stressful situation was increased by prenatal GC (Brabham et al. 2000; Hauser et al. 2009; Oliveira et al. 2006; Bakker et al. 1995; Hougaard et al. 2005; Muneoka et al. 1997; Shoener et al. 2006). Two studies reported a reduction of GR in the hippocampus in response to prenatal GC exposure (Brabham et al. 2000; Welberg et al. 2001), which could mediate the increased reactivity to the stressor via a reduction of the sensitivity of hippocampus to the negative feedback of GC. In guinea pigs, prenatal synthetic GC exposure had a sexual dimorphic effect on HPA axis activity. In males, the HPA axis response to stress was increased, whereas in females, the effect was modulated by the hormonal cycle. In the follicular early luteal phase it resulted in increased basal and challenged HPA axis activity, but this was reversed in the late luteal phase (Liu et al. 2001). In a long-term follow-up study of lamb exposed to prenatal GC, Sloboda and colleagues reported that the HPA axis exhibited no changes at 6 months, increased basal and challenged activity at 1 year and decreased basal and challenged activity at 3 years of age (Sloboda et al. 2000, 2002, 2007). This clearly underlines the importance of long-term follow-up studies as it suggests that prenatal GC might have effects that are age-dependent. Both in rhesus and vervet monkeys, challenged HPA axis activity was increased following prenatal synthetic GC treatment (Uno et al. 1994; de Vries et al. 2007), and this effect was also observed in basal activity in rhesus monkey. However, there was no evidence of impact on HPA axis activity in marmoset monkey offspring exposed to prenatal GC (Hauser et al. 2007). In conclusion, the challenged HPA axis activity is increased in most animal models after prenatal GC treatment; a similar increase in activity was reported in young infants (Tegethoff et al. 2009). The findings on basal activity seem to be more variable among studies. There is a strong interaction between the effects of prenatal GC treatment on HPA axis activity and the sex of the subjects in rats (Brabham et al. 2000; Hauser et al. 2009) and in guinea pigs (Liu et al. 2001). Although a similar sexual dimorphism was not reported in primate studies, it is noteworthy here to highlight that most primate studies only used males or were not able to account for the sex due to the limited number of subjects in these studies. In two studies both male and female marmoset monkeys were assessed for HPA axis activity and no differences were reported (Hauser et al. 2007, 2008).

A few studies focused on neurotransmitter changes in rodents. A general increase in serotoninergic function can be assumed, as prenatal synthetic GC promotes serotonin transporter expression in brainstem (Slotkin et al. 1996) and reduces serotonin turnover in hypothalamus, neocortex, hippocampus and midbrain as well as increases serotonin in hypothalamus and midbrain (Muneoka et al. 1997). Kreider et al. (2005) reported an increased cholinergic synaptic activity in the hippocampus of male rats. Prenatal GC results in an increase of tyrosine hydroxylase immunopositive cell counts in the substantia nigra pars compacta and an increase of dopamine in the dorsolateral striatum (McArthur et al. 2005), suggesting that this treatment results in an increased dopaminergic activity.

Finally, several studies investigated behavioural long-term effects of prenatal synthetic GC. In rodents, this treatment decreases locomotor activity and increases anxiety (Oliveira et al. 2006; Welberg et al. 2001), but there are also reports of increased locomotor activity (Muneoka et al. 1997) as well as reduced anxiety (Velisek 2006). The impact of prenatal GC treatment on learning was assessed most frequently in the Morris water maze (MWM). The MWM is the most commonly used behavioural task to assess spatial learning in rodents. It consists of a circular pool filled with cold water, in which a hidden escape platform is positioned. Learning is assessed by the reduction of latency to find the platform. Prenatal GC was reported to yield a general impairment of performance in this task (Brabham et al. 2000; Emgard et al. 2007; Kreider et al. 2005), although one study failed to replicate this effect (Oliveira et al. 2006). Noteworthy here, a study using cross-fostering showed that this reported deficit in spatial learning might not be mediated by the prenatal GC treatment but by the associated increased maternal care (Hauser et al. 2009). Cross-fostering consists of fostering treated and control pups to treated and control surrogate mother, resulting in litters composed of half-treated half-control pups reared either by treated or control dams. Learning was also assessed following prenatal GC in the eight-arm radial maze, a task in which the animal has to get a reward from each arm, re-entry in an arm already visited being an error. Prenatal GC treatment resulted in a quicker learning in males and in performance impairment in females (Kreider et al. 2005). In a set of two studies, we evaluated the possible association between prenatal GC exposure and symptoms of two psychiatric diseases, schizophrenia and depression. Rat offspring exposed to prenatal GC did not exhibit any alteration of prepulse inhibition or latent inhibition (Hauser et al. 2006), which respectively accounts for schizophrenia-induced disruption in sensory motor gating (Braff et al. 1992) and the ability to ignore irrelevant stimuli (Baruch et al. 1988). We also failed to obtain evidence of depressive-like behavioural performance in paradigm taxing processes affected by depression, namely anhedonia and behavioural despair (Hauser et al. 2009). Anhedonia was assessed using the progressive ratio schedule of reinforcement, a task in which the subject has to increase progressively the workload required to obtain a reward, with a reduction of maximum workload reached modelling anhedonia. Behavioural despair was assessed using the Porsolt forced swim task and in unconditioned stimulus pre-exposure in active avoidance. In both tasks the subject is exposed to an inescapable aversive situation (cold water in Porsolt and foot shocks in active avoidance), with reduced latency to stop escape attempts modelling behavioural despair. These findings confirmed and extended the existing report of unaffected performance in the Porsolt forced swim test following prenatal GC (Oliveira et al. 2006). In primates, we assessed the impact of prenatal GC on motor learning using an adaptation of Whishaw skilled reaching task for rats and discrimination learning and motivation for a palatable reward using the Cambridge Neuropsychological Test Automated Battery (CANTAB) system (Hauser et al. 2008). In the skilled reaching task, the subject has to reach through a narrow opening for a palatable reward. The CANTAB apparatus consists of a programmable touch screen. We assessed discrimination learning by rewarding only one out of two presented stimuli; once good performance was reached the rewarded and unrewarded stimuli were reversed to assess reversal learning. Motivation was assessed in a progressive ratio schedule of reinforcement. Marmosets exposed to GC during the last third of pregnancy showed no improvement of performance in the skilled reaching task with experience and an improvement of reversal learning, but no changes in motivation and no other effects were observed.

9.4 Mediation of Effects of Prenatal Synthetic GC Exposure in Adulthood

The first mediator that comes to mind for long-term effects of prenatal synthetic GC exposure is a major alteration of development (e.g. reduction of organ size, change in the differentiation of specific subpopulation of cells). Prenatal synthetic GC has only shown a transient reduction of birth weight in rodents (Hauser et al. 2006; Kreider et al. 2005; McArthur et al. 2005; Muneoka et al. 1997; Burlet et al. 2005), an effect that was not visible in primates (Hauser et al. 2008) except in studies using especially long treatments and/or high concentration of GC (Johnson et al. 1981; Novy and Walsh 1983). In addition, these studies with longer treatment duration reported decreased hippocampus size as well as reduced number of pyramidal and granule neurons. This indicates that the effects of prenatal GC treatment on physical growth and CNS development follow a dose-dependent curve. A similar dose-dependent effect of prenatal GC treatment on HPA axis activity was reported in human babies (Tegethoff et al. 2009). In clinical studies, the general outcome of prenatal synthetic treatment is also a transient reduction of birth weight, which is recovered after few days (Crowther et al. 2011; French et al. 1999). Considering these transient effects and the absence of other major defect at birth, major developmental effects resulting in abnormal organs following prenatal synthetic GC treatment are unlikely with the dose range used in clinics or in most animal studies. The reduced birth weight reported rather seems to represent the known effects of GC, namely growth reduction in favour of an increased maturation. Thus the low dose and short treatment used in clinics do not result in major alteration of development. Interestingly, increasing dose or duration of treatment was shown to have dramatic impact on survival in animal models. In a pilot study in rats, we observed that increasing the typical prenatal synthetic GC doses used in this species by twofold resulted in major developmental problem, with most of the litter being either stillborn or having major malformation leading to early life death (Hauser et al. unpublished results). A similar report was published in primates reporting an increased number of stillbirths following high-dose prenatal GC treatment (Novy and Walsh 1983).

The next most plausible mediator candidate is prenatal or foetal programming; it is the phenomenon by which a specific adulthood phenotype is set up based on foetal environment. Barker (Hales and Barker 2001) made the first proposition of prenatal programming by associating prenatal undernutrition with several adulthood diseases (including but not limited to metabolic syndrome, type 2 diabetes) in his thrifty phenotype hypothesis. In this hypothesis, prenatal undernutrition is perceived by the foetus, which consequentially adapts its development (e.g. via increased food storage in fat tissue). Barker highlighted that such beneficial strategy would be harmful to the offspring if their future environment was not under food restriction. The HPA axis has also been proposed to undergo prenatal programming. It is known that a stressful environment (pathogens, nutrient deprivation, high predation, etc.) results in increased HPA activity in the pregnant mother. In response to that, the foetus development is adapted and the offspring exhibit increased basal and challenged HPA axis activity, decreased GR mRNA and protein levels as well as behavioural fear and anxiety (Seckl 2001, 2004). This phenotype is more adapted to a threatening environment, and it increases the offspring survival rate despite its cost. As stated by Barker, a major drawback of prenatal programming is that it requires long-term prediction based on the information perceived by the foetus to reliably represent the environment in which the adult offspring will live. The effect of prenatal GC treatment can be interpreted in the frame of this theory as follows: prenatal GC is perceived by the developing foetus as indicator of a stressful environment (due to its activation of the HPA axis); however, this signal is a poor predictor of adulthood environment; thus, the infant will have a maladaptive phenotype in the form of an overreactive HPA axis in a normal environment that will yield no benefits or even be harmful.

The mediation of prenatal programming has yet to be understood; however, several indices suggest that epigeneticss play a major role. Epigenetics is the modification of genetic information over and above alteration in nucleotide sequence. Its control of gene expression is mediated by DNA methylation and/or modification of chromatin packaging (Wolffe 1998). Recent studies support the association between perinatal stress and epigenetics programming of element of the stress response. The laboratory of Meaney presented evidence associating early life variation in maternal care and epigenetics (Weaver et al. 2004). They reported that offspring of dam exhibiting high maternal care had reduced GR exon 1_7 methylation and hippocampal GR expression. Confirming the epigenetics nature of this effect, cross-fostering of these pups resulted in a phenotype that was dependent of the maternal care and not of the genetic parent. Although not in the central nervous system, another study by Thomassin et al. (2001) is supporting the association between prenatal GC treatment and epigenetics. They reported that prenatal levels of GC were modulating the demethylation of the glucocorticoid receptor responsive unit of the tyrosine aminotransferase gene promoter in rat liver. These studies provide the first evidences for a mechanism that could mediate the integration of foetal environment in genetic expression. Thus it seems that prenatal synthetic GC exposure and increased maternal behaviour both lead to epigenetics modification of gene expression in adulthood, not only in the brain but also in other organs. Mediation by epigenetics prenatal programming might therefore be the best way to explain the integration of environmental information and the consequent adapted foetal development, such as the thrifty phenotype in response to undernutrition or the increased stress response observed after prenatal synthetic GC.

9.5 Other Factors Modulating Prenatal GC Treatment Effects

Gender of the exposed foetus and the maternal care it received in early life have repeatedly been reported to modulate the effects of prenatal GC treatment. In the rat prenatal synthetic GC literature, most studies were performed using solely male subjects, while in other species the inclusion of both males and females is more common. Differences between genders are important considering the fact that each gender undergoes a specific prenatal development path from as early as the first week of life in humans. This differential development between male and female continues postnatally and peaks at puberty, when endocrine and physical differences are accentuated between genders. Thus, for studies trying to achieve highest possible translational value of their findings, the use of both males and females is essential. Considering that there are important endocrine, neurologic and behavioural differences between sexes, the effect of prenatal synthetic GC on any of these could be different between the two genders. The inclusion of both genders in studies requires an increase in number of experimental subject to achieve the necessary statistical power. This can be complicated with primate, which is the reason why in most studies, experimenters either only used males or did not use the sex as a factor in their analyses. In rat studies using both genders, experimenters reported a clear sexual dimorphism of prenatal GC treatment, with increased HPA axis reactivity being restricted to males (Brabham et al. 2000; Hauser et al. 2009). In guinea pig, the effect of prenatal synthetic GC exposure was an increased HPA axis response in males and oestrus cycle dependant in females (Liu et al. 2001). The sexual dimorphism observed in the effects of prenatal GC exposure can have a wide number of mediators; this was not yet fully investigated. An obvious mediator is the sex hormones themselves, particularly when considering the overlap between their activity pathway and the one of GC and the possibility of GR to form heterodimer with sexual hormone receptors.

Prenatal exposure to synthetic GC has been shown to result in altered maternal behaviour in rodents (Brabham et al. 2000; Hauser et al. 2009) and in altered infant home cage behaviours in primates (Hauser et al. 2008). Manipulations affecting maternal behaviour (e.g. early deprivation or early manipulation) have been shown to alter HPA axis activity as well as behavioural response in various tasks in adult rodents of the first as well as the second generation (Pryce and Feldon 2003; Iqbal et al. 2012) and primates (Pryce et al. 2011). It is therefore possible that part of the effects attributed to prenatal synthetic GC treatment could originate from the altered maternal behaviour. To be able to dissociate direct effects of prenatal synthetic GC treatment from indirect effects mediated via altered maternal behaviour, the best strategy is to use a cross-fostering design. Despite the increase in number of animals needed, the possibility to dissociate effects mediated via the treatment from those due to altered maternal behaviour is a considerable advantage in terms of data interpretation. An example where such refinement proved to be critical is the impact of prenatal synthetic GC on spatial learning. In the MWM, prenatal GC exposure was reported to result in a decrease of performance (Emgard et al. 2007). However this finding was reinterpreted after a study using a cross-fostering design reported that the decreased performance was due to the alteration of maternal behaviour associated with this treatment (Hauser et al. 2009).

9.6 Conclusions

There is no doubt that the use of prenatal GC is necessary in diagnosed preterm delivery, as this treatment clearly improves survival rate of newborn. Nevertheless, it is important to understand all the possible long-term effects such a treatment could yield and thereby develop better strategy to accommodate them. GCs are the key hormones of the HPA axis and inhibit growth; it is therefore not surprising that most clinical studies focused on the HPA axis and on physical growth of infants exposed to prenatal GC treatment. The general finding in this regard was that prenatal GC treatment results in a transient sensitisation of the HPA axis and a transient reduction of infant weight. The transient character of these effects is questioned by studies performed in animals, especially considering the life-long sensitisation of the HPA axis reported in rats. A long-term follow-up study in lamb showed that the impact of prenatal GC on HPA axis was age dependent. In regard of body weight, although most animal studies reported only a transient effect of prenatal GC treatment, there is at least one report of long-lasting effects in rats (Hauser et al. 2006). This highlights the importance to obtain a long-term follow-up study in humans, as the effect on HPA axis and/or body weight could remain silent until a certain age. On the background of these physiological changes, there were only few long-term effects of prenatal GC on cognition and behavioural performance in various tasks. This absence of long-term behavioural effects is positive considering the worries regarding side effects of this treatment in clinics. A very interesting finding obtained in rodent studies is that some of the effects traditionally attributed to prenatal GC treatment were shown to be mediated by the increased maternal care of dams receiving GC. First of all, this finding highlights the importance to use a cross-fostering design when investigating any perinatal treatment, to be able to dissociate the direct effect from those mediated via the mother. Based on these observed long-term effect of prenatal GC exposure mediated via alteration of maternal behaviour in rats, clinical investigation focusing on such possibilities in humans are highly warranted.

Epigenetics programming seems to be the mediator of both prenatal synthetic GC exposure and alteration of maternal behaviour. Accordingly, prenatal and early life programming could be reunited into one longer process: developmental epigenetics programming. Using this mechanism, growing organisms could adapt their development strategy according to modifications of their environment perceived either directly or indirectly. This hypothesis provides an elegant common mechanism to integrate information from the various life stages to have a continuous development adapted to the environment. While, natural selection works over generations and selects phenotypes according to long-term variation of the environment, developmental epigenetics programming provides a faster mechanism to adapt to rapid environmental changes. Because any environment will indeed present both fast and slow changes, it would be expected that organisms developed mechanisms to integrate both of them in their development to achieve the most adapted phenotype. The quickness of the developmental epigenetics programming is also responsible for its drawback, namely the risk of maladaptation, when a signal is not reliably representing the environment in which the developing organism will be in its adult life. Thus while developmental epigenetics programming compensates natural selection to accommodate fast environmental changes, natural selection slower process might compensate for the risk of maladaptation existing in developmental epigenetics programming. These two mechanisms would thereby act in synergy to achieve the best phenotype according to long-term selection and short-term adaptation.

References

Bakker JM et al (1995) Effects of short-term dexamethasone treatment during pregnancy on the development of the immune system and the hypothalamo-pituitary adrenal axis in the rat. J Neuroimmunol 63(2):183–191

Bartholome B et al (2004) Membrane glucocorticoid receptors (mGCR) are expressed in normal human peripheral blood mononuclear cells and up-regulated after in vitro stimulation and in patients with rheumatoid arthritis. FASEB J 18(1):70–80

Baruch I, Hemsley DR, Gray JA (1988) Differential performance of acute and chronic schizophrenics in a latent inhibition task. J Nerv Ment Dis 176(10):598–606

Beato M, Sanchez-Pacheco A (1996) Interaction of steroid hormone receptors with the transcription initiation complex. Endocr Rev 17(6):587–609

Brabham T et al (2000) Effects of prenatal dexamethasone on spatial learning and response to stress is influenced by maternal factors. Am J Physiol Regul Integr Comp Physiol 279(5):R1899–R1909

Braff DL, Grillon C, Geyer MA (1992) Gating and habituation of the startle reflex in schizophrenic patients. Arch Gen Psychiatry 49(3):206–215

Breuner CW, Orchinik M (2002) Plasma binding proteins as mediators of corticosteroid action in vertebrates. J Endocrinol 175(1):99–112

Brown RW et al (1996) Cloning and production of antisera to human placental 11 beta-hydroxysteroid dehydrogenase type 2. Biochem J 313(pt 3):1007–1017

Burlet G et al (2005) Antenatal glucocorticoids blunt the functioning of the hypothalamic-pituitary-adrenal axis of neonates and disturb some behaviors in juveniles. Neuroscience 133(1):221–230

Buttgereit F, Scheffold A (2002) Rapid glucocorticoid effects on immune cells. Steroids 67(6):529–534

Cannon WB (1939) The wisdom of the body. Peter Smith Publisher, New York

Coe CL, Lubach GR (2005) Developmental consequences of antenatal dexamethasone treatment in nonhuman primates. Neurosci Biobehav Rev 29(2):227–235

Crowther CA et al (2011) Repeat doses of prenatal corticosteroids for women at risk of preterm birth for improving neonatal health outcomes. Cochrane Database Syst Rev 2011(6):CD003935

Davis EP, Waffarn F, Sandman CA (2011) Prenatal treatment with glucocorticoids sensitizes the hpa axis response to stress among full-term infants. Dev Psychobiol 53:175–183

de Kloet ER, Reul JM, Sutanto W (1990) Corticosteroids and the brain. J Steroid Biochem Mol Biol 37(3):387–394

de Vries A et al (2007) Prenatal dexamethasone exposure induces changes in nonhuman primate offspring cardiometabolic and hypothalamic-pituitary-adrenal axis function. J Clin Invest 117(4):1058–1067

Diaz R, Brown RW, Seckl JR (1998) Distinct ontogeny of glucocorticoid and mineralocorticoid receptor and 11beta-hydroxysteroid dehydrogenase types I and II mRNAs in the fetal rat brain suggest a complex control of glucocorticoid actions. J Neurosci 18(7):2570–2580

Emgard M et al (2007) Prenatal glucocorticoid exposure affects learning and vulnerability of cholinergic neurons. Neurobiol Aging 28(1):112–121

Foix-L'Helias L et al (2008) Impact of the use of antenatal corticosteroids on mortality, cerebral lesions and 5-year neurodevelopmental outcomes of very preterm infants: the EPIPAGE cohort study. BJOG 115(2):275–282

French NP et al (1999) Repeated antenatal corticosteroids: size at birth and subsequent development. Am J Obstet Gynecol 180(1 pt 1):114–121

Gayrard V, Alvinerie M, Toutain PL (1996) Interspecies variations of corticosteroid-binding globulin parameters. Domest Anim Endocrinol 13(1):35–45

Gitau R et al (2001) Fetal hypothalamic-pituitary-adrenal stress responses to invasive procedures are independent of maternal responses. J Clin Endocrinol Metab 86(1):104–109

Groeneweg FL et al (2011) Rapid non-genomic effects of corticosteroids and their role in the central stress response. J Endocrinol 209(2):153–167

Grossmann C et al (2004) Transactivation via the human glucocorticoid and mineralocorticoid receptor by therapeutically used steroids in CV-1 cells: a comparison of their glucocorticoid and mineralocorticoid properties. Eur J Endocrinol 151(3):397–406

Hales CN, Barker DJ (2001) The thrifty phenotype hypothesis. Br Med Bull 60:5–20

Hauser J, Feldon J, Pryce CR (2006) Prenatal dexamethasone exposure, postnatal development, and adulthood prepulse inhibition and latent inhibition in Wistar rats. Behav Brain Res 175(1):51–61

Hauser J et al (2007) Effects of prenatal dexamethasone treatment on postnatal physical, endocrine, and social development in the common marmoset monkey. Endocrinology 148(4):1813–1822

Hauser J et al (2008) Effects of prenatal dexamethasone treatment on physical growth, pituitary-adrenal hormones, and performance of motor, motivational, and cognitive tasks in juvenile and adolescent common marmoset monkeys. Endocrinology 149(12):6343–6355

Hauser J, Feldon J, Pryce CR (2009) Direct and dam-mediated effects of prenatal dexamethasone on emotionality, cognition and HPA axis in adult Wistar rats. Horm Behav 56(4):364–375

Hougaard KS et al (2005) Prenatal stress may increase vulnerability to life events: comparison with the effects of prenatal dexamethasone. Brain Res Dev Brain Res 159(1):55–63

Iqbal M et al (2012) Transgenerational effects of prenatal synthetic glucocorticoids on hypothalamic-pituitary-adrenal function. Endocrinology 153(7):3295–3307

Johnson JW et al (1981) Long-term effects of betamethasone on fetal development. Am J Obstet Gynecol 141(8):1053–1064

Kreider ML et al (2005) Gestational dexamethasone treatment elicits sex-dependent alterations in loc.motor activity, reward-based memory and hippocampal cholinergic function in adolescent and adult rats. Neuropsychopharmacology 30(9):1617–1623

Liggins GC, Howie RN (1972) A controlled trial of antepartum glucocorticoid treatment for prevention of the respiratory distress syndrome in premature infants. Pediatrics 50(4):515–525

Liu L, Li A, Matthews SG (2001) Maternal glucocorticoid treatment programs HPA regulation in adult offspring: sex-specific effects. Am J Physiol Endocrinol Metab 280(5):E729–E739

Malkoski SP, Dorin RI (1999) Composite glucocorticoid regulation at a functionally defined negative glucocorticoid response element of the human corticotropin-releasing hormone gene. Mol Endocrinol 13(10):1629–1644

Mathews TJ, MacDorman MF (2011) Infant mortality statistics from the 2007 period linked birth/infant death data set. Natl Vital Stat Rep 59(6):1–30

Mazumder P et al (2008) Single versus multiple courses of antenatal betamethasone and neonatal outcome: a randomized controlled trial. Indian Pediatr 45(8):661–667

McArthur S et al (2005) Altered mesencephalic dopaminergic populations in adulthood as a consequence of brief perinatal glucocorticoid exposure. J Neuroendocrinol 17(8):475–482

McEwen BS, Wingfield JC (2003) The concept of allostasis in biology and biomedicine. Horm Behav 43(1):2–15

McKay LI, Cidlowski JA (1998) Cross-talk between nuclear factor-kappa B and the steroid hormone receptors: mechanisms of mutual antagonism. Mol Endocrinol 12(1):45–56

McKay LI, Cidlowski JA (1999) Molecular control of immune/inflammatory responses: interactions between nuclear factor-kappa B and steroid receptor-signaling pathways. Endocr Rev 20(4):435–459

McKay LI, Cidlowski JA (2000) CBP (CREB binding protein) integrates NF-kappaB (nuclear factor-kappaB) and glucocorticoid receptor physical interactions and antagonism. Mol Endocrinol 14(8):1222–1234

Muneoka K et al (1997) Prenatal dexamethasone exposure alters brain monoamine metabolism and adrenocortical response in rat offspring. Am J Physiol 273(5 pt 2):R1669–R1675
Murphy VE et al (2007) Metabolism of synthetic steroids by the human placenta. Placenta 28(1):39–46
N.I.H. Consensus (1994) Effect of corticosteroids for fetal maturation on perinatal outcomes. NIH Consens Statement 12(2):1–24
N.I.H. Consensus (2001) Antenatal Corticosteroids Revisited: Repeat Courses. NIH Consens Statement 17(2): 1–10
Noorlander CW et al (2006) Ontogeny of hippocampal corticosteroid receptors: effects of antenatal glucocorticoids in human and mouse. J Comp Neurol 499(6):924–932
Novy MJ, Walsh SW (1983) Dexamethasone and estradiol treatment in pregnant rhesus macaques: effects on gestational length, maternal plasma hormones, and fetal growth. Am J Obstet Gynecol 145(8):920–931
Oliveira M et al (2006) Induction of a hyperanxious state by antenatal dexamethasone: a case for less detrimental natural corticosteroids. Biol Psychiatry 59(9):844–852
Ou XM et al (2001) Heterodimerization of mineralocorticoid and glucocorticoid receptors at a novel negative response element of the 5-HT1A receptor gene. J Biol Chem 276(17):14299–14307
Owen D, Matthews SG (2003) Glucocorticoids and sex-dependent development of brain glucocorticoid and mineralocorticoid receptors. Endocrinology 144(7):2775–2784
Pascual-Le Tallec L, Lombes M (2005) The mineralocorticoid receptor: a journey exploring its diversity and specificity of action. Mol Endocrinol 19(9):2211–2221
Peets EA, Staub M, Symchowicz S (1969) Plasma binding of betamethasone-3H, dexamethasone-3H, and cortisol-14C—a comparative study. Biochem Pharmacol 18(7):1655–1663
Pryce CR (2008) Postnatal ontogeny of expression of the corticosteroid receptor genes in mammalian brains: inter-species and intra-species differences. Brain Res Rev 57(2):596–605
Pryce CR, Feldon J (2003) Long-term neurobehavioural impact of the postnatal environment in rats: manipulations, effects and mediating mechanisms. Neurosci Biobehav Rev 27(1–2):57–71
Pryce CR et al (2011) The developmental impact of prenatal stress, prenatal dexamethasone and postnatal social stress on physiology, behaviour and neuroanatomy of primate offspring: studies in rhesus macaque and common marmoset. Psychopharmacology 214(1):33–53
Roberts D, Dalziel S (2006) Antenatal corticosteroids for accelerating fetal lung maturation for women at risk of preterm birth. Cochrane Database Syst Rev 2006;3:CD004454
Rosenfeld P et al (1993) Ontogeny of corticosteroid receptors in the brain. Cell Mol Neurobiol 13(4):295–319
Schwab M, Klotz U (2001) Pharmacokinetic considerations in the treatment of inflammatory bowel disease. Clin Pharmacokinet 40(10):723–751
Seckl JR (2001) Glucocorticoid programming of the fetus; adult phenotypes and molecular mechanisms. Mol Cell Endocrinol 185(1–2):61–71
Seckl JR (2004) Prenatal glucocorticoids and long-term programming. Eur J Endocrinol 151(3):U49–U62
Shoener JA, Baig R, Page KC (2006) Prenatal exposure to dexamethasone alters hippocampal drive on hypothalamic-pituitary-adrenal axis activity in adult male rats. Am J Physiol Regul Integr Comp Physiol 290(5):1366–1373
Sloboda DM, Newnham JP, Challis JR (2000) Effects of repeated maternal betamethasone administration on growth and hypothalamic-pituitary-adrenal function of the ovine fetus at term. J Endocrinol 165(1):79–91
Sloboda DM et al (2002) The effect of prenatal betamethasone administration on postnatal ovine hypothalamic-pituitary-adrenal function. J Endocrinol 172(1):71–81
Sloboda DM et al (2007) Prenatal betamethasone exposure results in pituitary-adrenal hyporesponsiveness in adult sheep. Am J Physiol Endocrinol Metab 292(1):E61–E70
Slotkin TA et al (1996) Programming of brainstem serotonin transporter development by prenatal glucocorticoids. Brain Res Dev Brain Res 93(1–2):155–161
Song IH, Buttgereit F (2006) Non-genomic glucocorticoid effects to provide the basis for new drug developments. Mol Cell Endocrinol 246(1–2):142–146

Speirs HJ, Seckl JR, Brown RW (2004) Ontogeny of glucocorticoid receptor and 11beta-hydroxysteroid dehydrogenase type-1 gene expression identifies potential critical periods of glucocorticoid susceptibility during development. J Endocrinol 181(1):105–116

Tegethoff M, Pryce C, Meinlschmidt G (2009) Effects of intrauterine exposure to synthetic glucocorticoids on fetal, newborn, and infant hypothalamic-pituitary-adrenal axis function in humans: a systematic review. Endocr Rev 30(7):753–789

Thomassin H et al (2001) Glucocorticoid-induced DNA demethylation and gene memory during development. EMBO J 20(8):1974–1983

Uno H et al (1994) Neurotoxicity of glucocorticoids in the primate brain. Horm Behav 28(4):336–348

Velisek L (2006) Prenatal exposure to betamethasone decreases anxiety in developing rats: hippocampal neuropeptide y as a target molecule. Neuropsychopharmacology 31(10):2140–2149

Weaver IC et al (2004) Epigenetics programming by maternal behavior. Nat Neurosci 7(8):847–854

Welberg LA, Seckl JR, Holmes MC (2001) Prenatal glucocorticoid programming of brain corticosteroid receptors and corticotrophin-releasing hormone: possible implications for behaviour. Neuroscience 104(1):71–79

Westphal U (1983) Steroid-protein interaction: from past to present. J Steroid Biochem 19(1A):1–15

Wolffe AP (1998) Packaging principle: how DNA methylation and histone acetylation control the transcriptional activity of chromatin. J Exp Zool 282(1–2):239–244

Wright AP et al (1993) Structure and function of the glucocorticoid receptor. J Steroid Biochem Mol Biol 47(1–6):11–19

Chapter 10
Early Developmental Trajectories of Brain Development: New Directions in the Search for Early Determinants of Health and Longevity

F. Cirulli and A. Berry

Abstract Conditions experienced early in life can have enduring consequences. Results from epidemiological studies and basic research agree that individual differences in both physical and mental health (cognitive, social, and emotional development as well as metabolic asset) can be determined by the early environment. Both prenatal life and the early postnatal periods are crucial times when adverse experiences including psychological or toxic stress can have major impact on developing systems. The next step in research is to identify the mechanisms underlying such programming. Changes in the effectors of stress responses during critical developmental stages may favor vulnerability to obesity, mental health, and neurodegeneration. Such broad spectrum of effects may explain the comorbidity often found between different pathologies, which can greatly affect longevity and the quality of life during aging. In addition to genetic susceptibility, epigenetic processes—which rely upon permanent changes in gene expression—could underlie such long-term effects and offer promise for environmental or pharmacological interventions.

10.1 Introduction

Developing biological systems are strengthened by positive early experiences, which lay the groundwork for a lifelong health. Converging evidence, from neuroscience research, genomics, and the behavioral and social sciences, opens new

F. Cirulli, PhD (✉) • A. Berry
Behavioural Neuroscience Section, Department of Cell Biology and Neurosciences, Istituto Superiore di Sanità, Viale Regina Elena 299, 00161 Rome, Italy
e-mail: francesca.cirulli@iss.it

G. Laviola and S. Macrì (eds.), *Adaptive and Maladaptive Aspects of Developmental Stress*, Current Topics in Neurotoxicity 3, DOI 10.1007/978-1-4614-5605-6_10,

perspectives on health promotion and disease prevention. We are now well aware that experiences early in life—including adverse experiences—can produce important physiological changes which become embedded biological traces that persist into adulthood and can lead to increased vulnerability to both physical and mental health affecting longevity and healthspan at adult age. Genes and experiences interact to determine an individual's vulnerability to early adversity, and for children experiencing severe traumatic events, environmental influences can be as powerful as genetic predispositions in their impact on the odds of having chronic health problems later in life. Thus health promotion and disease prevention also depend upon our ability to strengthen the foundations of health and mitigate the adverse impacts of toxic stress in the prenatal and early childhood periods (Cirulli et al. 2009; Hunter 2012; Shonkoff et al. 2009; van Os et al. 2012).

In most mammals, development is a gradual process occurring before and after birth entailing a continuous accumulation of small changes. This process, resulting from the interaction between the individual genetic asset and the pre- and/or postnatal environment, is functional to the generation of a range of phenotypes suitable for different environments promoting the successful perpetuation of the genotype and affecting both the existence of the single individual and population biology (Bateson 2001; Bateson et al. 2004; D'Udine and Alleva 1983). During development, the organism is characterized by an elevated plasticity which allows the individual to adjust to changes but, at the same time, provides the substrate for increased vulnerability to later disease, resulting overall in a double-edged sward. The overall successful or detrimental result of the developmental program will depend upon the genetic background of the organism and on the stability of environmental conditions with respect to those which contributed to prime developmental trajectories (Bateson et al. 2004).

Knowledge of biological mechanisms occurring during the early stages of life, including pre- and perinatal phases, can allow understanding/predicting changes in brain function which might affect health, including mental health, throughout a life course (Bateson et al. 2004). An extensive and growing body of research demonstrates multiple linkages between childhood adversity and health impairments in the adult years. The "Adverse Childhood Experiences (ACE) Study" is one of the largest investigations ever conducted on the links between childhood maltreatment and later-life health status and well-being. Over, 17,000 patients participated to the study on voluntary bases undergoing a comprehensive physical examination providing detailed information about their childhood experience of abuse, neglect, and family dysfunction. Data from this study show a strong association among multiple occurrences of traumatic or abusive childhood events and a wide variety of conditions later in life, including cardiovascular disease (CVD), chronic lung disease, cancer, depression, alcoholism, and drug abuse (Felitti et al. 1998). Individuals reporting more ACE also had substantially greater risks for life-threatening psychiatric disorders overlapping mental health problems obesity, physical inactivity, and smoking (Anda et al. 2006; Felitti et al. 1998). Other longitudinal studies have reported similar associations between early stressful life events and adult diseases (Schilling et al. 2007). In all cases, the pattern has been the same: the greater the number of adverse experiences in childhood, the greater the likelihood of health problems later in life.

Research on the biology of adversity illustrates how the body's physiological balance is lost following exposure to chronic stress (or "allostatic load") (McEwen 1998). The activation of stress management systems results in a tightly integrated repertoire of responses involving the secretion of stress hormones, increased heart rate and blood pressure, elevation in blood sugar and inflammatory protein levels, redirection of blood flow to the brain, and increased vigilance behavior. The discrete activation of these systems represents a "positive stress response" which helps the organism to face an acute threat. Under circumstances of chronic or overwhelming adversity—and in the absence of social support—stress can become harmful for the individual (Shonkoff et al. 2009). It is possible to hypothesize that adversity early in life can affect the experience-dependent maturation of the cortico-limbic and neuroendocrine systems underlying emotional functioning, leading to increased stress responses at adulthood (Heim and Nemeroff 2001; Meaney 2001; Schore 2000; Seckl and Meaney 2004; Tronick and Reck 2009). Depressed patients with a history of childhood abuse are indeed characterized by a hyperactive hypothalamic–pituitary–adrenal (HPA) axis, a major component of the stress response (Heim and Nemeroff 2001). In addition, childhood abuse or neglect has been associated with abnormalities in brain regions involved in emotional disorders including an overall volume loss in hippocampus, corpus callosum, and prefrontal cortex; altered symmetry in cortical regions; and reduced neuronal density and integrity in the anterior cingulate (Bremner et al. 1997; Carrion et al. 2001; De Bellis et al. 2002; Driessen et al. 2000; Stein et al. 1997; Teicher et al. 2004).

In order to describe how vulnerability can become "embedded" into animal physiology, a "two-hit model" has been hypothesized. According to such model, genetic or environmental factors disrupt early development leading to a long-term vulnerability to a "second hit" that then leads to the onset of disease. The signaling pathways involved in cellular differentiation and maintenance could be targets for a "first hit" during early development. These same pathways may be targets for a "second hit" of stress occurring in the adolescent or adult organism. Thus, if the same pathways in both the developing and the mature organism are targets of stress, we can hypothesize an integration between genetic, developmental, and environmental factors that contribute to vulnerability and pathogenesis to diseases at adulthood (Maynard et al. 2001; Norman and Malla 1993; Pani et al. 2000).

10.2 Prenatal Stress

Stress has been defined as any change of the internal or external milieu perturbing the maintenance of homeostasis; in complex organisms it involves a coordinated set of intercellular signals and physiological and behavioral responses that result in avoidance (when possible) or adaptation to the stressful stimulus. The activation of the neuroendocrine system (HPA axis) in response to stress, in the short run, is essential for adaptation, maintenance of homeostasis, and survival (allostasis = maintaining stability through changes). Yet, over longer time intervals, it imposes a cost (allostatic load) that might affect growth, metabolism, reproduction, inflammatory/immune, and neuroendocrine responses (de Kloet et al. 2005; Maccari and Morley-Fletcher 2007; McEwen 1998; Seckl 2004).

During perinatal phases, stress and environmental adversities are transmitted to the fetus/newborn and act on developmental trajectories of specific tissues. These mechanisms that occur during sensitive periods of individual development may affect tissue organization and function. Since different organs show different sensitivity at different times, the effects of adversities on an animal's biology will be tissue, time, and challenge specific (Harris and Seckl 2011). Thus, the allostatic load experienced early during life has the potential to pervasively impact on the adult physiology, setting the stage for the emergence of phenotypes vulnerable to the onset of many and different pathologies during adult life, including CVD, type 2 diabetes (T2D), emotional disorders, and psychopathology (just to mention a few) which can greatly affect health and the quality of life of the adult age.

Human epidemiological studies consistently provide evidence for an association between environmental challenges during pregnancy, altered fetal growth and development, and the occurrence of pathological conditions later in life (Barker et al. 1993; Seckl 1998).

In the late 1980s Barker and colleagues observed that the regions in England showing the highest rates of low birth weight-related infant mortality were also characterized by the highest rates of mortality from CVD. These observations led to the hypothesis that those babies who survived, despite their reduced body weight, became at risk of CVD later in life (Barker et al. 1989, 1993; Barker and Osmond 1986). Interestingly, the relationship between reduced birth weight and the development of diseases at adult age (e.g., hypertension or T2D) appears to be independent of lifestyle-related risk factors (e.g., obesity, smoking, alcohol consumption, and social class) (Harris and Seckl 2011; Leon et al. 1996; Levine et al. 1994; Osmond et al. 1993). In addition, studies on twins have shown that the genetic background can only partially account for this association (Baird et al. 2001; Gluckman and Hanson 2004) suggesting that a certain part of the vulnerability to adult diseases might certainly find its roots during early life stages as a result of diversions or reprogramming in the developmental trajectories. Indeed, reduced weight at birth has been also associated with anxiety disorders, cognitive disabilities, depression, schizophrenia, attention-deficit/hyperactivity disorder (ADHD), and antisocial behaviors (Famularo and Fenton 1994; Khashan et al. 2008; Raikkonen and Pesonen 2009; Raikkonen et al. 2008; Wust et al. 2005). A number of maternal complications have been specifically related to negative fetal outcomes, including preeclampsia, depression, diabetes, infection/inflammation, and obesity (Zammit et al. 2009). Congenital abnormalities, including cardiovascular and neural tube defects, are also more common in offspring of diabetic or overweight mothers, potentially due to the teratogenic effects of hyperglycemia and relaxin (Hawdon 2011). Maternal psychosocial stress has been found to result in altered fetal weight; insulin resistance; altered metabolic, immune, and endocrine function; and decreased cognitive performance (Entringer et al. 2010). These data on human cohorts have been strengthened by studies performed on animal models (mostly rodents and monkeys) of prenatal stress ranging from restraint stress in the third week of pregnancy to disruption of maternal metabolic pathways through chronic undernutrition or high-fat diet-induced obesity. Effects of these early challenges result overall in increased anxiety and depressive-like behavior, impaired cognitive abilities, and increased markers of brain inflammation and oxidative stress in the offspring (Bilbo and Tsang 2010; Darnaudery and Maccari 2008; Maccari and Morley-Fletcher 2007; Vallee et al. 1997, 1999). In particular, exposure to prenatal stress in rodents results in increased responsiveness of the HPA axis to stress (Maccari et al. 1995; Morley-Fletcher et al. 2003; Vallee et al. 1997). In addition, reduced levels of both mineralocorticoid (MR) and glucocorticoid (GR) receptors (the main receptors for the adrenal stress hormones cortisol and corticosterone—CORT) are found in the hippocampus of adult offspring, revealing a possible mechanism for the deficit of HPA axis feedback processes (Maccari et al. 1995). Moreover, prenatal restraint stress (PRS) accelerates age-related alterations in the HPA axis also eliminating the characteristic hyporesponsiveness in newborn rats (Henry et al. 1994). A number of results suggest that the HPA alterations induced by prenatal stress may vary according to gender and to the nature or the intensity of the stressor. The hyperactivity of the HPA axis observed in both male and female rats exposed to prenatal stress is accompanied by increased anxiety-like behavior in adult males (Poltyrev et al. 1996; Vallee et al. 1997; Viltart et al. 2006) and by an increased behavioral response to novelty in both males and females (Fride et al. 1986; Louvart et al. 2005; Thompson 1957; Vallee et al. 1997; Wakshlak and Weinstock 1990). Interestingly, the effects of prenatal stress on anxiety are less marked in females (Zagron and Weinstock 2006) and possibly even opposite to those found in males, as suggested by the decreased anxiety-like behavior observed in prenatally stressed females (Cirulli et al. unpublished results). By contrast, depression-like behavior has been observed in both male (Morley-Fletcher et al. 2003, 2004) and female rats (Alonso et al. 1991). Circadian rhythmicity of CORT secretion is also changed by prenatal stress which might be mediated, at least in part, by a reduction in hippocampal MR/GR at specific times of the day (Koehl et al. 1999). The altered glucocorticoids (GCs) secretion and behavioral response induced by early environmental manipulations seem to occur earlier than the cognitive alterations that are mainly observed later in life, and these cognitive alterations could be one consequence of early HPA axis hyperactivity (Vallee et al. 1999; Viltart et al. 2006). Interestingly, the response of adult female PRS rats to early stress is less evident, and despite showing hyperactivity in the HPA axis, females are characterized by reduced anxiety-like behavior and improved learning, suggesting an early effect of stress on hormonal systems with an impact on their "organizational effects." Glucocorticoid hormones are also involved in the regulation of circulating glucose levels, body weight, and appetitive behavior; peripheral glucose utilization decreases following GC administration (Long et al. 1940; McMahon et al. 1988) or stressful events (Munck 1971). In the brain, GCs exert an inhibitory function on glucose uptake (Horner et al. 1990; Kadekaro et al. 1988). Indeed, Vallée and coworkers showed that the offspring of mothers which underwent restraint stress between the 14th and the 21st days of pregnancy were characterized by decreased body weight and food consumption in addition to increased basal glucose levels at adult age (Vallee et al. 1996) demonstrating a pervasive effect of prenatal stress not only on emotional but also on metabolic development of the organism.

As for the mechanisms, it is worth noticing that exposure to a wide range of insults during gestation appears to converge to similar effects on fetal growth, neurodevelopment, and metabolism (Zeltser and Leibel 2011). Indeed this "funneling effect" should not be surprising given the pervasive effects of stress on the functionality of the HPA axis, which represents a main link between central nervous system (CNS) adaptations to environmental stressors and peripheral–behavioral and endocrine–metabolic responses (Cirulli and Alleva 2009; McEwen et al. 1988, 1992; Sapolsky et al. 1986). The interface of the mother–fetus communication is the placenta, which has an important role in directing stress signals from the mother to the fetus through the expression of transporters that regulate the flux of glucose, amino acids, vitamins, and ions required for growth and development (Bale et al. 2010; Fowden et al. 2009). The placenta is a temporary organ that performs the functions of several adult organs for the growing fetus such as those related to metabolism, respiration, excretion, and endocrine function. It is characterized by high expression levels of the enzyme 11β-hydroxysteroid dehydrogenase type 2 (HSD2) which catalyze the rapid inactivation of GCs, acting as shield from the maternal GCs (Edwards et al. 1993; Meaney et al. 2007). However, this barrier is apparently incomplete, allowing a moderate amount of maternal GCs to reach the fetus (Benediktsson et al. 1997). Glucocorticoids are essential for the fetus to develop, as also suggested by the lethal phenotype of GR knockout mice (Cole et al. 1995). They bind to GR and MR, widely expressed in the fetal tissues and in the placenta, which act as transcription factors to alter gene expression. These hormones promote lung maturation (Ward 1994) and brain development by initiating terminal maturation, remodeling of axons and dendrites, and affecting cell survival (Meyer 1983; Yehuda et al. 1989). Thus, even modest perturbations in the levels of HSD2 enzyme might have a profound impact on fetal GC exposure and on the overall fetal development. Both in human and in animal studies, it has been shown that controlled fluctuations in the expression levels or in the functionality of HSD2 naturally occur during pregnancy to allow the correct amount of GCs to reach the fetus at the proper time for organ maturation. Interestingly, metabolic (Bertram et al. 2001; Langley-Evans 1997) and emotional stressors (Mairesse et al. 2007) have the equal potential to modify HSD2 functionality resulting in high levels of GCs reaching the fetus, which retard growth and program disease susceptibility (Benediktsson et al. 1997; Edwards et al. 1993; Seckl 1998). Thus, hormonal signals of maternal status, including GCs, insulin-like growth factors (IGFs), insulin, and leptin, are sensed by the placenta and transmitted to the fetus predominantly through effects on placental function. Notwithstanding this evidence, environmental adversities may not act alone in determining the developmental origin of health and diseases (DOHaDs). In fact, evidence from rodents' models has clearly shown that the genetic asset plays a critical role. For example, rats bred for high (HAB) or low (LAB) anxiety are able to differently modulate placental HSD2 in response to restraint stress during pregnancy, the latter showing increased levels. As a result, the offspring of LAB rats appear overall protected from the detrimental effects of high maternal GC exposure (Lucassen et al. 2009). In addition, HSD2 knockout mice show reduced birth weight only on C57Bl/6J but not on a 129× MF1 background (Holmes et al. 2006; Kotelevtsev et al. 1999).

Taken together, results from these studies clearly suggest that the interaction between the individual maternal genetic asset and environmental challenges, such as prenatal stress, might lead to a complex remodeling of the developing organism providing an "allostatic load" which results in lifelong changes. Such early life challenges have the potential to pervasively impact adults' physical and mental health affecting the quality of life especially in fragile times as during aging.

10.3 Postnatal Stress

Beyond the prenatal phases, childhood and adolescence also represent sensitive times of great brain plasticity and of physical and emotional development. Most of the studies in this field of research are currently committed to identify main aspects of adverse early experience associated with alterations in specific output systems and, consequently, the presence of windows of opportunity for targeted interventions to prevent or reverse dysfunctions. In addition, the identification of gene polymorphisms contributing to the genetic individual's variability in susceptibility to adverse health outcome at adult age appears promising (Sanchez et al. 2001).

Strong evidence suggest that stress experienced during early childhood may have long-lasting and often irreversible effects on emotion, behavior, growth, metabolism, reproductive, immune, and cardiovascular function and may also affect brain morphology and neurochemistry in addition to CNS expression levels of genes that have been related to anxiety and mood disorders (Charmandari et al. 2003; Pervanidou and Chrousos 2007). In particular, a clear association between anxiety, mood disorders, and HPA axis dysfunctions has been found. Children with a history of exposure to chronic uncontrolled stress often show elevated peripheral cortisol levels, especially in the evening, i.e., during the circadian hormonal trough, and elevated catecholamine concentrations. De Bellis and coworkers have reported increased 24 h urinary concentrations of catecholamines and their metabolites in sexually abused girls with depression and suicidal behavior, much more than in matched controls (De Bellis et al. 2002). The same group of children exhibited reduced evening corticotrophin-releasing hormone (CRH)-stimulated plasma corticotropin concentrations compared with matched non-abused symptom-free girls. These neuroendocrine changes reflect a potential mechanism of how chronic stress can lead to chronic hormonal disturbances with potential clinical and metabolic consequences. In fact, stress-related chronic alterations in cortisol secretion in children may also affect the final stature and body composition, as well as cause early onset obesity, metabolic syndrome, and T2D. Thus, the understanding of stress mechanisms leading to metabolic abnormalities in early life may lead to more effective prevention and intervention strategies of obesity-related health problems (Pervanidou and Chrousos 2011).

During the early postnatal phases the brain is experience-seeking and provided by a considerable plasticity which allows a fine-tuning between the external environment and the developing organism. Since the early work of Seymour Levine, an impressive amount of research has clearly shown that stressful experiences exert powerful effects on brain and body development. These effects can last throughout the entire lifespan influencing brain function and increasing the risk for depression and anxiety disorders. Studies performed in altricial rodents (e.g., mice and rats) have clearly shown that maternal care is crucial for an adequate development of the pups, representing the most relevant source of early stimulation. Thus modifications of the maternal environment may result in long-term changes in the pattern of neuroendocrine and emotional/behavioral responses later in life (Cirulli and Alleva 2009).

Wild rodents having pups are often forced to leave the nest for variable periods (minutes to hours) of time for foraging. This pattern of maternal attendance to the nest has been modeled in the laboratory settings by early handling (H), which consists of removing the pups from the mother and their cage and placing them in individual compartments for up to 15 min until weaning (Levine 1957). Animals handled during infancy show important changes in the functionality of the HPA axis in a way such that the ability of the adult organism to respond, cope, and adapt to novel and/or stressful stimuli is increased (Levine 1957; Meaney et al. 1991). For example, immediately after the exposure to an electric shock, H rats (tested at adulthood) show a faster peak in the release of the GC stress hormones and a rapid return to basal levels when compared to non-handled (NH) controls. The speed and short duration of the neuroendocrine responses characterizing the H subjects appear to be extremely adaptive preventing the organism to be exposed to high circulating GC levels that can result, especially under chronic stressful conditions, in neurotoxicity through different mechanisms (Lupien et al. 1998; McEwen 1998). These changes in neuroendocrine responses to stress are also accompanied by important changes in emotionality and in those neurotransmitter systems which regulate it such as the GABAergic one (Giachino et al. 2007).

The long-term effects of the H procedure appear to depend upon changes in the phenotype of those neurons involved in the stress response (Meaney et al. 1996). As an example, H subjects show an increased number of GR expression in the hippocampus, a brain region strongly implicated in GC feedback regulation (Meaney et al. 1989). In addition, CRH mRNA and protein levels are higher in NH compared with H animals already under basal conditions (about 2.5-fold) as well as in response to stress (Plotsky and Meaney 1993). It is important to point out that, depending on the duration of the dam-offspring separation, the effects on the HPA axis responses to stress are qualitatively different. Animals that have been exposed to repeated maternal separation of 180–360 min per day for the first 2 weeks of life show, at adulthood, increased plasma adrenocorticotrophin (ACTH) and CORT levels in response to a stressful challenge (when compared to NH controls), an effect opposite to that of the H procedure (Plotsky and Meaney 1993). This longer period of separation also results in decreased GR binding in both the hippocampus and the hypothalamus (Plotsky and Meaney 1993). Compared to handling, which imposes an external manipulation on the mother–infant relationship, social enrichment, in the form of communal rearing, in mice has very profound effects on animal's emotionality and the response to stress. These effects are also accompanied by important changes in central levels of brain-derived neurotrophic factor (BDNF). Overall, the availability of ever more sophisticated animal models represents a fundamental tool to translate basic research data into appropriate interventions for humans raised under traumatic or impoverished situations (Cirulli et al. 2010).

10.4 Neurobiological Determinants: Role of Neurotrophins

As mentioned above, early life adverse events might set the stage for the onset of emotional and psychiatric disorders at adult age, which have been often associated to important changes in the functionality of the neuroendocrine system and of the HPA axis in particular (Heim and Nemeroff 1999, 2001; Holsboer 2000). Thus, although it appears even more clear that the interaction among genetic, developmental, and environmental factors can produce long-lasting changes in HPA system physiology and emotional behavior, the molecular mechanisms underlying such effects are far from being completely elucidated (Meaney and Szyf 2005). Neurotrophins, such as nerve growth factor (NGF) and BDNF, play a pivotal role in brain development and plasticity representing good candidates for mediating some of the effects triggered by early experiences on brain function. In fact, these trophic factors are involved in synaptic and morphological plasticity with maximal levels at times of neuronal growth, differentiation, and synaptogenesis (Thoenen 1995). A growing body of evidence suggest that changes in BDNF signaling in different brain areas of the adult individual may be involved in the pathophysiology of psychiatric disorders, such as depression (Altar 1999; Berton et al. 2006; Castren et al. 2007; Duman and Monteggia 2006; Sen et al. 2008). In fact, though it does not control mood, BDNF plays an important functional role in the modulation of those networks which determine how a plastic change influences mood (Castren et al. 2007). Interestingly, it has been hypothesized that successful antidepressant treatments might act by promoting activity-dependent neuronal plasticity through the activation of BDNF systems, possibly inducing proliferative or survival effects on neural stem cells (Castren 2005).

Since NGF and BDNF play a pivotal role in shaping brain function, a pathological alteration in their activation early during perinatal phases could exert long-lasting effects on synaptic plasticity, impairing the ability of the organism to cope with novel/stressful situation, leading to psychopathology (Zubin and Spring 1977).

It has been already shown that manipulations of the early environment in rats can affect the expression of these neurotrophic factors both during development and at adult age (Branchi et al. 2006; Cirulli et al. 1998, 2000, 2003; Liu et al. 1997; Roceri et al. 2004; Sale et al. 2004). More in detail, daily maternal separations of 3 h performed over the first 14 postnatal days affect BDNF mRNA levels in limbic regions of rats (Roceri et al. 2004). BDNF gene expression is increased soon after (on postnatal day 17) maternal deprivation stress in the prefrontal cortex and hippocampus, while at adulthood, a long-lasting decrease in its expression has been found in the same brain area (Roceri et al. 2004). These changes are associated to reduced HPA axis responses to chronic swimming stress (Roceri et al. 2004). Thus, variations in the expression of BDNF may contribute to the generation of individual differences in stress neurocircuitry, providing a substrate for altered susceptibility to the onset of depressive disorders at adult age (Castren et al. 2007; Cirulli and Alleva 2009; Cirulli et al. 2009; Hunnerkopf et al. 2007). It is worth noticing that early social stimulation provided by experimental paradigms such as that of communal nesting (CN) might also affect the levels of neurotrophins (Branchi et al. 2006) in a way such that NGF and BDNF protein levels are markedly increased in the hippocampus and hypothalamus of CN mice compared to NH mice. According to the literature (Castren 2005), an increase in neurotrophin levels in the hippocampus might account for changes in the neurogenesis rate in the dentate gyrus. In fact, CN mice are characterized by an increased cell survival, in line with the hypothesis that BDNF signaling is required mainly for the long-term survival and less for proliferation of newborn brain cells (Sairanen et al. 2005).

10.5 Epigenetic Marks of Early Experiences

Epigenetics—also referred to as the influences on phenotype operating above the level of the genetic code—has been defined as "the study of heritable changes other than those in the DNA sequence that encompass two major modifications of DNA or chromatin: DNA methylation, the covalent modification of cytosine, and post-translational modification of histones including methylation, acetylation, phosphorylation and sumoylation" (Callinan and Feinberg 2006). In addition, recent evidence suggests that noncoding RNAs (and particularly microRNAs) contribute to the stock of epigenetic mechanisms that are found in major diseases and that can occur at critical developmental times (Esteller 2011).

The principles and concepts of epigenetics have been applied to the DOHaD, a hypothesis on fetal programming of diseases (Barker 1998). If some of the mechanisms for developmental plasticity described in the DOHaD approach are epigenetic, and disease-related outcomes are related to disruptions of epigenetic processes elicited by the fetal environment, then this emerging field may provide explanatory mechanisms that underlie some of the enduring effects of adverse fetal, infant, and childhood environments. Two types of genes are modified epigenetically (i.e., are epigenetically liable): imprinted genes and genes with metastable epialleles. Imprinted genes are those in which specifically either the maternally derived or the paternally derived allele is suppressed, thereby rendering them functionally haploid (i.e., with parent-of-origin monoallelic expression). In other non-imprinted genes, one or both alleles are regulated epigenetically, and these metastable epialleles result in varying levels of gene expression. New data show that a similar process is associated with human suicide (McGowan et al. 2008) and that marks of childhood adversity are found in adult human brains (McGowan et al. 2009; Szyf 2012). Although direct epidemiological evidence of an involvement of epigenetic dysregulation in human CVD, T2D, and obesity is still rare, compared with the field of cancer, this may be due to temporal and tissue specificity. Nonetheless, there is a strong epigenetic basis for the DOHaD model of disease pathogenesis based on animal studies, which (for example) show that minor alterations in the maternal diet during pregnancy can produce lasting changes in the physiology and metabolism of offspring (Gluckman and Hanson 2004).

There are numerous examples of the contribution of early environment to epigenetic alterations in stress responsiveness. Using rodent models, the group of Michael Meaney and Moshe Szyf investigated epigenetic effects of maternal care. Increased maternal care through licking and grooming increases hippocampal expression of the glucocorticoid receptor GR mRNA and protein, reduces hypothalamic CRH, and decreases HPA response to stress (Francis and Meaney 1999; Liu et al. 1997; Meaney and Szyf 2005). This work showed a direct relationship between maternal behavior and DNA methylation in the rat hippocampal GR gene. Further studies have then shown that central infusion of methionine (a methyl donor) is able to reverse such early effects, suggesting that the inherently stable epigenomic marks established by behavioral programming at a critical period early in life are potentially reversible later in life (Weaver et al. 2004). The maternal separation model of early life stress has also been demonstrated to have an epigenetic impact. Low maternal behavior produces HPA hyperactivity in adults, partly by increasing expression of the ACTH gene, in the anterior pituitary. As ACTH release is driven by the release of arginine-vasopressin gene (AVP) and CRH from the hypothalamus into the pituitary portal circulation, Murgatroyd and collaborators examined the effect of maternal separation on AVP and CRH gene expression (Murgatroyd et al. 2009). Early life maternal separation has been shown to induce DNA hypomethylation in the arginine AVP in neurons of the hypothalamic paraventricular nucleus that was sustained into adulthood (Murgatroyd et al. 2009). They found that maternal separation increased AVP but not CRH expression in the hypothalamus and that the increase was associated with DNA hypomethylation in an AVP enhancer region. This region appears to be the major binding site for the methyl CpG-binding domain protein MeCP2 that plays also a role in the regulation of the expression of stress-responsive genes such as BDNF (Martinowich et al. 2003). Another early life stress model, using stressed and abusive mothers (obtained, e.g., by limiting dams' ability to build the nest during postpartum days—see Ivy et al. (2008) and also Roth and Sullivan (2005)), showed that pups reared under these conditions are characterized by reduced levels of BDNF expression in the prefrontal cortex, which correlated with DNA hypermethylation at the activity-dependent exon IV promoter. The investigators were able to reverse this effect by infusing the DNA methylation inhibitor zebularine (Roth et al. 2009). This provides a biological basis for speculations about the effects of poverty on early experience and how exposure to abuse, family strife, emotional neglect, and harsh discipline may have epigenetic effects that produce individual differences in neural and endocrine response to stress. This series of events may ultimately increase the susceptibility to common adult disorders such as depression and anxiety, drug abuse, and diabetes, heart disease, and obesity. Investigating the effects of early life stress on the epigenome and adult behavioral phenotype may lead not only to a better understanding of the risk architecture of major diseases but could also allow preventive measures in risk populations and new diagnostic and, potentially, therapeutic approaches since, in contrast to genetic variations, epigenetic effects on the transcriptome are potentially reversible in adulthood.

Acknowledgements This work was supported by EU (FP7) Project DORIAN "Developmental Origin of Healthy and Unhealthy Aging: The Role of Maternal Obesity" (grant n. 278603) and by ERA net-NEURON "Poseidon" and RF-2009-1498890. The authors are grateful to Sara Capoccia e Veronica Bellisario for their skillful help in retrieving and selecting bibliographic entries.

References

Alonso SJ et al (1991) Effects of maternal stress during pregnancy on forced swimming test behavior of the offspring. Physiol Behav 50:511–517

Altar CA (1999) Neurotrophins and depression. Trends Pharmacol Sci 20:59–61

Anda RF et al (2006) The enduring effects of abuse and related adverse experiences in childhood. A convergence of evidence from neurobiology and epidemiology. Eur Arch Psychiatry Clin Neurosci 256:174–186

Baird J et al (2001) Testing the fetal origins hypothesis in twins: the Birmingham twin study. Diabetologia 44:33–39

Bale TL et al (2010) Early life programming and neurodevelopmental disorders. Biol Psychiatry 68:314–319

Barker DJ (1998) In utero programming of chronic disease. Clin Sci (Lond) 95:115–128

Barker DJ, Osmond C (1986) Infant mortality, childhood nutrition, and ischaemic heart disease in England and Wales. Lancet 1:1077–1081

Barker DJ et al (1989) Weight in infancy and death from ischaemic heart disease. Lancet 2:577–580

Barker DJ et al (1993) Fetal nutrition and cardiovascular disease in adult life. Lancet 341:938–941

Bateson P (2001) Fetal experience and good adult design. Int J Epidemiol 30:928–934

Bateson P et al (2004) Developmental plasticity and human health. Nature 430:419–421

Benediktsson R et al (1997) Placental 11 beta-hydroxysteroid dehydrogenase: a key regulator of fetal glucocorticoid exposure. Clin Endocrinol (Oxf) 46:161–166

Berton O et al (2006) Essential role of BDNF in the mesolimbic dopamine pathway in social defeat stress. Science 311:864–868

Bertram C et al (2001) The maternal diet during pregnancy programs altered expression of the glucocorticoid receptor and type 2 11beta-hydroxysteroid dehydrogenase: potential molecular mechanisms underlying the programming of hypertension in utero. Endocrinology 142:2841–2853

Bilbo SD, Tsang V (2010) Enduring consequences of maternal obesity for brain inflammation and behavior of offspring. FASEB J 24:2104–2115

Branchi I et al (2006) Early social enrichment augments adult hippocampal BDNF levels and survival of BrdU-positive cells while increasing anxiety- and "depression"-like behavior. J Neurosci Res 83:965–973

Bremner JD et al (1997) Magnetic resonance imaging-based measurement of hippocampal volume in posttraumatic stress disorder related to childhood physical and sexual abuse–a preliminary report. Biol Psychiatry 41:23–32
Callinan PA, Feinberg AP (2006) The emerging science of epigenomics. Hum Mol Genet 15(1):R95–R101
Carrion VG et al (2001) Attenuation of frontal asymmetry in pediatric posttraumatic stress disorder. Biol Psychiatry 50:943–951
Castren E (2005) Is mood chemistry? Nat Rev Neurosci 6:241–246
Castren E et al (2007) Role of neurotrophic factors in depression. Curr Opin Pharmacol 7:18–21
Charmandari E et al (2003) Pediatric stress: hormonal mediators and human development. Horm Res 59:161–179
Cirulli F, Alleva E (2009) The NGF saga: from animal models of psychosocial stress to stress-related psychopathology. Front Neuroendocrinol 30:379–395
Cirulli F et al (1998) Early maternal separation increases NGF expression in the developing rat hippocampus. Pharmacol Biochem Behav 59:853–858
Cirulli F et al (2000) NGF expression in the developing rat brain: effects of maternal separation. Brain Res Dev Brain Res 123:129–134
Cirulli F, Berry A, Alleva E (2003) Early disruption of the mother-infant relationship: effects on brain plasticity and implications for psychopathology. Neurosci Biobehav Rev 27:73–82
Cirulli F et al (2009) Early life stress as a risk factor for mental health: role of neurotrophins from rodents to non-human primates. Neurosci Biobehav Rev 33:573–585
Cirulli F et al (2010) Early life influences on emotional reactivity: evidence that social enrichment has greater effects than handling on anxiety-like behaviors, neuroendocrine responses to stress and central BDNF levels. Neurosci Biobehav Rev 34:808–820
Cole TJ et al (1995) Targeted disruption of the glucocorticoid receptor gene blocks adrenergic chromaffin cell development and severely retards lung maturation. Genes Dev 9:1608–1621
Darnaudery M, Maccari S (2008) Epigenetic programming of the stress response in male and female rats by prenatal restraint stress. Brain Res Rev 57:571–585
De Bellis MD et al (2002) Brain structures in pediatric maltreatment-related posttraumatic stress disorder: a sociodemographically matched study. Biol Psychiatry 52:1066–1078
de Kloet ER et al (2005) Stress and the brain: from adaptation to disease. Nat Rev Neurosci 6:463–475
Driessen M et al (2000) Magnetic resonance imaging volumes of the hippocampus and the amygdala in women with borderline personality disorder and early traumatization. Arch Gen Psychiatry 57:1115–1122
D'Udine B, Alleva E (eds) (1983) Early experience and sexual preferences in rodents. Cambridge University Press, Cambridge
Duman RS, Monteggia LM (2006) A neurotrophic model for stress-related mood disorders. Biol Psychiatry 59:1116–1127
Edwards CR et al (1993) Dysfunction of placental glucocorticoid barrier: link between fetal environment and adult hypertension? Lancet 341:355–357
Entringer S et al (2010) Attenuation of maternal psychophysiological stress responses and the maternal cortisol awakening response over the course of human pregnancy. Stress 13:258–268
Esteller M (2011) Non-coding RNAs in human disease. Nat Rev Genet 12:861–874
Famularo R, Fenton T (1994) Early developmental history and pediatric posttraumatic stress disorder. Arch Pediatr Adolesc Med 148:1032–1038
Felitti VJ et al (1998) Relationship of childhood abuse and household dysfunction to many of the leading causes of death in adults. The Adverse Childhood Experiences (ACE) Study. Am J Prev Med 14:245–258
Fowden AL et al (2009) Placental efficiency and adaptation: endocrine regulation. J Physiol 587:3459–3472
Francis DD, Meaney MJ (1999) Maternal care and the development of stress responses. Curr Opin Neurobiol 9:128–134

Fride E et al (1986) Effects of prenatal stress on vulnerability to stress in prepubertal and adult rats. Physiol Behav 37:681–687
Giachino C et al (2007) Maternal deprivation and early handling affect density of calcium binding protein-containing neurons in selected brain regions and emotional behavior in periadolescent rats. Neuroscience 145:568–578
Gluckman PD, Hanson MA (2004) Developmental origins of disease paradigm: a mechanistic and evolutionary perspective. Pediatr Res 56:311–317
Harris A, Seckl J (2011) Glucocorticoids, prenatal stress and the programming of disease. Horm Behav 59:279–289
Hawdon JM (2011) Babies born after diabetes in pregnancy: what are the short- and long-term risks and how can we minimise them? Best Pract Res Clin Obstet Gynaecol 25:91–104
Heim C, Nemeroff CB (1999) The impact of early adverse experiences on brain systems involved in the pathophysiology of anxiety and affective disorders. Biol Psychiatry 46:1509–1522
Heim C, Nemeroff CB (2001) The role of childhood trauma in the neurobiology of mood and anxiety disorders: preclinical and clinical studies. Biol Psychiatry 49:1023–1039
Henry C et al (1994) Prenatal stress increases the hypothalamo-pituitary-adrenal axis response in young and adult rats. J Neuroendocrinol 6:341–345
Holmes MC et al (2006) The mother or the fetus? 11beta-hydroxysteroid dehydrogenase type 2 null mice provide evidence for direct fetal programming of behavior by endogenous glucocorticoids. J Neurosci 26:3840–3844
Holsboer F (2000) The corticosteroid receptor hypothesis of depression. Neuropsychopharmacology 23:477–501
Horner HC et al (1990) Glucocorticoids inhibit glucose transport in cultured hippocampal neurons and glia. Neuroendocrinology 52:57–64
Hunnerkopf R et al (2007) Interaction between BDNF Val66Met and dopamine transporter gene variation influences anxiety-related traits. Neuropsychopharmacology 32(12):2552–2560
Hunter RG (2012) Epigenetic effects of stress and corticosteroids in the brain. Front Cell Neurosci 6:18
Ivy AS et al (2008) Dysfunctional nurturing behavior in rat dams with limited access to nesting material: a clinically relevant model for early-life stress. Neuroscience 154:1132–1142
Kadekaro M et al (1988) Local cerebral glucose utilization is increased in acutely adrenalectomized rats. Neuroendocrinology 47:329–334
Khashan AS et al (2008) Higher risk of offspring schizophrenia following antenatal maternal exposure to severe adverse life events. Arch Gen Psychiatry 65:146–152
Koehl M et al (1999) Prenatal stress alters circadian activity of hypothalamo-pituitary-adrenal axis and hippocampal corticosteroid receptors in adult rats of both gender. J Neurobiol 40:302–315
Kotelevtsev Y et al (1999) Hypertension in mice lacking 11beta-hydroxysteroid dehydrogenase type 2. J Clin Invest 103:683–689
Langley-Evans SC (1997) Hypertension induced by foetal exposure to a maternal low-protein diet, in the rat, is prevented by pharmacological blockade of maternal glucocorticoid synthesis. J Hypertens 15:537–544
Leon DA et al (1996) Failure to realise growth potential in utero and adult obesity in relation to blood pressure in 50 year old Swedish men. BMJ 312:401–406
Levine S (1957) Infantile experience and resistance to physiological stress. Science 126:405
Levine RS et al (1994) Blood pressure in prospective population based cohort of newborn and infant twins. BMJ 308:298–302
Liu D et al (1997) Maternal care, hippocampal glucocorticoid receptors, and hypothalamic-pituitary-adrenal responses to stress. Science 277:1659–1662
Long C et al (1940) The adrenal cortex and carbohydrate metabolism. Endocrinology 26:309–344
Louvart H et al (2005) Prenatal stress affects behavioral reactivity to an intense stress in adult female rats. Brain Res 1031:67–73

Lucassen PJ et al (2009) Prenatal stress reduces postnatal neurogenesis in rats selectively bred for high, but not low, anxiety: possible key role of placental 11beta-hydroxysteroid dehydrogenase type 2. Eur J Neurosci 29:97–103

Lupien SJ et al (1998) Cortisol levels during human aging predict hippocampal atrophy and memory deficits. Nat Neurosci 1:69–73

Maccari S, Morley-Fletcher S (2007) Effects of prenatal restraint stress on the hypothalamus-pituitary-adrenal axis and related behavioural and neurobiological alterations. Psychoneuroendocrinology 32(suppl 1):S10–S15

Maccari S et al (1995) Adoption reverses the long-term impairment in glucocorticoid feedback induced by prenatal stress. J Neurosci 15:110–116

Mairesse J et al (2007) Maternal stress alters endocrine function of the feto-placental unit in rats. Am J Physiol Endocrinol Metab 292:E1526–E1533

Martinowich K et al (2003) DNA methylation-related chromatin remodeling in activity-dependent BDNF gene regulation. Science 302:890–893

Maynard TM et al (2001) Neural development, cell-cell signaling, and the "two-hit" hypothesis of schizophrenia. Schizophr Bull 27:457–476

McEwen BS (1998) Protective and damaging effects of stress mediators. N Engl J Med 338:171–179

McEwen BS et al (1988) Glucocorticoid receptors and behavior: implications for the stress response. Adv Exp Med Biol 245:35–45

McEwen BS et al (1992) Paradoxical effects of adrenal steroids on the brain: protection versus degeneration. Biol Psychiatry 31:177–199

McGowan PO et al (2008) Promoter-wide hypermethylation of the ribosomal RNA gene promoter in the suicide brain. PLoS One 3:e2085

McGowan PO et al (2009) Epigenetic regulation of the glucocorticoid receptor in human brain associates with childhood abuse. Nat Neurosci 12:342–348

McMahon M et al (1988) Effects of glucocorticoids on carbohydrate metabolism. Diabetes Metab Rev 4:17–30

Meaney MJ (2001) Maternal care, gene expression, and the transmission of individual differences in stress reactivity across generations. Annu Rev Neurosci 24:1161–1192

Meaney MJ, Szyf M (2005) Maternal care as a model for experience-dependent chromatin plasticity? Trends Neurosci 28:456–463

Meaney MJ et al (1989) Neonatal handling alters adrenocortical negative feedback sensitivity and hippocampal type II glucocorticoid receptor binding in the rat. Neuroendocrinology 50:597–604

Meaney MJ et al (1991) The effects of neonatal handling on the development of the adrenocortical response to stress: implications for neuropathology and cognitive deficits in later life. Psychoneuroendocrinology 16:85–103

Meaney MJ et al (1996) Early environmental regulation of forebrain glucocorticoid receptor gene expression: implications for adrenocortical responses to stress. Dev Neurosci 18:49–72

Meaney MJ et al (2007) Epigenetic mechanisms of perinatal programming of hypothalamic-pituitary-adrenal function and health. Trends Mol Med 13:269–277

Meyer JS (1983) Early adrenalectomy stimulates subsequent growth and development of the rat brain. Exp Neurol 82:432–446

Morley-Fletcher S et al (2003) Prenatal stress in rats predicts immobility behavior in the forced swim test. Effects of a chronic treatment with tianeptine. Brain Res 989:246–251

Morley-Fletcher S et al (2004) Chronic treatment with imipramine reverses immobility behaviour, hippocampal corticosteroid receptors and cortical 5-HT(1A) receptor mRNA in prenatally stressed rats. Neuropharmacology 47:841–847

Munck A (1971) Glucocorticoid inhibition of glucose uptake by peripheral tissues: old and new evidence, molecular mechanisms, and physiological significance. Perspect Biol Med 14:265–269

Murgatroyd C et al (2009) Dynamic DNA methylation programs persistent adverse effects of early-life stress. Nat Neurosci 12:1559–1566

Norman RM, Malla AK (1993) Stressful life events and schizophrenia II: conceptual and methodological issues. Br J Psychiatry 162:166–174
Osmond C et al (1993) Early growth and death from cardiovascular disease in women. BMJ 307:1519–1524
Pani L et al (2000) The role of stress in the pathophysiology of the dopaminergic system. Mol Psychiatry 5:14–21
Pervanidou P, Chrousos GP (2007) Post-traumatic Stress Disorder in children and adolescents: from Sigmund Freud's "trauma" to psychopathology and the (Dys)metabolic syndrome. Horm Metab Res 39:413–419
Pervanidou P, Chrousos GP (2011) Metabolic consequences of stress during childhood and adolescence. Metabolism 61:611–619
Plotsky PM, Meaney MJ (1993) Early, postnatal experience alters hypothalamic corticotropin-releasing factor (CRF) mRNA, median eminence CRF content and stress-induced release in adult rats. Brain Res Mol Brain Res 18:195–200
Poltyrev T et al (1996) Role of experimental conditions in determining differences in exploratory behavior of prenatally stressed rats. Dev Psychobiol 29:453–462
Raikkonen K, Pesonen AK (2009) Early life origins of psychological development and mental health. Scand J Psychol 50:583–591
Raikkonen K et al (2008) Depression in young adults with very low birth weight: the Helsinki study of very low-birth-weight adults. Arch Gen Psychiatry 65:290–296
Roceri M et al (2004) Postnatal repeated maternal deprivation produces age-dependent changes of brain-derived neurotrophic factor expression in selected rat brain regions. Biol Psychiatry 55:708–714
Roth TL, Sullivan RM (2005) Memory of early maltreatment: neonatal behavioral and neural correlates of maternal maltreatment within the context of classical conditioning. Biol Psychiatry 57:823–831
Roth TL et al (2009) Epigenetic mechanisms in schizophrenia. Biochim Biophys Acta 1790:869–877
Sairanen M et al (2005) Brain-derived neurotrophic factor and antidepressant drugs have different but coordinated effects on neuronal turnover, proliferation, and survival in the adult dentate gyrus. J Neurosci 25:1089–1094
Sale A et al (2004) Enriched environment and acceleration of visual system development. Neuropharmacology 47:649–660
Sanchez MM et al (2001) Early adverse experience as a developmental risk factor for later psychopathology: evidence from rodent and primate models. Dev Psychopathol 13:419–449
Sapolsky RM et al (1986) The neuroendocrinology of stress and aging: the glucocorticoid cascade hypothesis. Endocr Rev 7:284–301
Schilling EA et al (2007) Adverse childhood experiences and mental health in young adults: a longitudinal survey. BMC Public Health 7:30
Schore AN (2000) Attachment and the regulation of the right brain. Attach Hum Dev 2:23–47
Seckl JR (1998) Physiologic programming of the fetus. Clin Perinatol 25:939–962; vii
Seckl JR (2004) Prenatal glucocorticoids and long-term programming. Eur J Endocrinol 151(suppl 3):U49–U62
Seckl JR, Meaney MJ (2004) Glucocorticoid programming. Ann N Y Acad Sci 1032:63–84
Sen S et al (2008) Serum brain-derived neurotrophic factor, depression, and antidepressant medications: meta-analyses and implications. Biol Psychiatry 64:527–532
Shonkoff JP et al (2009) Neuroscience, molecular biology, and the childhood roots of health disparities: building a new framework for health promotion and disease prevention. JAMA 301:2252–2259
Stein MB et al (1997) Hippocampal volume in women victimized by childhood sexual abuse. Psychol Med 27:951–959
Szyf M (2012) The early-life social environment and DNA methylation. Clin Genet 81:341–349

Teicher MH et al (2004) Childhood neglect is associated with reduced corpus callosum area. Biol Psychiatry 56:80–85

Thoenen H (1995) Neurotrophins and neuronal plasticity. Science 270:593–598

Thompson WR (1957) Influence of prenatal maternal anxiety on emotionality in young rats. Science 125:698–699

Tronick E, Reck C (2009) Infants of depressed mothers. Harv Rev Psychiatry 17:147–156

Vallee M et al (1996) Long-term effects of prenatal stress and handling on metabolic parameters: relationship to corticosterone secretion response. Brain Res 712:287–292

Vallee M et al (1997) Prenatal stress induces high anxiety and postnatal handling induces low anxiety in adult offspring: correlation with stress-induced corticosterone secretion. J Neurosci 17:2626–2636

Vallee M et al (1999) Long-term effects of prenatal stress and postnatal handling on age-related glucocorticoid secretion and cognitive performance: a longitudinal study in the rat. Eur J Neurosci 11:2906–2916

van Os J et al (2012) The environment and schizophrenia. Nature 468:203–212

Viltart O et al (2006) Prenatal stress alters Fos protein expression in hippocampus and locus coeruleus stress-related brain structures. Psychoneuroendocrinology 31:769–780

Wakshlak A, Weinstock M (1990) Neonatal handling reverses behavioral abnormalities induced in rats by prenatal stress. Physiol Behav 48:289–292

Ward RM (1994) Pharmacologic enhancement of fetal lung maturation. Clin Perinatol 21:523–542

Weaver IC et al (2004) Epigenetic programming by maternal behavior. Nat Neurosci 7:847–854

Wust S et al (2005) Birth weight is associated with salivary cortisol responses to psychosocial stress in adult life. Psychoneuroendocrinology 30:591–598

Yehuda R et al (1989) Enhanced brain cell proliferation following early adrenalectomy in rats. J Neurochem 53:241–248

Zagron G, Weinstock M (2006) Maternal adrenal hormone secretion mediates behavioural alterations induced by prenatal stress in male and female rats. Behav Brain Res 175:323–328

Zammit S et al (2009) Investigating whether adverse prenatal and perinatal events are associated with non-clinical psychotic symptoms at age 12 years in the ALSPAC birth cohort. Psychol Med 39:1457–1467

Zeltser LM, Leibel RL (2011) Roles of the placenta in fetal brain development. Proc Natl Acad Sci U S A 108:15667–15668

Zubin J, Spring B (1977) Vulnerability—a new view of schizophrenia. J Abnorm Psychol 86:103–126

Chapter 11
Adaptive Regulations in Developing Rodents Following Neonatal Challenges

Laurence Coutellier

Abstract Developmental phenotypic plasticity is a mechanism by which events early in life program brain for a pattern of neuroendocrine and behavioral responses in later life. The goal of this chapter is to give the reader a better understanding of what is developmental phenotypic plasticity and how it can lead to adaptive phenotypes in adulthood. Experimental evidences from rodents show that early experiences influence long-term development of behavioral, neuroendocrine, and cognitive functions. Different factors have been suggested to mediate the effects of neonatal conditions on offspring development, but their exact contribution as well as their interaction still needs to be clarified. Several studies demonstrated the important role of maternal behavior in mediating the effects of neonatal challenges on adaptive regulations in the developing rodents. It has been suggested that there is an inverse relationship between the amount of active maternal care received by the offspring and their later reactivity to stressful or challenging events. However, other studies found a dissociation between the level of maternal care and offspring phenotype. These results suggest that aside from the level of maternal care, non-maternal factor (gender, neonatal glucocorticoid levels) contributes to the adjustment of offspring phenotype to early environmental cues. Altogether, rodents-based evidence suggests that developmental plasticity is a very complex phenomenon mediated by multiple factors that interact one to each other. Ultimately, these researches will help to better understand how the conditions from the neonatal environment affect brain development and can lead to adaptive phenotypes in adulthood.

L. Coutellier (✉)
Behavioral and Functional Neuroscience Laboratory, School of Medicine,
Stanford University, 1050 A Arastradero Road, Palo Alto, CA 94304, USA
e-mail: laurence.coutellier@stanford.edu

G. Laviola and S. Macrì (eds.), *Adaptive and Maladaptive Aspects of Developmental Stress*, Current Topics in Neurotoxicity 3, DOI 10.1007/978-1-4614-5605-6_11,

11.1 Introduction

The discovery of the double helix structure of DNA by Watson and Crick in 1953 allowed scientists to explain all biological phenomena by the trio DNA–RNA–protein. However, the idea that an organism's phenotype is entirely explained by its genome has been revised. Indeed, "the concept of phenotypic plasticity has allowed researchers to go beyond the nature-nurture dichotomy to gain deeper insights into how organisms are shaped by the interaction of genetic and ecological factors" (Pigliucci 2001). For instance, the characteristics of the rearing environment affect the development of offspring leading to a wide variety of adult phenotypes within populations. This is referred to as "developmental phenotypic plasticity" or "programming." This mechanism has been shown to exist in many living organisms, including plants, insects, and mammals (Tollrian 1995; Agrawal 1999; Bateson et al. 2004). It allows organisms to develop a phenotype that will help them cope with the characteristics of their (future) environment. In this framework, the developing organism is not seen as a passive entity subjected to its environment, but rather it plays an active role based on specific characteristics of the environment. When environing conditions remain constant throughout the entire life span, this phenomenon may provide the organism with a better adaptive response to the future requirements and preserve health and survival. However, if the initial conditions are suddenly changed, the adaptive processes may be disrupted and predispose to certain diseases.

This programming phenomenon is particularly relevant in philopatric species such as rats and mice, in which the offspring inhabit the parental niche. Early environmental cues "inform" the developing offspring of the characteristics of their future environment, and the offspring regulate their phenotype accordingly. Laboratory rodents (e.g., rats and mice) became the preferred animal model for the study of developmental phenotypic plasticity. The pioneering work of Levine, Denenberg, and their colleagues, and later Meaney, Plotsky, and colleagues has shown that even quite subtle alterations of a rat's experience during the early postnatal period can have long-lasting consequences for defensive behavior, emotiveness, and stress responsiveness (for reviews, see Francis and Meaney 1999; Meaney 2001; Champagne et al. 2003; Cirulli et al. 2003). The dynamic interplay between the early environment, brain development, and stress reactivity later in life has since been extensively studied. Artificial manipulations of mother/offspring relationships during the early postnatal period of rats and mice suggest that subtle variations in the postnatal environment can have marked and persistent effects on rodent brain systems involved in coping with environmental challenges (e.g., Meaney et al. 1991; Liu et al. 1997). It was also suggested that there is an inverse relationship between the amount and quality of maternal care received during infancy and offspring stress reactivity in adulthood. This was called the maternal mediation hypothesis (Smotherman and Bell 1980; see also Würbel 2001). This hypothesis provides a mechanism by which variations in maternal care may adaptively mediate phenotypic plasticity of stress systems (e.g., hypothalamic–pituitary–adrenal [HPA] axis) of offspring. However, the maternal mediation hypothesis has recently been exposed to scrutiny. The main concern is the emphasis placed on the role of active maternal care as opposed to other contributing factors (e.g., neonatal corticosteroid levels, offspring gender—Macrì et al. 2004, 2009; Tang et al. 2006, 2011; Coutellier et al. 2008b, 2009). It has been suggested that maternal care cannot be considered the sole and unique factor mediating developmental phenotypic adjustments in rodents. Other factors are now being studied to understand how they can favor phenotypic plasticity in response to neonatal environmental cues.

The main goal of this chapter is to better understand this "programming" or "developmental phenotypic plasticity" as a phenomenon by which neonatal cues change the organism's development. The aim is to give the reader a clear picture of how environmental challenges, acting in a critical period in life, affect the adult rodent phenotype and how it helps organisms to adapt to their environment in adulthood. First we will present how developmental challenges modulate individual adaptation to the future environment. We will then discuss the adaptive significance of developmental phenotypic plasticity and briefly explain how this process might become maladaptive. Finally, we will discuss the factors contributing to this programming phenomenon.

11.2 Individual Adaptation Following Developmental Challenges

Developmental phenotypic plasticity has been extensively studied in rodents. The early postnatal environment of the offspring has been challenged using a wide variety of manipulations ranging from mother–pup separation, exposure to stressors (e.g., predator odors), or variability in the foraging environment. The adult phenotype of the offspring has been analyzed to determine how these challenges affect the animals' capacity to cope with environmental challenges later in life. Early studies investigating the link between neonatal experiences and the regulation of adult stress and fear responses revealed that mild experimental challenges may help the organism to cope with acute stressors in adult life (Levine et al. 1957). Levine et al. (1957) showed that brief (3 min) daily mother–infant separations during the first 3 weeks of life have persistent effects on the rat pups' pituitary adrenal axis, as exhibited by reduced adrenal gland weight 24 h after a saline injection. This result has been replicated many times, and brief (3–15 min) daily mother–offspring separation (early handling, EH) during the first 1 or 2 weeks of life is consistently associated with reduced HPA axis and fear responses in the adult offspring (e.g., Meaney et al. 1991; Lui et al. 1997; Macrì et al. 2004), enhanced spatial working memory, and a greater competitive ability to obtain limited food rewards in the presence of a conspecific (Tang et al. 2006). This lent support to the hypothesis that early moderate environmental challenges (such as brief periods of maternal separation) affect the development of the offspring in a way that allow them to efficiently cope with acute stressors or challenges during their adult life.

Because of the artificial nature of these neonatal manipulations, new experimental designs using more natural settings have been recently developed and have confirmed further the importance of early environmental cues on the developing pups. For instance, dam–offspring rat dyads were exposed to different levels of foraging conditions (Macrì and Würbel 2007). Food availability was varied in space and time. Specifically, one group of dams had access to food ad libitum in the home cage (minimal challenges); one group had food ad libitum in an exploration cage located at some distance from the nest (mild challenges). Pups raised by dams with access to food away from the nest cage showed reduced fearfulness and HPA reactivity when adult (Macrì and Würbel 2007). Similarly, in mice, mild challenges in the maternal environment induced by a predator odor result in adult mice that are less fearful (Coutellier et al. 2008a) and that demonstrate better cognitive abilities (Coutellier and Würbel 2009). Altogether, a large amount of studies demonstrate that increasing levels of neonatal challenges result in adjustments in the offspring phenotype allowing them to effectively respond to acute stressors in adulthood. This efficient ability to maintain homeostasis despite the exposure to multiple stressors throughout life has been referred to as "resilience" (stability throughout development—McEwen 1998; Feder et al. 2009).

11.3 Adaptive Significance of Developmental Phenotypic Plasticity

An important question is whether developmental phenotypic plasticity is an adaptive process. Living under conditions of high environmental demand represents an important cost for the organism because behavioral and endocrine stress systems need to be constantly activated. Thus, if an individual develops in a way to reduce its stress and fear responses in adulthood, it will minimize the cost of living in such a highly demanding environment. In this case, developmental plasticity can favor an individual if the conditions in which it develops are similar to the ones it is going to face in adulthood. It looks like developmental plasticity is an adaptive phenomenon since it induces attributes that help the organism to cope with the characteristics of the environment in which it lives (Gluckman et al. 2005).

However, in the case of dissociation between environmental cues during early development and the characteristics of the future habitat, phenotypic mismatch may occur: the adult phenotype does not match the needs of the habitat which may be costly in terms of both survival and reproductive success (Bateson et al. 2004). Studies in rodents demonstrate that exposure to adverse conditions during the neonatal period alters the programming of the neuroendocrine and neuroimmune systems. For instance, long periods (3–4 h) of daily maternal separations (MS) during the first 1–3 weeks of life induce enhanced HPA axis responses and increased fearfulness in adulthood (Plotsky and Meaney 1993; Huot et al. 2004), increased corticotropin-releasing factor (CRF) expression in the hypothalamus and reduced cortical glucocorticoid receptor (GR) expression (Huot et al. 2004), altered cognitive abilities

(Huot et al. 2002; Aisa et al. 2009), dysregulation of the serotonergic and cholinergic systems (Aisa et al. 2009), altered immune response to infection (Meagher et al. 2010), and increased vulnerability to influenza virus infection in lung (Avitsur et al. 2006) and to stroke (Craft et al. 2006). In this chapter we are not aiming at explaining in depth the adverse consequences of early postnatal stress. For this particular thematic we are directing the reader to Chap. 10. However, it is clear that a large number of experimental studies support the view that exposure to adversity during the neonatal period affects brain morphology, neurochemistry, and expression levels of genes in the central nervous system. This is associated with an increased vulnerability to psycho- and physiopathologies at adulthood. These aberrant phenotypes are clearly maladaptive under all environmental conditions (Gluckman et al. 2005).

Thus, the adaptive value of a phenotype depends on how accurately the factors or cues that mediate developmental plasticity predict the future environment (Windig et al. 2004). However, labeling a trait as an adaptation implies that it has evolved in response to a specific form of selection or that there is a cause-and-effect relationship between the trait and the environment in which it is found. For adaptive plasticity to evolve there must be a trade-off among traits that cause one phenotype to have higher fitness in one environment and an alternative phenotype to have higher fitness in another environment (Doughty and Reznick 2004). To assess the relative fitness of diverse phenotypes, offspring should be investigated in environments that differ in how demanding or challenging they are for the animals. For instance, components of reproductive success (i.e., Darwinian fitness), such as the age at which sexual maturity occurs, fecundity, or survival rate, might help to assess the adaptive value of different phenotypes depending on the environment to which they are exposed. More "naturalistic" experimental designs have been recently developed to better address this point (Macrì and Würbel 2007; Coutellier et al. 2008a, b, 2009; Coutellier and Würbel 2009).

11.4 Factors Mediating Environment-Dependent Regulations of Offspring Phenotypes

In the previous paragraphs we reviewed experimental evidence demonstrating that early experiences influence long-term development of behavioral, neuroendocrine, and cognitive functions in rodents. These phenotypic regulations are likely adaptive because they help organisms cope with the characteristics and challenges of the environment in which they live. However, the exact mechanism underlying this programming phenomenon remains unclear. Several maternal and non-maternal neonatal factors have been shown to influence offspring development, but the precise contribution and interaction of these factors still need to be clarified.

1. Maternal behavior as a mediator of the effects of neonatal challenges on offspring development

(a) Qualitative aspects of maternal care

Due to the altricial nature of rodents, pups stay in the confines of a safe and stable nest while brain plasticity is at its highest. Their mother is, however, directly exposed to their future environment during her foraging trips. Since she is the only connection the pups have to their future environment, it could be an adaptive response if their brain systems involved in coping with environmental challenges were modulated by the mother's behavior. In fact, many studies have shown that environment-dependent variations in maternal behavior are associated with variations in the development of the offspring's phenotype. Furthermore, postnatal environmental manipulations that were shown to affect adult behavior and brain morphology were also shown to alter maternal behavior. For instance, conditions such as EH (e.g., Meaney et al. 1991), environmental enrichment (e.g., Chapillon et al. 1999; Coutellier et al. 2008b), cross-fostering (e.g., Anisman et al. 1998), exposure to predator odor (e.g., McLeod et al. 2007; Coutellier et al. 2008a), or variability in food availability (e.g., Léonhardt et al. 2007; Macrì and Würbel 2007; Coutellier et al. 2008b, 2009) have all been shown to increase the level of active maternal care and to down regulate the fear and stress reactivity of the adult offspring. Increased active maternal care may be a protective response by the dams toward their pups under environmental conditions that may be perceived as threatening. In comparison, low challenging conditions such as non-handling (NH, i.e., leaving rat pups completely undisturbed during the first 2 postnatal weeks) or easy access to food (Macrì and Würbel 2007; Coutellier et al. 2008b, 2009) are associated with lower levels of active maternal care and higher HPA axis reactivity and fearfulness in the offspring. Interestingly, it was also found that naturally high levels of active maternal care in NH rats are associated with reduced stress and fear responses in the adult offspring (Lui et al. 1997; Caldji et al. 1998). These results provide support to the maternal mediation hypothesis exposed by Smotherman and Bell (1980). It is proposed that the effects of the early environment characteristics on offspring phenotype could in fact be mediated indirectly via its effect on maternal care (Richards 1966; Denenberg et al. 1969; Lui et al. 1997; Caldji et al. 1998; Macrì et al. 2004 also see reviews by Meaney 2001; Macrì and Würbel 2006) and, more generally, that developmental plasticity of fear and stress responses in rodents is maternally mediated (Bell et al. 1974; Smotherman et al. 1977).

The mechanism by which active maternal care affects the expression of the HPA axis in offspring has been proposed to be based on epigenetic processes. Weaver et al. (2004) found that the level of active maternal care affects the offspring's epigenome at the glucocorticoid receptor (GR) gene promoter in the hippocampus: offspring receiving high levels of active maternal care showed reduced DNA methylation of the exon 17 region of the promoter, which was associated with an increased expression of GR in the hippocampus and, therefore, enhanced negative feedback sensitivity to glucocorticoids. These findings provide a mechanism by which variations in maternal care may adaptively mediate phenotypic plasticity of the HPA system.

(b) Other maternal factors: nest attendance

Aside from the quality of maternal care received by the pups (active vs. passive nursing), other aspects of maternal behavior have been recognized to influence the development of offspring. Various studies have highlighted the importance of maternal presence. Specifically, Moriceau and Sullivan (2006) demonstrated in rats that maternal presence controls pups' learning abilities in an odor-shock conditioning paradigm through modulation of pup corticosterone. Macrì et al. (2004) showed that despite receiving similar levels of active nursing, offspring reared in the condition of EH or MS showed different HPA responses to restraint and behavioral fear response. The authors demonstrated that the temporal distribution of maternal care by EH and MS dams was significantly different and might contribute to the difference observed in offspring. Similar findings in mice have demonstrated that offspring reared by mothers with access to environmental enrichment have reduced behavioral fearfulness in comparison to offspring reared by unenriched dams, despite receiving a similar amount of maternal care (Coutellier et al. 2008b). It was observed that enriched dams spend less time in their nest; this lower level of nest attendance might mediate the effects observed on the offspring. This idea has been further supported by a multiple regression analysis showing an inverse relationship between the time the mother spent away from the nest and adult offspring fearfulness. These experimental evidences suggest that not only the quality of maternal care influences the development of the offspring in rodents but also the presence of the mother in the nest.

2. Non-maternal factors contribute to the effects of neonatal challenges on developing rodents

(a) Evidence for the implication of non-maternal factors

We previously described a series of studies demonstrating the importance of maternal behavior in mediating the effects of neonatal challenges on the adaptive regulations in developing rodents. Another line of evidence in rats and mice indicates that early environmental effects on stress reactivity and fearfulness cannot be fully explained by variations in maternal behavior (Macrì et al. 2004; Macrì and Würbel 2006; Tang et al. 2006; Coutellier et al. 2008a, b). For example, Macrì et al. (2004) demonstrated that both long (MS) and brief (EH) periods of maternal separation early in life result in high levels of active maternal care, but while EH offspring were less fearful and had a downregulated HPA axis, MS offspring were found more fearful and highly stress reactive. Thus, indistinguishable maternal styles resulted in differential adult offspring, supporting the idea that maternal care is not the unique factor mediating the effects of the early environment on offspring development. Another set of evidence supports this idea. Highly challenging foraging conditions (variable foraging demand—VFD) and moderately challenging foraging conditions (high foraging demand—HFD) in mice have both been associated with increased active maternal care. However, VFD male offspring were found more fearful than HFD offspring (Coutellier et al. 2009). Thus, while offspring received a similar level of active maternal care, their adult phenotype differed significantly. This finding indicates that a high level of active maternal care is not the only parameter that leads to a downregulation of the fear and stress systems of the offspring and may not be sufficient in explaining developmental phenotypic plasticity. Tang et al. (2006, 2011) further developed this idea by using a split-litter methodology: half of the litter was exposed for 3 min daily to a novel environment and the other half remained in the home cage. The dam was removed prior to the separation and was returned to the home cage upon litter reunion to insure that all pups received the same amount of maternal care. They showed that pups that were separated from their mother and placed in a novel environment for brief periods have a more effective HPA axis when confronted to unexpected stressors and better spatial memory during adulthood compared to pups that were left in the home cage. These changes occurred in spite of both groups of pups receiving the same amount of active maternal care throughout their lactation period. The dissociation between the level of maternal care and offspring phenotype suggests that other factors than maternal factors contribute to the adjustment of offspring phenotype to early environmental cues.

(b) Corticosteroids as a possible contributing factor to individual adaptations to developmental challenges

A possible contributing factor to the adaptive development of rodents is the level of neonatal corticosteroids. Challenging and stressful environmental conditions are known to induce an increase in the circulating plasma level of stress hormones (e.g., corticosterone in rodents—Sapolsky 2004). This hormone could be passed from the mother to the offspring via milk during suckling episodes. Studies in rats and mice demonstrated that when lactating dams are supplemented with corticosterone in the drinking water, pups show high concentrations of plasmatic corticosterone (Catalani et al. 1993; Macrì et al. 2009). Therefore, because maternal corticosterone varied according to environmental conditions and because offspring's HPA axis is sensitive to the level of maternal corticosterone, it is likely that maternal corticosteroids contribute to offspring development.

Experimental evidence supports the idea that corticosterone levels during the neonatal period modulate individual development. Specifically, low levels of neonatal corticosterone have been associated with increased glucocorticoid receptors in the hippocampus, with reduced stress reactivity and improved cognitive abilities in rats (Catalani et al. 1993, 2000) and mice (Macrì et al. 2009). Interestingly, while low doses of neonatal corticosterone result in individual adaptation, high doses seem to lead to negative outcomes. Offspring exposed to elevated corticosterone levels during their development are found to have increased HPA-axis activity and behavioral anxiety (Brummelte et al. 2006; Macrì et al. 2007), reduced hippocampal levels of brain-derived neurotrophic factor (BDNF), and cell proliferation in the dentate gyrus (Brummelte et al. 2006; Macrì et al. 2009). These evidences suggest that exposure to different levels of corticosteroid during the neonatal period might interact with the level of active maternal care affecting offspring development.

(c) Gender as an important factor influencing the trajectory of effects of early life challenges on the development of rodents

Most research aiming to unravel the link between early environmental and maternal cues and the adult offspring phenotype has focused on males. Males have been favored because the estrous cycle is known to affect both the behavioral and physiological responses of females (e.g., Romeo et al. 2003). As a consequence, the conclusions made are based mainly on data coming from only male offspring, preventing scientists to determine whether neonatal variations affect offspring in a sex-specific manner. This point is very important especially if experimental neonatal manipulations in rodents are used to model early life events and to examine developmental hypotheses on the etiology of human psychopathologies such as anxiety, schizophrenia, and depression. These diseases, and other mental and neurological disorders, are highly sex-specific in human. Experimental studies using rodents as a model should take into account possible sex-specific effects.

Studies analyzing the effects of neonatal cues on both male and female offspring consistently demonstrate sex-specific results. For instance, Barha et al. (2007) showed that, in rats, the level of maternal licking/grooming affects working memory and stress reactivity more in female than in male offspring. Similarly, Noschang et al. (2010) found that females subjected to an EH paradigm showed impairments in spatial learning when compared to a non-handled group, while this effect was not observed in males. This cognitive impairment in females was associated with a decrease in nitric oxide production, an important cellular messenger molecule. Desbonnet et al. (2008) showed increased CRF immunoreactivity and increased colocalization of c-Fos and CRF following stress in the hypothalamus of maternally separated females but not males. Gender differences were also observed in mice. Coutellier et al. (2008a) and Coutellier and Würbel (2009) demonstrated that variations in the postnatal foraging maternal environment of mice affect female's corticosterone response to an isolation/novelty stressor and behavioral fearfulness, whereas there were no effects on male offspring. Gross et al. (2012) found that the level of expression of the glycoprotein Reelin, a master molecule for development and differentiation of the hippocampus, was increased in EH males when compared to NH males, while this difference was not observed in females.

Taken together, these findings suggest a significant role of the offspring's sex on the development of neurobehavioral and neurocognitive functions. They indicate that gender is an important factor influencing the trajectory of neonatal challenges effects on the development of rodents. These evidences point to the necessity of including both sexes in the analysis of early life influences on phenotypic plasticity. More work is needed for a better understanding of the mechanisms underlying sex-specific effects on adaptive regulations in developing rodents.

11.5 Conclusion

This chapter focused on the phenomenon by which early life events, even seemingly minor ones, program rodent's brain for a pattern of neuroendocrine and behavioral responses in later life. Early experiences are capable of enhancing or suppressing the expression of certain genetic traits and, by doing so, may change the outcome for behavioral and cognitive performance in adulthood (de Kloet et al. 2005). Many experimental evidences support this idea and show that developmental phenotypic plasticity is a powerful phenomenon that helps the organisms to live according to the characteristics of their environment.

The exact mechanism by which this plasticity occurs remains unclear. Many factors seem to interact and contribute to the development of the offspring's phenotype. There is increasing support for a broader view of the factors mediating adaptive regulations in developing rodents, than the one-factor maternal mediation hypothesis once proposed. However, the exact contribution of each of these factors as well as their interactions remains to be determined. Recently, new paradigms have been developed (Macrì and Wurbel 2007; Coutellier et al. 2008a, b, 2009; Coutellier and Würbel 2009) helping to answer these questions. The ultimate goal is to understand how the conditions from the neonatal environment affect brain development and can lead to adaptive or abnormal phenotypes in adulthood.

References

Agrawal AA (1999) Induced responses to herbivory in wild radish: effects on several herbivores and plant fitness. Ecology 80:1713–1723

Aisa B, Gil-Bea FJ, Marcos B, Tordera R, Lasheras B, Del Rio J, Ramirez MJ (2009) Neonatal stress affects vulnerability of cholinergic neurons and cognition in the rat: involvement of the HPA axis. Psychoneuroendocrinology 34:1495–1505

Anisman H, Zaharia MD, Meaney MJ, Merali Z (1998) Do early-events permanently alter behavioral and hormonal responses to stress? Int J Dev Neurosci 16:149–164

Avitsur R, Hunzeker J, Sheridan JF (2006) Role of early stress in the individual differences in host response to viral infection. Brain Behav Immun 20:339–348

Barha CK, Pawluski JL, Galea LAM (2007) Maternal care affects male and female offspring working memory and stress reactivity. Physiol Behav 92:939–950

Bateson P, Barker D, Clutton-Brock T, Deb D, D'Udine B, Foley RA, Gluckman P, Godfrey K, Kirkwood T, Lahr MM, McNamara J, Metcalfe NB, Monoghan P, Spencer HG, Sultan SE (2004) Developmental plasticity and human health. Nature 430:419–421

Bell RW, Nitschke W, Bell NJ, Zachman TA (1974) Early experience, ultrasonic vocalizations, and maternal responsiveness in rats. Dev Psychobiol 7:235–242

Brummelte S, Pawluski JL, Galea LA (2006) High post-partum levels of corticosterone given to dams influence postnatal hippocampal cell proliferation and behavior of offspring: a model of post-partum stress and possible depression. Horm Behav 50:370–382

Caldji C, Tannenbaum B, Sharma S, Francis D, Plotsky PM, Meaney MJ (1998) Maternal care during infancy regulates the development of neural systems mediating the expression of fearfulness in the rat. Proc Natl Acad Sci U S A 95:5335–5340

Catalani A, Marinelli M, Scaccianoce S, Nicolai R, Muscolo LA, Porcu A, Koranyi L, Piazza PV, Angelucci L (1993) Progeny of mothers drinking corticosterone during lactation has lower stress-induced corticosterone secretion and better cognitive performance. Brain Res 624:209–215

Catalani A, Casolini P, Scaccianoce S, Patacchioli FR, Spinozzi P, Angelucci L (2000) Maternal corticosterone during lactation permanently affects brain corticosteroid receptors, stress response and behavior in rat progeny. Neuroscience 100:319–325

Champagne FA, Francis DD, Mar A, Meaney MJ (2003) Variations in maternal care in the rat as a mediating influence for the effects of environment on development. Physiol Behav 79:359–371

Chapillon P, Manneche C, Belzung C, Caston J (1999) Rearing environmental enrichment in two inbred strains of mice. 1. Effects on emotional reactivity. Behav Genet 29:41–46

Cirulli F, Berry A, Alleva E (2003) Early disruption of the mother–infant relationship: effects on brain plasticity and implications for psychopathology. Neurosci Biobehav Rev 27:73–82

Coutellier L, Würbel H (2009) Early environmental cues affect object recognition memory in adult female but not male C57Bl/6 mice. Behav Brain Res 203:312–315

Coutellier L, Friedrich AC, Failing K, Marashi V, Würbel H (2008a) Effects of rat odour and shelter on maternal behavior in C57BL/6 dams and on fear and stress responses in their adult offspring. Physiol Behav 94:393–404

Coutellier L, Friedrich AC, Failing K, Würbel H (2008b) Variations in the postnatal maternal environment in mice: effects on maternal behavior and behavioral and endocrine responses in the adult offspring. Physiol Behav 93:395–407

Coutellier L, Friedrich AC, Failing K, Marashi V, Würbel H (2009) Effects of foraging demand on maternal behavior and adult offspring anxiety and stress response in C57Bl/6 mice. Behav Brain Res 196:192–199

Craft TK, Zhang N, Glasper ER, Hurn PD, Devries AC (2006) Neonatal factors influence adult stroke outcome. Psychoneuroendocrinology 31:601–613

De Kloet ER, Sibug RM, Helmerhorst FM, Schmidt MV (2005) Stress, genes and the mechanism of programming the brain for later life. Neurosci Biobehav Rev 29:271–281

Denenberg VH, Rosenberg KM, Paschke R, Zarrow MX (1969) Mice reared with rat aunts: effects on plasma corticosterone and open field activity. Nature 221:73–74

Desbonnet L, Garrett L, Daly E, McDermott KW, Dinan TG (2008) Sexually dimorphic effects of maternal separation stress on corticotrophin-releasing factor and vasopressin systems in the adult rat brain. Int J Dev Neurosci 26:259–268

Doughty P, Reznick DN (2004) Patterns and analysis of adaptive phenotypic in animals. In: Dewitt TJ, Scheiner SM (eds) Phenotypic plasticity—functional and conceptual approaches. Oxford University Press, Oxford, UK, pp 126–150

Feder A, Nestler EJ, Charney DS (2009) Psychobiology and molecular genetics of resilience. Nat Rev Neurosci 10:446–457

Francis DD, Meaney MJ (1999) Maternal care and the development of stress response. Curr Opin Neurobiol 9:128–134

Gluckman PD, Hanson MA, Spencer HG, Bateson P (2005) Environmental influences during development and their later consequences for health and disease: implications for the interpretation of empirical studies. Proc Biol Sci 272:671–677

Gross CM, Flubacher A, Tinnes S, Heyer A, Scheller M, Herpfer I, Berger M, Frotscher M, Lieb K, Haas CA (2012) Early life stress stimulated hippocampal reelin gene expression in a sex-specific manner: evidence for corticosterone-mediated action. Hippocampus 22(3):409–420

Huot RL, Plotsky PM, Lenox RH, McNamara RK (2002) Neonatal maternal separation reduces hippocampal mossy fiber density in adult Long Evans rats. Brain Res 950:52–63

Huot RL, Gonzales ME, Ladd CO, Thrivikraman KV, Plotsky PM (2004) Foster litters prevent hypothalamic-pituitary-adrenal axis sensitization mediated by neonatal maternal separation. Psychoneuroendocrinology 29:279–289

Léonhardt M, Matthews SG, Meaney MJ, Walker C-D (2007) Psychological stressors as a model of maternal adversity: diurnal modulation of corticosterone responses and changes in maternal behavior. Horm Behav 51:77–88

Levine S, Alpert M, Lewis GW (1957) Infantile experience and the maturation of the pituitary adrenal axis. Science 126:1347

Liu D, Diorio J, Tannenbaum B, Caldji C, Francis D, Freedman A, Sharma S, Pearson D, Plotsky PM, Meaney MJ (1997) Maternal care, hippocampal glucocorticoid receptors and hypothalamic-pituitary-adrenal responses to stress. Science 277:1659–1661

Macrì S, Würbel H (2006) Developmental plasticity of HPA and fear responses in rats: a critical review of the maternal mediation hypothesis. Horm Behav 50:667–680

Macrì S, Würbel H (2007) Effects of variations in postnatal maternal environment on maternal behavior and fear and stress responses in rats. Anim Behav 73:171–184

Macrì S, Mason GJ, Würbel H (2004) Dissociation in the effects of neonatal maternal separations on maternal care and the offspring's HPA and fear responses in rats. Eur J Neurosci 20:1017–1024

Macrì S, Granstrem O, Shumilina M, Antunes Gomes Dos Santos FJ, Berry A, Saso L, Laviola G (2009) Resilience and vulnerability are dose-dependently related to neonatal stressors in mice. Horm Behav 56:391–398

McEwen BS (1998) Stress, adaptation, and disease. Allostasis and allostatic load. Ann N Y Acad Sci 840:33–44

McLeod J, Sinal CJ, Perrot-Sinal TS (2007) Evidence for non-genomic transmission of ecological information via maternal behavior in female rats. Genes Brain Behav 6:19–29

Meagher MW, Sieve AN, Johnson RR, Satterlee D, Belyavski M, Mi W, Prentice TW, Welsh TH Jr, Welsh CJ (2010) Neonatal maternal separation alters immune, endocrine and behavioral responses to acute Theiler's virus infection in adult mice. Behav Genet 40:233–249

Meaney MJ (2001) Maternal care, gene expression and the transmission of individual differences in stress across generations. Annu Rev Neurosci 24:1161–1192

Meaney MJ, Mitchell JB, Aitken DH, Bhatnagar S, Bodnoff SR, Iny LJ, Sarrieau A (1991) The effects of neonatal handling on the development of the adrenocortical response to stress: implications to neuropathology and cognitive deficits in later life. Psychoneuroendocrinology 16:85–103

Moriceau S, Sullivan RM (2006) Maternal presence serves as a switch between learning fear and attraction in infancy. Nat Neurosci 9:1004–1006

Noschang CG, Krolow R, Fontella FU, Arcego DM, Diehl LA, Weis SN, Arteni NS, Dalmaz C (2010) Neonatal handling impairs spatial memory and leads to altered nitric oxide production and DNA breaks in a sex specific manner. Neurochem Res 35(7):1083–1091

Pigliucci M (2001) Phenotypic plasticity: beyond nature and nurture. The Johns Hopkins University Press, Baltimore, MD, 344p

Plotsky PM, Meaney MJ (1993) Early, postnatal experience alters hypothalamic corticotropin-releasing factor (CRF) mRNA, median eminence CRF content and stress-induced release in adult rats. Brain Res Mol Brain Res 18:195–200

Richards MPM (1966) Infantile handling in rodents: a reassessment in the light of recent studies of maternal behavior. Anim Behav 14:582

Romeo RD, Mueller A, Sisti HM, Ogawa S, McEwen BS, Brake WG (2003) Anxiety and fear in adult male and female C57BL/6 mice are modulated by maternal separation. Horm Behav 43:561–567

Sapolsky RM (2004) Why zebras don't get ulcers: the acclaimed guide to stress, stress-related diseases, and coping, vol 2. Henry Holt & Company, New York

Smotherman WP, Bell RW (1980) Maternal mediation of early experience. In: Bell RW, Smotherman WP (eds) Maternal influences and early behavior. Spectrum, New York, pp 201–210

Smotherman WP, Brown CP, Levine S (1977) Maternal responsiveness following differential pup treatment and mother–pup interactions. Horm Behav 8:242–253

Tang AC, Akers KG, Reeb BC, Romeo RD, McEwen BS (2006) Programming social, cognitive, and neuroendocrine development by early exposure to novelty. Proc Natl Acad Sci U S A 103:15716–15721

Tang AC, Reeb-Sutherland BC, Yang Z, Romeo RD, McEwen BS (2011) Neonatal novelty-induced persistent enhancement in offspring spatial memory and the modulatory role of maternal self-stress regulation. J Neurosci 31:5348–5352

Tollrian R (1995) Predator-induced morphological defenses—costs, life history shifts, and maternal effects in *Daphnia pulex*. Ecology 76:1691–1705

Weaver IC, Cervoni N, Champagne FA, D'Alessio AC, Sharma S, Seckl JR, Dymov S, Szyf M, Meaney MJ (2004) Epigenetic programming by maternal behavior. Nat Neurosci 7:847–854

Windig JJ, de Kovel CGF, De Jong G (2004) Genetics and mechanics of plasticity. In: Dewitt TJ, Scheiner SM (eds) Phenotypic plasticity—functional and conceptual approaches. Oxford University Press, New York, pp 31–49

Würbel H (2001) Ideal homes? Housing effects on rodent brain and behavior. Trends Neurosci 24:207–211

Chapter 12
Adaptive and Maladaptive Regulations in Response to Environmental Stress in Adolescent Rodents

Simone Macrì and Giovanni Laviola

Abstract Adolescent mammals exhibit a plethora of age-specific behaviours that mark a discontinuity with earlier stages of life and prepare the individual to the challenges of adulthood. Stress sensitivity during adolescence is remarkably different from earlier and later maturational stages. Specifically, although adolescent and adult mammals mount a similar endocrine response to external stressors, such response is excessively prolonged in adolescents. This unique response profile may relate to an asynchronous developmental profile of the several components of the hypothalamic–pituitary–adrenocortical axis. Experiential factors encountered throughout this period are thus likely to persistently adjust long-term regulations to the adult environment. Such age-specific developmental plasticity entails both risks and opportunities. Whilst being sensitive to external stimulation favours the integration of external cues into the mature function, the latter may result pathologic if inappropriately stimulated. For example, precocious experience with drugs of abuse or exposure to adverse environments may favour the onset of conduct disorders (e.g. addiction, emotional disturbances). In this chapter we describe the inextricable link between adolescent plasticity and long-term individual regulations both in terms of predisposition to pathology and in terms of adaptive plasticity to the adult environment.

12.1 Introduction

The stress response system allows immediate individual allostatic responses to external threats and favours ultimate developmental adaptive regulations that "prepare" the organism to cope with future challenges (see Chap. 2). Whilst plastic

S. Macrì (✉) • G. Laviola
Section of Behavioral Neuroscience, Department of Cell Biology and Neurosciences, Istituto Superiore di Sanità, Viale Regina Elena 299 (Block 19, Floor D, Room 15), Rome 00161, Italy
e-mail: simone.macri@iss.it; giovanni.laviola@iss.it

G. Laviola and S. Macrì (eds.), *Adaptive and Maladaptive Aspects of Developmental Stress*, Current Topics in Neurotoxicity 3, DOI 10.1007/978-1-4614-5605-6_12,

regulations can occur throughout the entire lifetime, such plasticity, defined as the set of different phenotypes branching from the same genotype (West-Eberhard 2005), is not constant but fluctuates depending on the specific developmental stage. Beyond the earliest stages of neonatal life (including pregnancy and infancy, described in the previous chapters), the adolescent period—in mammals—is characterized by an elevated sensitivity to context. Thus, environmental influences encountered during adolescence are capable of influencing developmental trajectories and result in long-term adaptations in several behavioural and cognitive domains, and in their neurobiological determinants. The role of peripubertal influences in the regulation of developmental adjustments has been detailed in many studies encompassing different mammalian species (Leussis et al. 2008; Lurzel et al. 2011; Oldehinkel and Bouma 2011; Sachser et al. 2011; Spear 2000).

Adolescence is defined as the transition period between infancy and adulthood (Laviola et al. 2003; Laviola and Terranova 1998; Spear 2000), and its occurrence is generally associated with the hormonal events characterizing puberty. Although the latter is included in the course of adolescence, it does not coincide with it, whereby this life stage is characterized by a massive maturational spurt encompassing somatic growth, cognitive abilities, and emotional responses. All these events are paralleled by neurological and endocrine rearrangements concurring to dictate individual development. Specifically, clinical experimental data demonstrate that the brain encounters massive restructuring during this highly plastic developmental stage. In particular, Jay Giedd and collaborators showed that whereas cortical grey matter undergoes a regional-specific loss throughout adolescence, white matter progressively increases (Giedd et al. 1999). These variations in the neurobiological determinants of the growth spurt are associated with several behavioural idiosyncrasies: specifically, both in humans and in other mammals, adolescents are characterized by discontinuities in social interaction (shifting from interactions with parents and family members to intense bonds and interactions with peers), massive increases in sensation and novelty seeking, and in risk-taking behaviours (Adriani et al. 1998; Laviola and Adriani 1998; Sachser et al. 2011). Several authors identified, in rodents, specific age windows during which individual development resembles human adolescence across a spectrum of behavioural, neuroanatomical, neurochemical, and neuroendocrine parameters (Andersen 2003; Laviola et al. 2003; Spear 2000).

12.1.1 Acute Stressors Differentially Affect Adolescent and Adult Individuals

Several authors investigated the ontogenetic development of the systems involved in the sensitivity to external stressors. For example, Romeo and collaborators described the developmental variations occurring at the levels of the hypothalamic–pituitary–gonadal (HPG) axis (Romeo 2010; Romeo et al. 2002) and of the hypothalamic–pituitary–adrenocortical (HPA) axis in rodents (Romeo 2010). Specifically, they reported that HPA reactivity in the form of immediate corticosterone and ACTH reactivity in response to 30-min restraint stress is prolonged in adolescent rats compared to adults (Romeo et al. 2004). Specifically, in adult rats, restraint is associated with a peak in ACTH and corticosterone output few minutes after the onset of the stressor, which steadily declines within an additional half hour. Whereas adolescent rats mount a similar peak in corticosteroids following the onset of the stressor, they fail to show the rapid decline observed in adults (Romeo et al. 2004). Among the factors potentially contributing to this unique sensitivity to stress, a differential development of the several components of the HPA axis may play a remarkable role. Thus, age-specific maturation rate of the paraventricular nucleus of the hypothalamus (Romeo 2010), of the pituitary or the adrenal gland may all participate in these responses. Additionally, the expression of glucocorticoid and mineralocorticoid receptors exhibits age- and region-specific variations throughout development (Pryce 2008).

Whilst a detailed description of the fundamental mechanisms governing growth during adolescence falls beyond the scopes of this chapter, we here reiterate the existence of numerous biological events taking place during this developmental stage. Such biological turmoil entails risks and opportunities whereby experiential factors capable of interfering with developmental trajectories are likely to persistently deviate individual phenotypic adjustments. For example, Lenroot and collaborators analyzed the degree of heritability of different brain structures throughout development (Lenroot and Giedd 2008; Lenroot et al. 2009). Previous studies had already shown that several aspects of brain morphology are transmitted across generations (Sowell et al. 2001; Thompson et al. 2001a, b). Through a series of studies conducted in twins, Lenroot and colleagues showed that such heritability is also dependent on the specific brain structure and on the given developmental stage. For example, they observed that whilst the genetic influences on the development of primary motor and sensory cortices are noticeable early in life, the same natural predispositions exert their role on the regulation of brain structures associated with reasoning abilities during adolescence (Lenroot and Giedd 2008; Lenroot et al. 2009). These studies indicate that genes are differentially expressed depending on the specific time window and that they exert their maximal effects on higher-order brain functions during adolescence. Ultimately, elevated plasticity during adolescence appears naturally determined and potentially characterized by an elevated sensitivity to contextual influences. Thus, just as appropriate levels of environmental challenges may favour adaptive regulations, so also negative events may result in severely adverse consequences (see below).

In this chapter we will describe the multifaceted nature of the adolescent sensitivity to context. Specifically, we will examine a series of studies demonstrating that various forms of environmental challenges (ranging from mild to severe) regulate long-lasting individual adjustments; we will also discuss whether these adjustments shall be regarded as adaptive, maladaptive, or pathological.

In the following paragraphs we will describe (1) the rationale and the basic methodology behind the studies addressing the long-term consequences of stress during adolescence, (2) the main results obtained, (3) the general interpretations provided and (4) the alternative interpretations that might be proposed within an evolutionary/adaptive framework.

12.2 Persistent Effects of Stress During Adolescence: Rationale, Methodology, and Principal Outcomes

In the introductory section we mentioned a series of developmental milestones occurring throughout adolescence, and we implied that they conspire to render the individual particularly sensitive to environmental stimulation. In line with other scholars (Ellis et al. 2012), we believe that such sensitivity should be framed within an evolutionary adaptive context and that the consequences, positive or negative, depend on the specific environmental conditions. The bidirectional nature of such sensitivity has been elegantly conceptualized by Susan Andersen through the dichotomy: "point of vulnerability or window of opportunity?" (Andersen 2003). This definition—that received supplementary elaboration and experimental support in a series of studies (Adriani and Laviola 2004)—further implies that whilst some events may induce disease states in the long term, some others may favour resilience and compensatory adaptive processes. Specifically, whereas severely negative events may relate to psychiatric disturbances (e.g. early onset depression, first psychotic episodes, substance abuse and symptoms of attention deficit/hyperactivity), positive—moderately challenging—experiences may beget long-term advantages whereby they hinder disease progression or reduce symptoms' severity. Both positive and negative consequences of teenage stress rest upon the elevated, age-specific plasticity characterizing adolescent individuals.

12.2.1 Adolescence as a Point of Vulnerability

Despite such comprehensive heuristic perspective, most of the studies investigating the consequences of teenage stress predominantly focussed on detrimental consequences rather than on adaptive processes. We believe that such bias relates to the fact that several psychiatric disturbances emerge during or shortly after adolescence. For example, Andersen and Teicher (2008) reported that the risk to develop depression is dramatically increased in adolescence compared to infancy and childhood. Based on this evidence, they also described the neurobiological events characterizing adolescence and potentially predisposing individuals to depressive disorders. Beside alterations in hormonal regulation and neurochemical transmission, the latter are characterized by change of mood, sadness, and inability to experience pleasure (anhedonia). Specifically, they reported that different brain regions show asynchronous growth rates during adolescence and that this differential maturational rate may constitute a remarkable risk factor. For example, during adolescence, dopaminergic transmission is particularly elevated in the prefrontal cortex and downregulated in the dorsal and ventral striatum. Whilst the former is predominantly involved in the processing of contextual stimuli and in the mediation of individual reactivity to stressful events (Brenhouse et al. 2008), the latter is involved in the regulation of reward processes (e.g. natural rewards and drugs of abuse (Nestler and Carlezon 2006)). Thus, an imbalance between emotional processing and reward perception may render adolescence particularly sensitive to external stressors (a hallmark of depression) and insensitive to natural rewards (i.e. anhedonic). Within this framework, a series of studies demonstrated that, compared to adult individuals, adolescent rodents display a reduced sensitivity to the administration of dopaminergic agonists (Adriani and Laviola 2000; Laviola et al. 1995; Spear and Brake 1983). Laviola et al. (1995) observed that repeated exposure to cocaine had differential effects in adolescent and adult rats. Specifically, they observed that whilst adult rats showed a prominent sensitization towards the adverse consequences of cocaine (stereotypies, generally considered an indicator of poor welfare, likely mediated by an overstimulation of the striatum), adolescents failed to show such response. Conversely, adolescent rats showed sensitization to the locomotor-activating effects of cocaine (likely dependent on the activation of the nucleus accumbens). Thus, whereas adult rats, in response to a repeated psychostimulants administration, showed both positive and negative behavioural responses, adolescent rats failed to show negative reactions. This may offer support to the possibility that, during adolescence, the organism is naturally prone to develop drug-related disorders. Specifically, being sensitive to the positive (rewarding) consequences of psychoactive drug administration and much less sensitive to its negative effects increases the likelihood to bolster drug-seeking behaviour. This would ultimately relate to an increased vulnerability to later-onset adverse consequences of drugs of abuse. Andersen (2003) exemplified the link between adolescence-specific sensitivity to drugs of abuse and the subsequent emergence of reward-related deficits. Specifically, she proposed that whilst "the immature organism adapts by incorporating environmental information permanently into the mature structure and function [...], the mature organism compensates to accommodate changes in the environment" (Andersen 2003). In other words, exposure to drugs of abuse in adulthood may elicit a fluctuation in the individual homeostasis, likely to return to baseline following a given amount of time. Conversely, the same exposure during adolescence may induce long-lasting (mal)adaptive processes likely to interfere with natural developmental trajectories. This analysis further exemplifies the link between the underlying neurobiology and the observable sensitivity to context.

Beside depression, adolescence is frequently associated with the onset of additional psychiatric disorders like attention deficit/hyperactivity, borderline disorder and schizophrenia, which show a peak onset around this developmental stage (Rajji et al. 2009). Finally, a substantial bulk of epidemiological and experimental evidence indicates that adolescents are at high risk of developing substance use disorders (Adriani et al. 2006; Adriani and Laviola 2000; Spear and Varlinskaya 2005). The latter may also relate to the elevated sensation-seeking and risk-taking behaviour frequently exhibited by adolescent individuals.

12.2.2 Social Emotionality and Stress-Related Sensitivity

It is thus not surprising that many experimental studies attempted to induce phenotypic abnormalities, reminiscent of the aforementioned disorders, through the administration of different forms of stress during highly plastic developmental stages (e.g. infancy and adolescence). Recent studies investigating the effects of stress during adolescence on the development of emotional disturbances attempted to design stressful paradigms tailored to this specific life stage. In the light of the crucial role played by social experiences around puberty, many of these studies attempted to manipulate group size and composition (e.g. isolation, crowding, presence/absence of aggressive conspecifics). For example, Leussis et al. (2008) proposed an isolation paradigm in which rats are isolated from littermates between postnatal days 30–35 (mid-adolescence). Such isolation constitutes a fundamental stressor to adolescent rats whereby, during this period, they exhibit elevated levels of social interaction and play behaviour (Cirulli et al. 1996; Terranova et al. 1993, 1999). Specifically, periadolescent rodents normally exhibit elevated levels of playful behaviours, which progressively translate into adult-like forms of competitive behaviours (Terranova et al. 1993, 1998). Precluding the possibility to interact with others during this developmental stage has been shown to induce a plethora of behavioural abnormalities highly reminiscent of symptoms of human depression. For example, adolescent rats, isolated from their littermates for 6 days, showed a short-term reduction in coping abilities while facing adverse experimental conditions (Leussis and Andersen 2008). Specifically, rats exposed to this form of social stress showed reduced attempts to escape an inevitable foot shock and increased levels of immobility in the forced swim test (Porsolt et al. 1977a, b). In this paradigm, rats show a specific sequela of events: frantic movements against the sidewalls (escape attempts) followed by a period of swimming, which eventually precedes the attainment of an immobile posture. A number of studies demonstrated that this sequela can be modified through the administration of antidepressant compounds, which are capable of increasing the number of escape attempts at the expense of immobility. Thus, increased immobility is generally interpreted as an index of a depressive-like phenotype. Beside the alteration of emotional responses, social isolation during this developmental stage resulted in long-term alterations in the exhibition of social interaction with conspecifics and reproductive behaviour (Potegal and Einon 1989). Specifically, Toth et al. (2011) recently confirmed that isolation from weaning through adulthood resulted in long-term abnormalities in social aggressive patterns, deficiencies in social communication and increased defensive behaviours.

Resting on the natural predisposition of adolescent rats to engage in active playful social interaction, Bingham et al. (2011) attempted to investigate whether variations in the social *milieu* resulted in short- and long-term alterations in emotional responses. To this aim, they exposed adolescent and adult rats to a series of agonistic encounters and then measured (24–72 h after the last encounter) behavioural indices of anxiety- and depressive-related profiles in the defeated (socially stressed) individuals. The authors showed that social stress had differential effects if applied early in adolescence or in adulthood. Specifically, social defeat resulted in short-term increased indices of behavioural anxiety in adolescence and in reduced anxiety in adulthood. Additionally, the long-term consequences of peripubertal exposure to social stress are reminiscent of reduced anxiety (Bingham et al. 2011). In the same line, Romeo et al. (2006) reported that repeated physical stressors during adolescence differentially regulate individual HPA reactivity compared to the same stressors applied in adulthood. Specifically, whilst adult rats exposed to seven daily 30-min restraint sessions show a minimal corticosterone output in response to an additional restraint (tolerance), adolescent rats fail to show such habituation profile (Romeo et al. 2006). Thus, compared to adulthood, adolescence is characterized by differential stress reactivity both in baseline conditions (i.e. immediate responses to a single stressor differ between adolescents and adults, see above and Adriani and Laviola (2000)) and in response to repeated stressful experiences. Additionally, the consequences of external stressors may vary depending on whether they are encountered around puberty or later in life. Within this frame, Laviola and collaborators (Laviola et al. 2002) reported that a different manipulation of the social environment (crowding) had variable effects depending on whether it was applied in adolescence or in adulthood. Specifically, the authors kept adolescent and adult male mice in highly crowded (nine subjects per cage) or in standard rearing conditions and then evaluated individual corticosterone reactivity in response to a saline injection. Compared to standard rearing, adult mice kept under crowding conditions showed an increased corticosterone output in response to the injection. Adolescents kept under crowding conditions failed to show such increased elevation in corticosterone response compared to controls (Laviola et al. 2002).

12.2.3 Adolescence as a Window of Opportunity

Although studies investigating the negative consequences of stress during adolescence largely exceed those aimed at addressing the potentially positive influences exerted by stimulating environments, a number of scholars attempted to regard adolescence as a "window of opportunity" (Adriani and Laviola 2004; Andersen 2003). Within this realm, particular attention has been given to the possibility of providing stimulating environments to a developing individual in order to favour functional adjustments. The different methodologies aimed at providing moderately challenging environments to neonate rodents have already been described in different chapters of this book (see Chaps. 10 and 11). Here we will describe the rationale and some of the different methodologies adopted to provide adolescent rodents with enriched environments (EE), herein referred to as "a combination of complex inanimate and social stimulation" (Rosenzweig et al. 1978) generally constituted by large cages containing a running wheel and several toys that are regularly replaced by unfamiliar ones (Laviola et al. 2008). The original reports indicating that stimulating environments may impinge on individual development date back to the 1940s when Hebb (1947) observed that rats reared under enriched conditions showed improved problem-solving capabilities compared to conventionally housed rats. Following this pioneering report, many authors replicated these findings and demonstrated that EE was sufficient to persistently increase cell proliferation and neurogenesis in brain areas involved in memory formation and retention (van Praag et al. 1999). Additional studies extended the original observations to biological domains other than cognition

(Nithianantharajah and Hannan 2006). These observations were generally interpreted in the light of "positive" effects exerted by EE on individual development. Subsequently, several authors attempted to demonstrate that environmental stimulation may contrast the adverse consequences of previous "negative" experiences (Nithianantharajah and Hannan 2006). It is within this realm that EE started being regarded as a reciprocal treatment, a tool to investigate the possibility that stimulating experiential factors may hinder the progression of experimentally induced (i.e. genetic preparations, pharmacological treatments, environmental adversities) abnormal phenotypes (Laviola et al. 2008). For example, just as EE delayed the progression of motor symptoms in transgenic models of Huntington's disease (van Dellen et al. 2000) so also voluntary physical exercise reduced the severity of behavioural abnormalities in mouse models of Parkinson's disease (Tillerson et al. 2003). Beside these observations, several authors attempted to identify the molecular mechanisms mediating such EE-dependent recovery of brain functions. For example, Fischer et al. (2007) observed that EE reinstated learning behaviour and access to long-term memories in mice in which temporal neuronal loss was experimentally induced. The authors also showed that such recovery correlated with increased histone acetylation (Fischer et al. 2007), which is associated with increased gene transcription in several brain areas.

Within the context of emotional disturbances, EE has been extensively used as a tool to counteract the adverse consequences of precocious negative experiences impinging on HPA axis development (Laviola et al. 2008). A large body of scientific evidence demonstrates that adverse neonatal conditions favour the onset of psychiatric disorders in humans (Heim and Nemeroff 2001); by the same token, clinical data indicate that exposure to favourable stimuli can mitigate the symptoms (Hollon et al. 2002). Both the potentially pathological consequences of neonatal adversities and the beneficial effects of stimulating environments on emotional disturbances have been successfully translated in basic animal research. Specifically, countless studies (see Chaps. 8, 9, 10 and 11) demonstrated that, in rodents, perinatal stressors are capable of inducing abnormal phenotypes reminiscent of human emotional disturbances (for comprehensive reviews, see Pryce et al. (2005), Maccari and Morley-Fletcher (2007), Marco et al. (2011)). In order to evaluate whether stimulating environments may also hinder the progression of depressive-like abnormalities in rodents, several authors adopted EE paradigms as reciprocal treatments following perinatal stressors. To give but few examples, EE during adolescence has hampered the long-term consequences of prenatal stress on behavioural fearfulness, social interaction, endocrine stress reactivity, and immune responses (Laviola et al. 2004; Morley-Fletcher et al. 2003). Thus, developing rats that were stressed in the womb (three daily 45-min sessions of maternal restraint stress during the last third of pregnancy) showed increased corticosterone response to restraint stress and behavioural fearfulness, and reduced immune reactivity and playful behaviours; all these effects were efficiently counteracted by exposure to EE throughout adolescence (Laviola et al. 2004; Morley-Fletcher et al. 2003). Similarly, Francis et al. (2002) demonstrated that EE was able to counteract the long-term consequences of repeated daily 3-h maternal separations. Thus, whilst maternal separation resulted in long-term increased behavioural fearfulness and endocrine stress reactivity, EE was capable of eliminating these effects (Francis et al. 2002). Most of these studies ascribed these effects to the stimulating and potentially "beneficial" effects of enriched environments. As mentioned above, most studies unravelling the molecular determinants of EE-related behavioural and neuroendocrine variations rested upon the original overarching definition of enrichment (Rosenzweig et al. 1978). Such a broad definition may hinder some relevant aspects of EE. Specifically, whilst decomposing EE into its fundamental characteristics, we note that it may constitute a complex source of stress to rodents. Specifically, beside the long-term effects on individual adaptations, EE has been shown to favour aggressive behaviour, specifically in male mice (McQuaid et al. 2012). Compared to standard housing conditions, EE may also elicit repeated stress responses in laboratory mice. Although hypothetical, in line with what suggested for newborn pups (see Chap. 11), we propose that some of the effects of EE may also relate to the repeated stressful challenges to which subjects are exposed during a plastic stage of individual development. Specifically, mice capable of reinstating homeostasis in response to these repeated challenges may prepare to the multiple stressors to be encountered in adulthood. Future studies (potentially involving exogenous corticosterone administration during adolescence) are needed to clarify this aspect.

12.3 Conclusion

12.3.1 Long-Lasting Consequences of Teenage Stress Within an Adaptive Framework

The prevailing views associated with the studies described in the previous paragraphs are generally skewed towards a human-centred disease perspective. In other words, these studies are generally advocated to support the hypothesis that given events may favour pathology or protection (resilience) against negative outcomes. In order to substantiate these claims, laboratory animals are used under the assumption of a phylogenetic proximity between rodents and humans, thereby resting upon strong evolutionary considerations. Quite ironically, these considerations are sometimes neglected when it comes to discussing the results obtained in the light of the original hypotheses. Specifically, experimental findings garnered in this field are generally interpreted as proxies for symptoms of disease and framed within a good-to-bad scale. To give an example, increased HPA response to a given stressor is often considered to be negative and, conversely, reduced reactivity is generally considered positive. Yet, hardly ever is such isolated outcome measure framed within an evolutionary adaptive context and interpreted within the specific environment in which it is exhibited. In other words, in spite of evolutionary adaptive considerations (see Chap. 2) suggesting to interpret a given phenotype within the specific constraints in which it is exhibited, experimental psychobiology often labels a given parameter as good or bad. Maintaining the example of the HPA axis reactivity, life-history theories predict that the consequences of variable levels of arousal depend on the environment in which they are displayed. In other words, whilst under certain environmental conditions elevated HPA reactivity may beget advantages and reduced reactivity disadvantages in terms of lifetime fitness, under different environments the situation may be opposite. Few authors started addressing this possibility. For example, Norbert Sachser and his group, through a series of studies conducted in guinea pigs, provided a particularly cogent exemplification of this broad perspective (Sachser et al. 2011). Specifically, the authors conducted a series of experiments in which they reared adolescent guinea pigs under variable social conditions. In an early study (Sachser and Lick 1991) they reared adolescent male guinea pigs (approximately 1 month old) either in a mixed-sex colony or in a pair together with a single female. Same-sex couples of colony-reared and pair-reared adult male guinea pigs were then placed into an unfamiliar cage together with an unknown female. Under this testing condition, colony-reared individuals rapidly established dominance hierarchies without recurring to overt fights; additionally, neither the winner nor the loser showed significant elevations in plasma cortisol levels. Pair-reared individuals displayed a completely different response profile. Thus, they showed elevated levels of aggressive behaviours that often escalated into overt fights nearing severe injuries or death. Finally, they showed remarkable elevations in cortisol response (Sachser and Lick 1991). Additional studies further corroborated this evidence (Sachser 1993; Sachser et al. 1994). In an independent series of studies, the authors investigated the ability of guinea pigs reared under different social conditions (isolated, pair- or colony-reared) to enter a colony of unfamiliar individuals in adulthood. Similar to what observed in the previous studies, the authors showed that adult individuals that experienced a colony environment throughout adolescence displayed a rapid adaptation to the novel social environment. Such capability to rapidly infiltrate a group of unfamiliar individuals encompassed both the behavioural and the hormonal domains (Sachser 1993; Sachser et al. 1994). Isolated and pair-reared individuals (exposed to these treatments during adolescence) showed an opposite behavioural and endocrine profile in adulthood. Beside the originality of the experimental design, the authors also offered an innovative interpretation of their findings. Rather than limiting their analysis to rearing-dependent variations in social competences (the more social interactions encountered during development, the better social competences in adulthood and vice versa), they framed their observations within a broad evolutionary adaptive context. Within this realm, the authors proposed that adolescent guinea pigs used the information provided by their developmental environment to predict the challenges to be encountered in adulthood. Specifically, they proposed that whereas colony rearing during adolescence anticipated an adult environment rich of age-matched conspecifics, isolation rearing forecasted a niche in which opportunities for social interactions were scant (Sachser et al. 2011). Sachser and collaborators then attempted to map the observed phenotypes onto the specific requirements of the two predicted environments. Specifically, a developmental context in which the number of individuals is low may favour the maturation of an aggressive phenotype (ready to successfully handle the few social threats encountered), sustained by a readily reactive HPA axis (itself favouring aggressive behaviour). Conversely, an environment characterized by the presence of several conspecifics would demand a much less aggressive behavioural phenotype associated with a low-reactive HPA axis. This latter phenotype would permit low levels of aggressive behaviours in interactions with strangers and would facilitate "queuing" in the long path to achieve a higher rank in the dominance hierarchy (Sachser et al. 2011).

Although this theoretical framework still requires experimental corroboration, it offers a heuristic model against which experimental hypotheses can be tested and falsified.

Acknowledgements This work was supported by the grant "ECS-EMOTION" from the Department of Antidrug Policies c/o Presidency of the Council of Ministers, Italy to GL and SM. The funders had no role in study design, data collection and analysis, decision to publish, or preparation of the manuscript.

References

Adriani W, Laviola G (2000) A unique hormonal and behavioral hyporesponsivity to both forced novelty and d-amphetamine in periadolescent mice. Neuropharmacology 39:334–346

Adriani W, Laviola G (2004) Windows of vulnerability to psychopathology and therapeutic strategy in the adolescent rodent model. Behav Pharmacol 15:341–352

Adriani W, Chiarotti F, Laviola G (1998) Elevated novelty seeking and peculiar d-amphetamine sensitization in periadolescent mice compared with adult mice. Behav Neurosci 112:1152–1166

Adriani W, Deroche-Gamonet V, Le Moal M, Laviola G, Piazza PV (2006) Preexposure during or following adolescence differently affects nicotine-rewarding properties in adult rats. Psychopharmacology (Berl) 184:382–390

Andersen SL (2003) Trajectories of brain development: point of vulnerability or window of opportunity? Neurosci Biobehav Rev 27:3–18

Andersen SL, Teicher MH (2008) Stress, sensitive periods and maturational events in adolescent depression. Trends Neurosci 31:183–191

Bingham B, McFadden K, Zhang X, Bhatnagar S, Beck S, Valentino R (2011) Early adolescence as a critical window during which social stress distinctly alters behavior and brain norepinephrine activity. Neuropsychopharmacology 36:896–909

Brenhouse HC, Sonntag KC, Andersen SL (2008) Transient D1 dopamine receptor expression on prefrontal cortex projection neurons: relationship to enhanced motivational salience of drug cues in adolescence. J Neurosci 28:2375–2382

Cirulli F, Terranova ML, Laviola G (1996) Affiliation in periadolescent rats: behavioral and corticosterone response to social reunion with familiar or unfamiliar partners. Pharmacol Biochem Behav 54:99–105

Ellis BJ, Del Giudice M, Dishion TJ, Figueredo AJ, Gray P et al (2012) The evolutionary basis of risky adolescent behavior: implications for science, policy, and practice. Dev Psychol 48:598–623

Fischer A, Sananbenesi F, Wang X, Dobbin M, Tsai LH (2007) Recovery of learning and memory is associated with chromatin remodelling. Nature 447:178–182

Francis DD, Diorio J, Plotsky PM, Meaney MJ (2002) Environmental enrichment reverses the effects of maternal separation on stress reactivity. J Neurosci 22:7840–7843

Giedd JN, Blumenthal J, Jeffries NO, Castellanos FX, Liu H et al (1999) Brain development during childhood and adolescence: a longitudinal MRI study. Nat Neurosci 2:861–863

Hebb DO (1947) The effects of early experience on problem solving at maturity. Am Psychol 2:306–307

Heim C, Nemeroff CB (2001) The role of childhood trauma in the neurobiology of mood and anxiety disorders: preclinical and clinical studies. Biol Psychiatry 49:1023–1039

Hollon SD, Munoz RF, Barlow DH, Beardslee WR, Bell CC et al (2002) Psychosocial intervention development for the prevention and treatment of depression: promoting innovation and increasing access. Biol Psychiatry 52:610–630

Laviola G, Adriani W (1998) Evaluation of unconditioned novelty-seeking and d-amphetamine-conditioned motivation in mice. Pharmacol Biochem Behav 59:1011–1020

Laviola G, Terranova ML (1998) The developmental psychobiology of behavioural plasticity in mice: the role of social experiences in the family unit. Neurosci Biobehav Rev 23:197–213

Laviola G, Wood RD, Kuhn C, Francis R, Spear LP (1995) Cocaine sensitization in periadolescent and adult rats. J Pharmacol Exp Ther 275:345–357

Laviola G, Adriani W, Morley-Fletcher S, Terranova ML (2002) Peculiar response of adolescent mice to acute and chronic stress and to amphetamine: evidence of sex differences. Behav Brain Res 130:117–125

Laviola G, Macrì S, Morley-Fletcher S, Adriani W (2003) Risk-taking behavior in adolescent mice: psychobiological determinants and early epigenetic influence. Neurosci Biobehav Rev 27:19–31

Laviola G, Rea M, Morley-Fletcher S, Di Carlo S, Bacosi A et al (2004) Beneficial effects of enriched environment on adolescent rats from stressed pregnancies. Eur J Neurosci 20:1655–1664

Laviola G, Hannan AJ, Macrì S, Solinas M, Jaber M (2008) Effects of enriched environment on animal models of neurodegenerative diseases and psychiatric disorders. Neurobiol Dis 31(2):159–168

Lenroot RK, Giedd JN (2008) The changing impact of genes and environment on brain development during childhood and adolescence: initial findings from a neuroimaging study of pediatric twins. Dev Psychopathol 20:1161–1175

Lenroot RK, Schmitt JE, Ordaz SJ, Wallace GL, Neale MC et al (2009) Differences in genetic and environmental influences on the human cerebral cortex associated with development during childhood and adolescence. Hum Brain Mapp 30:163–174

Leussis MP, Andersen SL (2008) Is adolescence a sensitive period for depression? Behavioral and neuroanatomical findings from a social stress model. Synapse 62:22–30

Leussis MP, Lawson K, Stone K, Andersen SL (2008) The enduring effects of an adolescent social stressor on synaptic density, part II: poststress reversal of synaptic loss in the cortex by adinazolam and MK-801. Synapse 62:185–192

Lurzel S, Kaiser S, Sachser N (2011) Social interaction decreases stress responsiveness during adolescence. Psychoneuroendocrinology 36:1370–1377

Maccari S, Morley-Fletcher S (2007) Effects of prenatal restraint stress on the hypothalamus-pituitary-adrenal axis and related behavioural and neurobiological alterations. Psychoneuroendocrinology 32(Suppl 1):S10–S15

Marco EM, Macri S, Laviola G (2011) Critical age windows for neurodevelopmental psychiatric disorders: evidence from animal models. Neurotox Res 19:286–307

McQuaid RJ, Audet MC, Anisman H (2012) Environmental enrichment in male CD-1 mice promotes aggressive behaviors and elevated corticosterone and brain norepinephrine activity in response to a mild stressor. Stress 15:354–360

Morley-Fletcher S, Rea M, Maccari S, Laviola G (2003) Environmental enrichment during adolescence reverses the effects of prenatal stress on play behaviour and HPA axis reactivity in rats. Eur J Neurosci 18:3367–3374

Nestler EJ, Carlezon WA Jr (2006) The mesolimbic dopamine reward circuit in depression. Biol Psychiatry 59:1151–1159

Nithianantharajah J, Hannan AJ (2006) Enriched environments, experience-dependent plasticity and disorders of the nervous system. Nat Rev Neurosci 7:697–709

Oldehinkel AJ, Bouma EM (2011) Sensitivity to the depressogenic effect of stress and HPA-axis reactivity in adolescence: a review of gender differences. Neurosci Biobehav Rev 35:1757–1770

Porsolt RD, Bertin A, Jalfre M (1977a) Behavioral despair in mice: a primary screening test for antidepressants. Arch Int Pharmacodyn Ther 229:327–336

Porsolt RD, Le Pichon M, Jalfre M (1977b) Depression: a new animal model sensitive to antidepressant treatments. Nature 266:730–732

Potegal M, Einon D (1989) Aggressive behaviors in adult rats deprived of playfighting experience as juveniles. Dev Psychobiol 22:159–172

Pryce CR (2008) Postnatal ontogeny of expression of the corticosteroid receptor genes in mammalian brains: inter-species and intra-species differences. Brain Res Rev 57:596–605

Pryce CR, Ruedi-Bettschen D, Dettling AC, Weston A, Russig H et al (2005) Long-term effects of early-life environmental manipulations in rodents and primates: potential animal models in depression research. Neurosci Biobehav Rev 29:649–674

Rajji TK, Ismail Z, Mulsant BH (2009) Age at onset and cognition in schizophrenia: meta-analysis. Br J Psychiatry 195:286–293

Romeo RD (2010) Adolescence: a central event in shaping stress reactivity. Dev Psychobiol 52:244–253

Romeo RD, Richardson HN, Sisk CL (2002) Puberty and the maturation of the male brain and sexual behavior: recasting a behavioral potential. Neurosci Biobehav Rev 26:381–391

Romeo RD, Lee SJ, Chhua N, McPherson CR, McEwen BS (2004) Testosterone cannot activate an adult-like stress response in prepubertal male rats. Neuroendocrinology 79:125–132

Romeo RD, Bellani R, Karatsoreos IN, Chhua N, Vernov M et al (2006) Stress history and pubertal development interact to shape hypothalamic-pituitary-adrenal axis plasticity. Endocrinology 147:1664–1674

Rosenzweig MR, Bennett EL, Hebert M, Morimoto H (1978) Social grouping cannot account for cerebral effects of enriched environments. Brain Res 153:563–576

Sachser N (1993) The ability to arrange with conspecifics depends on social experiences around puberty. Physiol Behav 53:539–544

Sachser N, Lick C (1991) Social experience, behavior, and stress in guinea pigs. Physiol Behav 50:83–90

Sachser N, Lick C, Stanzel K (1994) The environment, hormones, and aggressive behaviour: a 5-year-study in guinea pigs. Psychoneuroendocrinology 19:697–707

Sachser N, Hennessy MB, Kaiser S (2011) Adaptive modulation of behavioural profiles by social stress during early phases of life and adolescence. Neurosci Biobehav Rev 35:1518–1533

Sowell ER, Thompson PM, Mattson SN, Tessner KD, Jernigan TL et al (2001) Voxel-based morphometric analyses of the brain in children and adolescents prenatally exposed to alcohol. Neuroreport 12:515–523

Spear LP (2000) The adolescent brain and age-related behavioral manifestations. Neurosci Biobehav Rev 24:417–463

Spear LP, Brake SC (1983) Periadolescence: age-dependent behavior and psychopharmacological responsivity in rats. Dev Psychobiol 16:83–109

Spear LP, Varlinskaya EI (2005) Adolescence. Alcohol sensitivity, tolerance, and intake. Recent Dev Alcohol 17:143–159

Terranova ML, Laviola G, Alleva E (1993) Ontogeny of amicable social behavior in the mouse: gender differences and ongoing isolation outcomes. Dev Psychobiol 26:467–481

Terranova ML, Laviola G, de Acetis L, Alleva E (1998) A description of the ontogeny of mouse agonistic behavior. J Comp Psychol 112:3–12

Terranova ML, Cirulli F, Laviola G (1999) Behavioral and hormonal effects of partner familiarity in periadolescent rat pairs upon novelty exposure. Psychoneuroendocrinology 24:639–656

Thompson PM, Cannon TD, Narr KL, van Erp T, Poutanen VP et al (2001a) Genetic influences on brain structure. Nat Neurosci 4:1253–1258

Thompson PM, Vidal C, Giedd JN, Gochman P, Blumenthal J et al (2001b) Mapping adolescent brain change reveals dynamic wave of accelerated gray matter loss in very early-onset schizophrenia. Proc Natl Acad Sci U S A 98:11650–11655

Tillerson JL, Caudle WM, Reveron ME, Miller GW (2003) Exercise induces behavioral recovery and attenuates neurochemical deficits in rodent models of Parkinson's disease. Neuroscience 119:899–911

Toth M, Mikics E, Tulogdi A, Aliczki M, Haller J (2011) Post-weaning social isolation induces abnormal forms of aggression in conjunction with increased glucocorticoid and autonomic stress responses. Horm Behav 60:28–36

van Dellen A, Blakemore C, Deacon R, York D, Hannan AJ (2000) Delaying the onset of Huntington's in mice. Nature 404:721–722

van Praag H, Kempermann G, Gage FH (1999) Running increases cell proliferation and neurogenesis in the adult mouse dentate gyrus. Nat Neurosci 2:266–270

West-Eberhard MJ (2005) Developmental plasticity and the origin of species differences. Proc Natl Acad Sci U S A 102(suppl 1):6543–6549

Chapter 13
Oxidative Stress and Hormetic Responses in the Early Life of Birds

David Costantini

Abstract Understanding the physiological processes underlying the long-term effects of early life environmental conditions on Darwinian fitness is of central importance to evolutionary ecology, biomedical research, and conservation science. In particular, the extent to which early stress exposure is detrimental to fitness may depend on its severity, with mild stress exposure actually having a stimulatory and, possibly, beneficial effect through a hormetic response to the stressful stimulus. Oxidative stress is an aspect of stress physiology that has received comparatively less attention than hormones when considering the long-term effects of early life conditions. We therefore need to combine hormesis and oxidative stress in order to better understand how the early environment can help shape a phenotype adapted to the conditions it is most likely to experience in its adult environment. This chapter aims to discuss how hormones and nutrients can shape the redox machinery of the embryo and its future susceptibility to oxidative stress; how hormesis might provide a mechanistic framework for interpreting the long-term effects of early exposure to various stressor magnitudes; and how to reconcile discrepancies among studies. Examples from avian research are especially emphasised throughout the chapter, notably because avian models may have many advantages over mammalian models when addressing the (mal)adaptive effects of early life experiences on adult phenotype.

D. Costantini (✉)
Animal Health and Comparative Medicine, Veterinary and Life Sciences,
Institute for Biodiversity, College of Medical, University of Glasgow,
Graham Kerr Building, Glasgow, G12 8QQ, UK
e-mail: davidcostantini@libero.it

G. Laviola and S. Macrì (eds.), *Adaptive and Maladaptive Aspects of Developmental Stress*, Current Topics in Neurotoxicity 3, DOI 10.1007/978-1-4614-5605-6_13,

13.1 Introduction

13.1.1 Early Life Conditions and Phenotypic Development

The adult phenotype is the result of a complex interplay between genetic background and pre- and post-natal developmental conditions. The up-regulation and down-regulation of genes are triggered by both social and non-social environmental stimuli experienced early in life—this is then expressed in individual morphology, physiology, and behaviour. Thus the environment is important; not only does it select for genetic variation, but it also contributes to the build-up of phenotypic variation (Gilbert 2001; Metcalfe and Monaghan 2001; West-Eberhard 2003; Monaghan 2008). A classic example of this complex interaction between genome and environment is demonstrated by the phenotypic plasticity concept, which refers to the capacity of a genotype to give rise to a variety of phenotypes, depending on the prevailing environmental conditions (Pigliucci 2001).

Recognising the importance of the continuous interaction between genes and environment that occurs during development has represented a valuable theoretical expansion of the Modern Synthesis, stuck for decades on the pivotal evolutionary role of genes (Pigliucci et al. 2006). At the same time, developmental systems are also guided by self-organisation properties of their components, without the contribution of any genetically determined programme (Gerhart and Kirschner 2007; Badyaev 2011). As a whole, it is becoming clear that a significant portion of genomic architecture, developmental processes and their interaction arise from processes other than natural selection, such as neutral evolution or compensatory rearrangements of anatomical traits or metabolic networks (Badyaev 2011). Moreover, it has become evident that a certain degree of the variation in brain development and behaviour, which occurs among individuals, may be built up by epigenetic modifications to the genome (e.g. chromatin structure, gene expression), which early life experiences might induce (Szyf et al. 2005; Curley and Branchi 2012). However, it is important to note that these theoretical steps forward (i.e. both the experimental identification and the description of mechanisms that regulate responses to social and non-social environmental stimuli, and how such responses shape adult phenotype, thus affect fitness) are still in their infancy.

Over the past few decades most studies have emphasised that the variation in maternal transmission patterns of hormones, especially androgens, can significantly contribute to phenotypic variation of offspring (Schwabl 1993; Groothuis et al. 2005; Carere and Balthazart 2007; Groothuis and Schwabl 2008). However, the adaptive role of maternal hormones and their part in phenotypic organisation have been questioned (Carere and Balthazart 2007). Moreover, sexual hormones may not change rapidly in response to environmental stimuli, and this may be a strong limitation when attempting to define a general mechanism of animal response to environment. Hence, more recently, attention has been shifted to how developmental stress experienced directly (e.g. heat stress, chilling, food shortage, parasites, toxicant exposure) or indirectly through parents (e.g. maternal transfer of nutrients in the egg, parental foraging effort) contributes to shape the phenotype. Several studies have proposed that glucocorticoids (i.e. hormones secreted when exposed to a stressor) passed from mother to offspring via the egg contents, placenta or milk may activate mechanisms underlying the link between pre- and post-natal environmental conditions (Marasco et al. 2012; Love and Williams 2008a, b; Henriksen et al. 2011; Schoech et al. 2011; Haussmann et al. 2012). In a recent study of wild European starlings (*Sturnus vulgaris*), induction of pre-natal stress through *in ovo* injection of corticosterone decreased stress responsiveness of nestlings at fledging (Love and Williams 2008a). The authors suggested that such plasticity in hypothalamic–pituitary–adrenal axis activity may be the result of a predictive adaptive response, that is, maternal stress hormones might provide offspring with a predictive signal of how their future environment *sensu lato* will be (Love and Williams 2008a, b; see also Gluckman and Hanson 2004, for predictive adaptive responses in human evolution). This form of maternal-induced developmental plasticity might be adaptive later in life under severe or unpredictable environmental conditions because a hyperresponsive stress system could be too costly to sustain under harsh conditions (Love and Williams 2008a). Therefore, it is the predictive nature of the environmental cue that might determine whether the developmental strategy will be adaptive or not.

13.1.2 Oxidative Stress in Early Life

Oxidative stress is an aspect of stress physiology that has received comparatively less attention than hormones when considering the long-term effects of early life conditions. It is a kind of cellular biochemical stress caused by an imbalance between pro-oxidants and antioxidant defences, leading to the generation of oxidative damage to biomolecules (Sies 1991; Halliwell and Gutteridge 2007), as well as disruption to redox signalling and balance and the antioxidant response (Jones 2006), and overoxidation of thiols (Sohal and Orr 2012). Oxidative stress can also be defined as the rate at which oxidative damage is generated (Costantini and Verhulst 2009). Implicit in this definition is that oxidative stress is a continuous variable that is unlikely ever to be zero since pro-oxidants are continually produced and some oxidative damage is always generated. Persistent oxidative stress may give rise to pathological conditions and is increasingly implicated in several human pathologies (over 150 disorders), cellular senescence, and ageing (Beckman and Ames 1998; Hulbert et al. 2007; Furness and Speakman 2008).

In the last few decades, we have seen a merger between the laboratory approaches to the study of oxidative stress mechanisms and the ecological principles. "Antioxidant ecology" and "oxidative stress ecology" are areas emerging from this marriage between mechanistic and functional perspectives (McGraw et al. 2010). These sister fields address the principle that organisms are engaged in a molecular struggle with pro-oxidant compounds produced from basic metabolic activities. These molecules have important functions in the body (e.g. cell signalling) but are also responsible for damaging cells. Animals use a large battery of antioxidant resources to offset deleterious effects of pro-oxidants (Pamplona and Costantini 2011). This is why ecologists have hypothesised that oxidative damage may be an important challenge for individual fitness, given its links with reproductive performance, growth patterns, cellular senescence, and survival (Alonso-Alvarez et al. 2006; Costantini and Dell'Omo 2006; Bize et al. 2008; Costantini 2008; Monaghan et al. 2009; Saino et al. 2011). However, the role of oxidative stress as a constraint or facilitator of developmental processes is virtually unexplored, especially in birds. Avian models may have many advantages over mammalian models when addressing the consequences of early life on adult phenotype (for a full discussion see Henriksen et al. 2011). As stressed by Henriksen et al. (2011), the avian egg can be experimentally manipulated independently of the mother, while the mammalian foetal environment is difficult to access and manipulate without interfering with the mother's physiology. Therefore, manipulation of the environment within the avian egg is an ideal way of addressing mechanisms underlying maternal effects on offspring development. Furthermore, the avian neuroendocrine system is very similar to that of mammals, hence theoretically allowing some degree of generalisation. Last but not least, birds show a large variation in developmental modes (the altricial–precocial developmental axis), making it possible to experimentally test the effects of developmental stress under a very large range of life-history strategies.

13.1.3 Stress Is Not Always Bad

A further point that warrants careful attention is the ongoing debate about whether or not the induction of stress during early development is detrimental for the organism. Although it is often assumed that exposure to stress in early life may jeopardise the survival of the individual, this assumption appears weak because priming effects of stress physiology induced by exposure to mild stressors might improve the capacity of the individual to withstand higher levels of stress later in life. Biphasic responses to a stressor with low dose stimulation and high dose inhibition of the stress response have been termed "hormetic responses" (Fig. 13.1; Forbes 2000; Parsons 2001; Costantini et al. 2010; Mattson and Calabrese 2010). Therefore, hormesis might provide a valuable mechanistic and theoretical framework for both developmental biologists and behavioural ecologists that are interested in the fitness consequences of early exposure to stressors.

13.1.4 Aims of the Chapter

This chapter aims to discuss how hormones and nutrients can shape the redox machinery of the embryo and its future susceptibility to oxidative stress; how hormesis might provide a mechanistic framework for interpreting the long-term effects of early exposure to various stressor magnitudes; and how to reconcile discrepancies among studies.

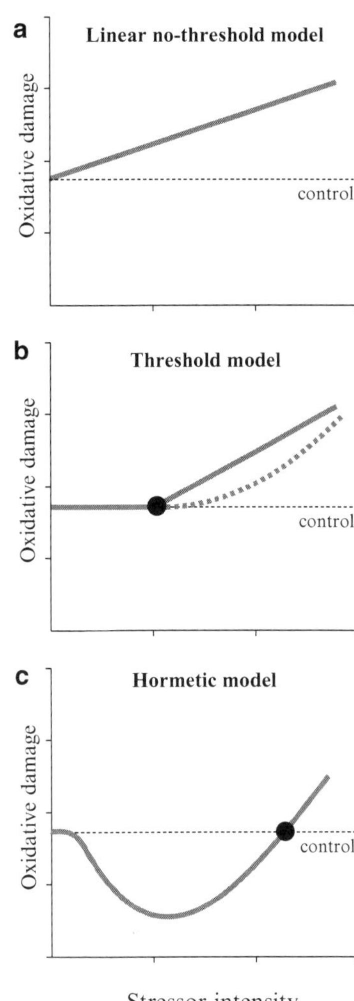

Fig. 13.1 Examples of common dose–response models. In these examples, oxidative damage is used as an end point. (**a**) In the *linear no-threshold model* oxidative damage increases linearly with the stressor intensity, with no threshold of activation of the physiological response. (**b**) In the *threshold model* the stressor has no biological effect until a threshold (indicated by the *black circle*) is reached, above which oxidative damage can increase linearly or nonlinearly. (**c**) In the *hormetic model* the response to the environmental stressor is biphasic, with oxidative damage decreasing and increasing at mild and high doses of the stressor, respectively. The zero equivalent point (*black circle*) refers to the area of the hormetic curve where oxidative damage equals that of the control (i.e. zero exposure)

13.2 Mechanisms: Interplay Among Hormones, Nutrients, Oxidative Stress, and Hormesis

13.2.1 Hormones

Maternal transfer of hormones and compounds with antioxidant properties to the egg can have significant effects on phenotypic development. The importance of this hormone deposition was first highlighted in the 1990s, when Schwabl (1993, 1996) demonstrated that female canaries (*Serinus canaria*) deposited testosterone in their eggs, depending on the laying order. Later-laid eggs contained more testosterone and hatched nestlings that begged more vigorously and grew faster than their siblings. Later, Schwabl et al. (1997) found that the pattern of testosterone deposition in eggs was completely reversed in the cattle egret (*Bubulcus ibis*). At the end of the 1990s, Royle et al. (1999) showed that two dietary compounds with antioxidant properties (tocopherols and carotenoids) could also be deposited differentially among eggs, depending on their laying sequence.

Although yolk testosterone might have a beneficial effect on offspring (Schwabl 1993), it can also induce costs, such as immune suppression, oxidative stress, and mortality (Sockman and Schwabl 2000). Given the link between oxidative stress and testosterone metabolites, and that the level of yolk testosterone in eggs within a clutch may be negatively correlated to that of yolk dietary antioxidants, Royle et al. (2001) hypothesised that these two maternal factors have complementary, but opposing effects (i.e. testosterone induces and dietary antioxidants mitigate oxidative stress). Royle et al. (2001) also suggested that the effect of testosterone on offspring hatched from last-laid eggs would be detrimental if food were scarce due to reduced availability of dietary antioxidants.

It is only relatively recently that experiments started to investigate these hypotheses, with the first three studies seeking to analyse the effects of androgens on oxidative state by injecting testosterone into the yolk. However, these studies produced conflicting results. Tobler and Sandell (2009) found that injecting eggs with 0.5 ng testosterone on day 3 of incubation reduced plasma non-enzymatic antioxidant capacity in zebra finch (*Taeniopygia guttata*) males, but not females, at 10 days old. No effects emerged when birds were 34 days old. At the same time, Galván and Alonso-Alvarez (2010) found no effects of injecting eggs of the same clutch with 26 ng testosterone on total red blood cell glutathione or plasma non-enzymatic antioxidant capacity in nestling great tits (*Parus major*). In contrast, Noguera et al. (2011) found that injecting third-laid eggs with 261 ng testosterone on the day of laying increased plasma non-enzymatic antioxidant capacity transiently and reduced plasma malondialdehyde (MDA, end product of lipid peroxidation) in yellow-legged gull (*Larus michahellis*) nestlings. There are several possible explanations for the observed differences between these studies: the dose of testosterone administered could have been in the stimulatory range of the dose–response curve for the gull, but in the detrimental range for the zebra finch (see paragraph on hormesis below; see also Fig. 13.2); the temporal window of manipulation of yolk androgens, which has a critical role for the response of the redox physiology (e.g. hyporesponsive and hyperresponsive periods), might have differed between studies; zebra finches have a higher pace of life than gulls—hence they might have invested less in antioxidant protection against the pro-oxidant effects of testosterone, but this does not explain why blood antioxidants were unaffected in great tits that have a high pace of life too; the age at which the birds' response to treatment was tested differed among studies—thus changes in blood redox state across development could have caused the discrepancies between studies; the distribution of testosterone across the yolk layers could have differed among studies, therefore exposing birds to different levels of androgens during different phases of embryonic development (e.g. Lipar et al. 1999); and lastly, zebra finches, great tits, and gulls may have differentially metabolised the injected testosterone, which is a precursor of the androgen

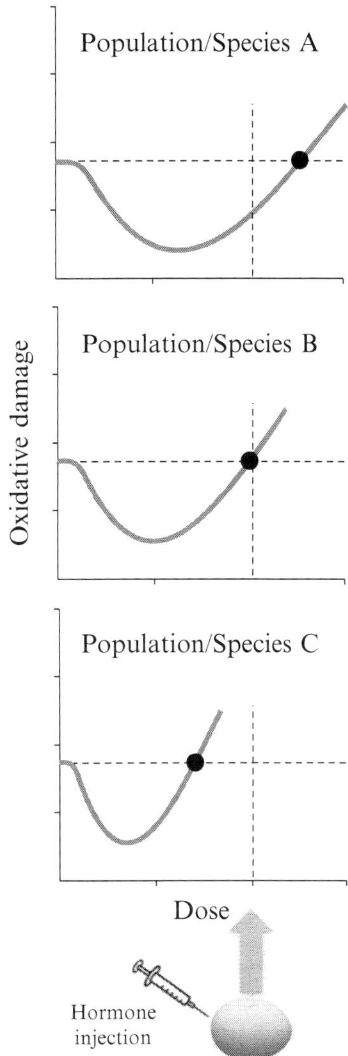

Fig. 13.2 The range of hormonal doses (or stressor intensity) over which the beneficial effects occur can vary among populations or species, possibly reflecting evolutionary adaptations. These differences indicate the importance of the hormetic zone (the area under the curve and above the zero equivalent point) when predicting the responses of a population or species to an environmental perturbation

dihydrotestosterone and the oestrogen estradiol (aromatisation: biosynthesis of oestrogens from testosterone catalysed by the enzyme aromatase)—the latter can have important pro-oxidant effects (Brambilla et al. 2003; Casagrande et al. 2011).

Following pioneering studies on mammals (e.g. Ward 1972; Dahlöf et al. 1977), recent avian studies reported the presence of stress hormones (corticosterone) in the egg (e.g. Downing and Bryden 2002; Eriksen et al. 2003; Hayward and Wingfield 2004; Love et al. 2005; Rubolini et al. 2005; Saino et al. 2005; Love and Williams 2008a, b) and suggested that the mother could affect offspring development through the transfer of these hormones into the egg. Henriksen et al. (2011) recently reviewed studies on pre-natal corticosterone-mediated effects in birds (see also Spencer et al. 2009; Schoech et al. 2011 for a review on post-natal corticosterone-mediated effects) and concluded that findings were inconsistent (Casagrande et al. 2012a, b). The discrepancy between results might have been due to the interaction of post-natal environmental variation, sex, age, development mode, and type of treatment; alternatively, maternal corticosterone could have affected offspring behaviour and physiology via alteration of another egg component (e.g. androgens, progesterone). Considering that corticosterone may affect free radical production, oxidative damage accumulations as well as non-enzymatic and enzymatic antioxidant levels (Costantini et al. 2011), maternal stress could influence the embryo's oxidative state (Haussmann et al. 2012). This could be very important because oxidative stress may reduce survival of embryos or interfere with development (Surai 2002).

13.2.2 Nutrients

Early nutrition (quality and quantity) can have significant downstream effects on adult physiology or life-history traits through its modulation of body composition. Although available evidence suggests that aerobic life has evolved by reducing the relative abundance of structural components that are highly susceptible to oxidative damage, there is still great variation in body composition between species (e.g. degree of unsaturation of cell membranes or amino acid composition of proteins; Hulbert et al. 2007; Pamplona and Barja 2007), whose causes remain poorly understood. The link between body composition and susceptibility to oxidative stress might therefore be ecologically and evolutionarily relevant. Diet is not only a source of molecular antioxidants as discussed above but also a source of nutrients, such as fatty acids and amino acids—these can expose an organism to different oxidative stress threats due to having differential molecular susceptibilities to oxidation. Animals can, however, actively regulate nutrient metabolism, allowing them to create different cell compositions. Discrimination of dietary fats might be especially important because n-3 polyunsaturated fatty acids are less resistant to oxidative damage than their n-6 counterparts. Such regulatory abilities of cell membrane composition may also be important when the female is depositing nutrients into the egg. For example, the n-6 polyunsaturate, arachidonic acid, forms between 8 and 19% (w/w) of the phospholipid fatty acids of egg yolk of the northern gannet *Morus bassanus*, the great skua *Catharacta skua*, the American white pelican *Pelecanus erythrorhynchos* and the double-crested cormorant *Phalacrocorax auritus* (Surai et al. 2001). In pelicans and cormorants, such yolk composition is consistent with the consumption of freshwater fish in which arachidonic acid may be a significant acyl constituent, but this finding is more difficult to explain for the gannet and skua because they largely feed on marine fish with a low arachidonic acid content. These results suggest that bird females may have an active control of fatty acid deposition in the egg; hence, they may directly influence the susceptibility of embryo tissues to oxidative stress (Surai et al. 1996).

13.2.3 Early Life in a Hormetic Framework

The Developmental Programming Hypothesis was proposed to explain the links between the effect of environmental stressors on females during pregnancy, the variation in patterns of growth and development associated with the quality of the maternal environment and the development of pathologies in offspring later in life (Barker et al. 1993; Seckl 1998). The idea behind this hypothesis is that females may prime offspring to prevalent environmental conditions that they are likely to experience as adults (environmental matching). An adaptationist view is clearly implicit in this model, that is, the developmental programming should confer adaptive Darwinian advantages to the offspring. However, this paradigm has been criticised because the characteristics of adult environment could change in an unpredictable way, hence resulting in a mismatch between this and the environment experienced during early development or by the mother. Individuals developmentally adapted to one environment may be at risk when exposed to another later in life and their fitness may be consequently reduced (Bateson et al. 2004). The Developmental Programming Hypothesis, therefore, might prove unreliable when trying to predict the effects of early life on adult condition. However, it has been suggested that it is the intensity of early stress experienced that is important because this will programme the degree of stress responsiveness in relation to environmental conditions. In this context, hormesis could provide a useful conceptual and mechanistic model to fill in the gaps of the Developmental Programming Hypothesis because it highlights the importance of stressor intensity so that its consequences for individual health and survival can be considered.

There is increasing evidence that some toxicants, nutrients or environmental stressors can have stimulatory or beneficial effects at low exposure levels, while being toxic at higher levels. Additionally, environmental priming of certain physiological processes can result in their improved functioning in later life, regardless of matching or mismatching. Both these phenomena have been called hormesis (Calabrese and Blain 2005), a term first coined by "Chester Southam and John Ehrlich in 1943".

Three general categories of dose–response relationships can be recognised. Generally speaking, in the case of toxicants, one might expect the negative effects of exposure to become more severe (and fitness to therefore decline) as exposure increases (linear no-threshold dose–response relationship; Fig. 13.1a), or there may be no noticeable effect until the dose reaches a threshold value (termed the "No Observed Adverse Effect Level" in the toxicological literature; threshold dose–response relationship, Fig. 13.1b), above which fitness declines linearly or non-linearly with dose. In a third case (hormesis), the dose–response relationship is biphasic, with low and high doses stimulating or inhibiting the stress response, respectively (Fig. 13.1c; Mattson and Calabrese 2010). It is important to highlight that stimulation of the physiological response might not necessarily be beneficial for the organism (e.g. if there is a trade-off between the stimulated trait and another).

The concept of hormesis is generally applicable to developmental biologists, behavioural ecologists and evolutionary biologists since it might provide an important link between the prevailing environmental circumstances and organism function. Hormesis might increase the stress tolerance and response effectiveness of an organism (homeodynamics *sensu* Yates 1994; Lloyd et al. 2001; Stebbing 2009) and generate epigenetic effects (Vaiserman 2010).

One type of hormesis that is relevant to this chapter is the "priming" or "conditioning" effect, whereby exposure to mild levels of a stressor increases the capacity to cope with subsequent exposure to higher levels of the same stressor, compared to individuals that were never exposed or were previously exposed to a high stress intensity. I term this *conditioning hormesis* according to the terminology proposed by Calabrese et al. (2007). In order to investigate this process, individuals are initially exposed to different levels of a stressor (including zero exposure), and then subsequently, the same individuals are exposed to high levels of the same stressor. The extent to which individuals can cope with the high-level exposure (i.e. the extent to which they show detrimental effects, such as tissue damage) is compared between the groups that experienced pre-exposure and those that experienced no exposure. Previous studies suggest that the magnitude of stimulation is rarely higher than twice the control group, and, in general, the maximum stimulation for hormesis is 30–60% higher than the control group (Calabrese 2010).

In ecological terms, conditioning or priming hormesis might translate to an increase in phenotypic plasticity, hence the ability of an organism or species to respond to environmental stimuli. In general, the phenotypic response to the current environment may depend on both its similarity to previous environments (mismatches leading to ecological limits on plasticity; Auld et al. 2010; Costantini et al. 2012), and on the nature of previous responses, since trait expression early in ontogeny may affect later trait expression (so-called plasticity-history limits; Auld et al. 2010). However, some evidence suggests that hormetic responses can be generalised across stressors so that conditions experienced in early life allow an individual to withstand other kinds of stressful conditions later in life (Le Bourg and Minois 1997; Bartling et al. 2003; Honma et al. 2003; Le Bourg et al. 2004; Le Bourg 2005). Hormesis thus has the potential to increase phenotypic plasticity, so having the opposite effects of both ecological and plasticity-history limits. Consequently, hormetic mechanisms could provide a fail-safe to buffer maladaptive effects of maternal investment when the maternal or early growth environment is not predictive of the adult environment.

The timing of initial exposure to a stressor may be important, for example, if there is a critical time window or developmental stage in which the priming effect can occur (i.e. sensitive periods in early life). Although studies show that stimulatory or beneficial effects generated by exposure to mild stress can emerge at any time throughout the individual's life, they also suggest that such effects might be stronger if exposure occurs early in life (Yahav and McMurtry 2001; Bartling et al. 2003; Honma et al. 2003; Le Bourg et al. 2004; Burger et al. 2007; Mangel 2008). In rats, exposure of pups to a mild stressor can dampen the stress response mediated by glucocorticoids when exposed to a stressor later in life. In contrast, exposure to strong stressors in early life can cause subsequent hypersensitivity to stressors, manifested in an oversecretion of glucocorticoids. Such hypersensitivity has been shown to have detrimental effects on health (Levine et al. 1967; Liu et al. 1997; Romero 2004; pp. 692–694 in Nelson 2005), such as excessive energy mobilisation (Wingfield et al. 1998), immunosuppression (e.g. Webster Marketon and Glaser 2008), oxidative stress (e.g. Costantini et al. 2011) or abandonment of reproductive activity (e.g. Wingfield and Hunt 2002). Thus a dampening in the production of glucocorticoids might be beneficial because it could limit the magnitude of these negative side effects. However, the negative fitness consequences of the priming process might only be evident if there is no subsequent exposure to a higher level of the stressor (Fig. 13.3). The existence of such a cost would explain why natural selection has favoured a system design that requires priming in order to work effectively. Evidently, the probability of encountering high levels of the stressor in the environment will influence the evolution of the response, as will the extent to which the first exposure carries costs. As illustrated by the rat example above, different levels of stress exposure in early life can have different outcomes. Recent work in rats suggests that exposure to relatively high stress levels (i.e. 24 h of maternal deprivation, a model for maternal neglect) can, in addition to negative effects, actually have beneficial effects on some aspects of brain function (particularly in the hippocampus), which are especially evident under conditions of high stress in adulthood (Oomen et al. 2010).

The application of the hormetic model may also be especially helpful when interpreting effects of diet during development, in terms of both energy and nutrient intake. Dietary energy restriction can, for example, increase levels of heat shock proteins (HSPs) in several different tissues (Heydari et al. 1993) and increase levels of other cytoprotective molecules, such as antioxidants, reducing levels of oxidative stress (Sanz et al. 2006; Ugochukwu and Figgers 2007). Hormetic effects induced by dietary energy restriction can also involve changes in systems that regulate cellular energy metabolism (Rodgers et al. 2005; Hyun et al. 2006; Liu et al. 2006), in particular through changes in mitochondrial activity (*mitohormesis*, Tapia 2006). Hormesis-like biphasic curves are also known to emerge in response to specific non-essential nutrients (Holmes et al. 1980; Hayes 2007), where fitness benefits and costs emerge at low and high intakes of a nutrient, respectively. It is important to note here that hormesis should not be confused with the so-called Bertrand's rule or essentiality, which is a dose–response relationship for essential substances. It can be characterised by negative Darwinian fitness consequences when there is not enough and when there is too much of these substances (Chapman 1998; Raubenheimer et al. 2005; Raubenheimer and Simpson 2009). This calls for an integration of the concepts of hormesis and essentiality when looking at the effects of nutrients on stress levels.

13.3 Conclusions

In order to understand the phenotypic and fitness consequences of early life experiences, the measurements of oxidative damage and antioxidant responses could provide informative tools that integrate well with classic studies of endocrine status dynamics. Oxidative stress might not only be a proxy variable that is useful in

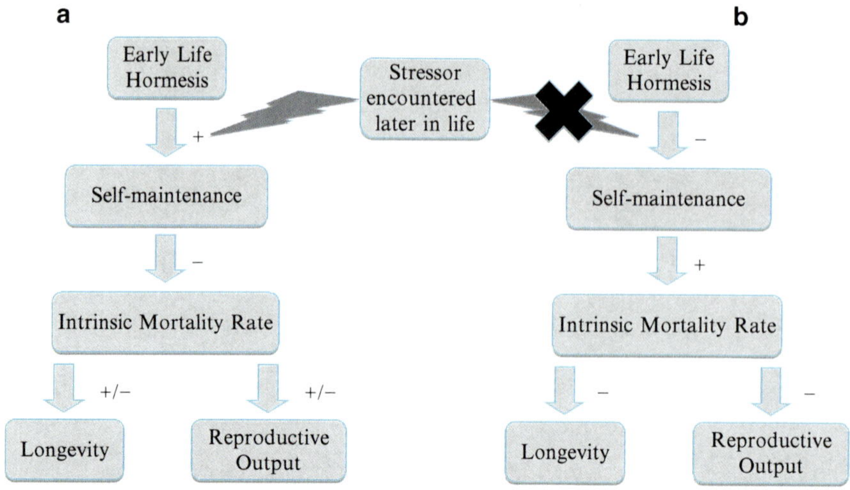

Fig. 13.3 Hypothesised scenarios representing downstream effects of early life hormesis on life-history strategies. + = increase; – = decrease. The long-term effects of priming or conditioning hormesis on the stress physiology machinery will depend on whether or not there will be subsequent exposure to a high level of the stressor. In fact, the priming process might have detrimental consequences for individual fitness if there is no subsequent exposure. The probability of encountering high levels of the stressor in the environment will influence the evolution of the response, as will the extent of the costs of the initial exposure. (**a**) Represents an example whereby the stressor will be encountered later in life. Hormesis will improve the capacity for self-maintenance and will have similar or opposite effects on longevity or reproduction. According to the literature, both longevity and reproductive output may be increased or are traded-off against each other. (**b**) Represents an example whereby the stressor is not encountered later in life and this can lead to reduced longevity and reproductive output

quantifying the physiological responses of the organism, but might also work as an inducer and guide development through the generation of constraints on phenotypic responses to environmental stimuli, redox cell signalling in regulatory processes and epigenetic effects. It is important to recognise that the stressor dose *sensu lato* is important and so is the timing of exposure (e.g. duration, sensitive periods of life), the environmental conditions experienced after the end of the developmental period and characteristics of the population or species (Fig. 13.2; see also Holmes et al. (1980) for an example of interspecies differences in nutritional hormesis). Obviously, the links between early life and phenotypic development can be studied from many different angles and under several theoretical underpinnings, but addressing these points under a hormetic framework could provide a novel tool for answering many questions. It might also allow discrepancies between studies to be reconciled, in modern behavioural toxicology and ecological and evolutionary developmental biology.

Acknowledgements I thank Shona Smith, Valeria Marasco, and two anonymous reviewers for valuable comments that helped me improve the presentation of the manuscript. During manuscript preparation, I was supported by a NERC postdoctoral fellowship (NE/G013888/1).

References

Alonso-Alvarez C, Bertrand S, Devevey G, Prost J, Faivre B, Chastel O, Sorci G (2006) An experimental manipulation of life-history trajectories and resistance to oxidative stress. Evolution 60:1913–1924

Auld JR, Agrawal AA, Relyea RA (2010) Re-evaluating the costs and limits of adaptive phenotypic plasticity. Proc Biol Sci 277:503–511

Badyaev AV (2011) Origin of the fittest: link between emergent variation and evolutionary change as a critical question in evolutionary biology. Proc Biol Sci 278:1921–1929

Barker DJ, Gluckman PD, Godfrey KM, Harding JE, Owens JA, Robinson JS (1993) Fetal nutrition and cardiovascular disease in adult life. Lancet 341:938–941

Bartling B, Friedrich I, Silber RE, Simm A (2003) Ischemic preconditioning is not cardioprotective in senescent human myocardium. Ann Thorac Surg 76:105–111

Bateson P, Barker D, Clutton-Brock T, Deb D, D'Udine B, Foley RA, Gluckman P, Godfrey K, Kirkwood T, Lahr MM, McNamara J, Metcalfe NB, Monaghan P, Spencer HG, Sultan SE (2004) Developmental plasticity and human health. Nature 430:419–421

Beckman KB, Ames BN (1998) The free radical theory of aging matures. Physiol Rev 78:547–581

Bize P, Devevey G, Monaghan P, Doligez B, Christe P (2008) Fecundity and survival in relation to resistance to oxidative stress in a free-living bird. Ecology 89:2584–2593

Brambilla G, Ballerini A, Civitareale C, Fiori M, Neri B, Cavallina R, Nardoni A, Giannetti L (2003) Oxidative stress as a bio-marker of estrogens exposure in healthy veal calves. Anal Chim Acta 483:281–288

Burger JM, Hwangbo DS, Corby-Harris V, Promislow DE (2007) The functional costs and benefits of dietary restriction in *Drosophila*. Aging Cell 6:63–71

Calabrese EJ (2010) Hormesis is central to toxicology, pharmacology and risk assessment. Hum Exp Toxicol 29:249–261

Calabrese EJ, Blain R (2005) The occurrence of hormetic dose responses in the toxicological literature, the hormesis database: an overview. Toxicol Appl Pharmacol 202:289–301

Calabrese EJ, Bachmann KA, Bailer AJ, Bolger PM, Borak J, Cai L, Cedergreen N, Cherian MG, Chlueh CC, Clarkson TW, Cook RR, Diamond DM, Doolittle DJ, Dorato MA, Duke SO, Feinendegen L, Gardner DE, Hart RW, Hastings KL, Hayes AW, Hoffmann GR, Ives JA, Jaworowski Z, Johnson TE, Jonas WB, Kaminski NE, Keller JG, Klaunig JE, Knudsen TB, Kozumbo WJ, Lettleri T, Liu SZ, Maisseu A, Maynard KI, Masoro EJ, McClellan RO, Mehendale HM, Mothersill C, Newlin DB, Nigg HN, Oehme FW, Phalen RF, Philbert MA, Rattan SIS, Riviere JE, Rodricks J, Sapolsky RM, Scott BR, Seymour C, Sinclair DA, Smith-Sonneborn J, Snow ET, Spear L, Stevenson DE, Thomas Y, Tubiana M, Williams GM, Mattson MP (2007) Biological stress response terminology: integrating the concepts of adaptive response and preconditioning stress within a hormetic dose–response framework. Toxicol Appl Pharmacol 222:122–128

Carere C, Balthazart J (2007) Sexual versus individual differentiation: the controversial role of avian maternal hormones. Trends Endocrinol Metab 18:73–80

Casagrande S, Dell'Omo G, Costantini D, Tagliavini J, Groothuis T (2011) Variation of a carotenoid-based trait in relation to oxidative stress and endocrine status during the breeding season in the Eurasian kestrel: a multi-factorial study. Comp Biochem Physiol A Mol Integr Physiol 160:16–26

Casagrande S, Costantini D, Groothuis TGG (2012a) Interaction between sexual steroids and immune response in affecting oxidative status of birds. Comparative Biochemistry and Physiology Part A 163:296–301

Casagrande S, Costantini D, Dell'Omo G, Tagliavini J, Groothuis TGG (2012b) Differential effects of testosterone metabolites oestradiol and dihydrotestosterone on oxidative stress and carotenoid-dependent colour expression in a bird. Behav Ecol Sociobiol 66: 1319–1331

Chapman PM (1998) New and emerging issues in ecotoxicology—the shape of testing to come? Australas J Ecotoxicol 4:1–7

Costantini D (2008) Oxidative stress in ecology and evolution: lessons from avian studies. Ecol Lett 11:1238–1251
Costantini D, Dell'omo G (2006) Effects of T-cell-mediated immune response on avian oxidative stress. Comp Biochem Physiol A Mol Integr Physiol 145:137–142
Costantini D, Verhulst S (2009) Does high antioxidant capacity indicate low oxidative stress? Funct Ecol 23:506–509
Costantini D, Metcalfe NB, Monaghan P (2010) Ecological processes in a hormetic framework. Ecol Lett 13:1435–1447
Costantini D, Marasco V, Møller AP (2011) A meta-analysis of glucocorticoids as modulators of oxidative stress in vertebrates. J Comp Physiol B 181:447–456
Costantini D, Monaghan P, Metcalfe N (2012) Early life experience primes resistance to oxidative stress. J Exp Biol 215:2820–2826
Curley JP, Branchi I (2012) Ontogeny of stable individual differences: gene-environment and epigenetic mechanisms. In: Carere C, Maestripieri D (eds) Animal personalities. Behavior, physiology, and evolution. The University of Chicago Press, Chicago
Dahlöf L-G, Hård E, Larsson K (1977) Influence of maternal stress on offspring sexual behaviour. Anim Behav 25:958–968
Downing JA, Bryden WL (2002) A non-invasive test of stress in laying hens. Rural Industries Research and Development Corporation (RIRDC), Barton
Eriksen MS, Haug A, Torjesen PA, Bakken M (2003) Prenatal exposure to corticosterone impairs embryonic development and increases fluctuating asymmetry in chickens (*Gallus gallus domesticus*). Br Poult Sci 44:690–697
Forbes VE (2000) Is hormesis an evolutionary expectation? Funct Ecol 14:12–24
Furness L, Speakman JR (2008) Energetics and longevity in birds. Age 30:75–87
Galvan I, Alonso-Alvarez C (2010) Yolk testosterone shapes the expression of a melanin-based signal in great tits: an antioxidant-mediated mechanism? J Exp Biol 213:3127–3130
Gerhart J, Kirschner M (2007) The theory of facilitated variation. Proc Natl Acad Sci USA 104:8582–8589
Gilbert SF (2001) Ecological developmental biology: developmental biology meets the real world. Dev Biol 233:1–12
Gluckman PD, Hanson MA (2004) The developmental origins of the metabolic syndrome. Trends Endocrinol Metab 15:183–187
Groothuis TGG, Schwabl H (2008) Hormone-mediated maternal effects in birds: mechanisms matter but what do we know of them? Philos Trans R Soc Lond B Biol Sci 363:1647–1661
Groothuis TGG, Müller W, von Engelhardt N, Carere C, Eising C (2005) Maternal hormones as a tool to adjust offspring phenotype in avian species. Neurosci Biobehav Rev 29:329–352
Halliwell B, Gutteridge JMC (2007) Free radicals in biology and medicine. Oxford University Press, Oxford
Haussmann MF, Longenecker AS, Marchetto NM, Juliano SA, Bowden RM (2012) Embryonic exposure to corticosterone modifies the juvenile stress response, oxidative stress and telomere length. Proc Biol Sci 279(1732):1447–1456
Hayes DP (2007) Nutritional hormesis. Eur J Clin Nutr 61:147–159
Hayward LS, Wingfield JC (2004) Maternal corticosterone is transferred to avian yolk and may alter offspring growth and adult phenotype. Gen Comp Endocrinol 135:365–371
Henriksen R, Rettenbacher S, Groothuis TGG (2011) Prenatal stress in birds: pathways, effects, function and perspectives. Neurosci Biobehav Rev 35:1484–1501
Heydari AR, Wu B, Takahashi R, Strong R, Richardson A (1993) Expression of heat shock protein 70 is altered by age and diet at the level of transcription. Mol Cell Biol 13:2909–2918
Holmes RS, Moxon LN, Parsons PA (1980) Genetic variability of alcohol dehydrogenase among Australian *Drosophila* species: correlation of ADH biochemical phenotype with ethanol resource utilization. J Exp Zool 214:199–204
Honma Y, Tani M, Yamamura K, Takayama M, Hasegawa H (2003) Preconditioning with heat shock further improved functional recovery in young adult but not in middle-aged rat hearts. Exp Gerontol 38:299–306

Hulbert AJ, Pamplona R, Buffenstein R, Buttemer WA (2007) Life and death: metabolic rate, membrane composition, and life span of animals. Physiol Rev 87:1175–1213

Hyun DH, Emerson SS, Jo DG, Mattson MP, de Cabo R (2006) Calorie restriction up-regulates the plasma membrane redox system in brain cells and suppresses oxidative stress during aging. Proc Natl Acad Sci USA 103:19908–19912

Jones DP (2006) Redefining oxidative stress. Antioxid Redox Signal 8:1865–1879

Le Bourg E (2005) Hormetic protection of *Drosophila melanogaster* middle-aged male flies from heat stress by mildly stressing them at young age. Naturwissenschaften 92:293–296

Le Bourg E, Minois N (1997) Increased longevity and resistance to heat shock in *Drosophila melanogaster* flies exposed to hypergravity. C R Acad Sci III 320:215–221

Le Bourg E, Toffin E, Massé A (2004) Male *Drosophila melanogaster* flies exposed to hypergravity at young age are protected against a non-lethal heat shock at middle age but not against behavioral impairments due to this shock. Biogerontology 5:431–443

Levine S, Haltmeyer GC, Karas GG (1967) Physiological and behavioral effects of infantile stimulation. Physiol Behav 2:55–59

Lipar JL, Ketterson ED, Nolan V Jr, Casto JM (1999) Egg yolk layers vary in the concentration of steroid hormones in two avian species. Gen Comp Endocrinol 115:220–227

Liu D, Diorio J, Tannenbaum B, Caldji C, Francis DD, Freedman A, Sharma S, Pearson D, Plotsky PM, Meaney MJ (1997) Maternal care, hippocampal glucocorticoid receptors, and hypothalamic-pituitary-adrenal responses to stress. Science 277:1659–1662

Liu D, Chan SL, de Souza-Pinto NC, Slevin JR, Wersto RP, Zhan M, Mustafa K, de Cabo R, Mattson MP (2006) Mitochondrial UCP4 mediates an adaptive shift in energy metabolism and increases the resistance of neurons to metabolic and oxidative stress. Neuromolecular Med 8:389–414

Lloyd D, Aon MA, Cortassa S (2001) Why homeodynamics, not homeostasis? Scientific World Journal 1:133–145

Love OP, Williams TD (2008a) Plasticity in the adrenocortical response of a free-living vertebrate: the role of pre- and post-natal developmental stress. Horm Behav 54:496–505

Love OP, Williams TD (2008b) The adaptive value of stress-induced phenotypes: effects of maternally derived corticosterone on sex-biased investment, cost of reproduction, and maternal fitness. Am Nat 172:E135–E149

Love OP, Chin EH, Wynne-Edwards KE, Williams TD (2005) Stress hormones: a link between maternal condition and sex-biased reproductive investment. Am Nat 166:751–766

Mangel M (2008) Environment, damage and senescence: modelling the life-history consequences of variable stress and caloric intake. Funct Ecol 22:422–430

Marasco V, Robinson J, Herzyk P, Spencer KA (2012) Pre- and post-natal stress in context: effects on the stress physiology in a precocial bird. J Exp Biol 215:3955–3964

Mattson MP, Calabrese EJ (2010) Hormesis: a revolution in biology, toxicology and medicine. Springer, New York

McGraw KJ, Cohen A, Costantini D, Horak P (2010) The ecological significance of antioxidants and oxidative stress: a marriage of functional and mechanistic approaches. Funct Ecol 24:947–949

Metcalfe NB, Monaghan P (2001) Compensation for a bad start: grow now, pay later? Trends Ecol Evol 16(5):254–260

Monaghan P (2008) Early growth conditions, phenotypic development and environmental change. Philos Trans R Soc Lond B Biol Sci 363:1635–1645

Monaghan P, Metcalfe NB, Torres R (2009) Oxidative stress as a mediator of life history tradeoffs: mechanisms, measurement and interpretation. Ecol Lett 12:75–92

Nelson RJ (2005) An introduction to behavioral endocrinology. Sinauer, Sunderland

Noguera JC, Alonso-Alvarez C, Kim SY, Morales J, Velando A (2011) Yolk testosterone reduces levels of oxidative damages during postnatal development. Biol Lett 7:93–95

Oomen CA, Soeters H, Audureau N, Vermunt L, van Hasselt FN, Manders EMM, Joels M, Lucassen PJ, Krugers H (2010) Severe early life stress hampers spatial learning and neurogenesis, but improves hippocampal synaptic plasticity and emotional learning under high-stress conditions in adulthood. J Neurosci 30:6635–6645

Pamplona R, Barja G (2007) Highly resistant macromolecular components and low rate of generation of endogenous damage: two key traits of longevity. Ageing Res Rev 6:189–210

Pamplona R, Costantini D (2011) Molecular and structural antioxidant defences against oxidative stress in animals. Am J Physiol Regul Integr Comp Physiol 301:R843–R863

Parsons PA (2001) The hormetic zone: an ecological and evolutionary perspective based upon habitat characteristics and fitness selection. Q Rev Biol 76:459–467

Pigliucci M (2001) Phenotypic plasticity: beyond nature and nurture. The Johns Hopkins University Press, Baltimore

Pigliucci M, Murren CJ, Schlichting CD (2006) Phenotypic plasticity and evolution by genetic assimilation. J Exp Biol 209:2362–2367

Raubenheimer D, Simpson SJ (2009) Nutritional pharmecology: doses, nutrients, toxins, and medicines. Integr Comp Biol 49:329–337

Raubenheimer D, Lee K-P, Simpson SJ (2005) Does Bertrand's rule apply to macronutrients? Proc Biol Sci 272:2429–2434

Rodgers JT, Lerin C, Haas W, Gygi SP, Spiegelman BM, Puigserver P (2005) Nutrient control of glucose homeostasis through a complex of PGC-1alpha and SIRT1. Nature 434:113–118

Romero LM (2004) Physiological stress in ecology: lessons from biomedical research. Trends Ecol Evol 19:249–255

Royle NJ, Surai PF, McCartney RJ, Speake BK (1999) Parental investment and egg yolk lipid composition in gulls. Funct Ecol 13:298–306

Royle NJ, Surai PF, Hartley IR (2001) Maternally derived androgens and antioxidants in bird eggs: complementary but opposing effects? Behav Ecol 12:381–385

Rubolini D, Romano M, Boncoraglio G, Ferrari RP, Martinelli R, Galeotti P, Fasola M, Saino N (2005) Effects of elevated egg corticosterone levels on behavior, growth, and immunity of yellow-legged gull *Larus michahellis* chicks. Horm Behav 47:592–605

Saino N, Romano M, Ferrari RP, Martinelli R, Møller AP (2005) Stressed mothers lay eggs with high corticosterone levels which produce low-quality offspring. J Exp Zool A Comp Exp Biol 303a:998–1006

Saino N, Caprioli M, Romano M, Boncoraglio G, Rubolini D, Ambrosini R, Bonisoli-Alquati A, Romano A (2011) Antioxidant defenses predict long-term survival in a passerine bird. PLoS One 6:e19593

Sanz A, Pamplona R, Barja G (2006) Is the mitochondrial free radical theory of aging intact? Antioxid Redox Signal 8:582–599

Schoech SJ, Rensel MA, Heiss RS (2011) Short- and long-term effects of developmental corticosterone exposure on avian physiology, behavioral phenotype, cognition, and fitness: a review. Curr Zool 57:514–530

Schwabl H (1993) Yolk is a source of maternal testosterone for developing birds. Proc Natl Acad Sci USA 90:11446–11450

Schwabl H (1996) Maternal testosterone in the avian egg enhances postnatal growth. Comp Biochem Physiol A Physiol 114:271–276

Schwabl H, Mock DW, Gieg JA (1997) A hormonal mechanism for parental favouritism. Nature 386:231

Seckl JR (1998) Physiologic programming of the fetus. Clin Perinatol 25:939–964

Sies H (1991) Oxidative stress II. Oxidants and antioxidants. Academic, London

Sockman KW, Schwabl H (2000) Yolk androgens reduce offspring survival. Proc Biol Sci 267:1451–1456

Sohal RS, Orr WC (2012) The redox stress hypothesis of aging. Free Radic Biol Med 52:539–555

Southam CM, Ehrlich J (1943) Effects of extracts of western red cedar heartwood on certain wood-decaying fungi in culture. Phytopathology 33:517–524

Spencer KA, Evans NP, Monaghan P (2009) Postnatal stress in birds: a novel model of glucocorticoid programming of the hypothalamic-pituitary-adrenal axis. Endocrinology 150:1931–1934

Stebbing ARD (2009) Interpreting 'dose-response' curves using homeodynamic data: with an improved explanation for hormesis. Dose Response 7:221–233

Surai P (2002) Natural antioxidants in avian nutrition and reproduction. Nottingham University Press, Nottingham

Surai PF, Noble RC, Speake BK (1996) Tissue-specific differences in antioxidant distribution and susceptibility to lipid peroxidation during development of the chick embryo. Biochim Biophys Acta 1304:1–10

Surai PF, Bortolotti GR, Fidgett AL, Blount JD, Speake BK (2001) Effects of piscivory on the fatty acid profiles and antioxidants of avian yolk: studies on eggs of the gannet, skua, pelican and cormorant. J Zool 255:305–312

Szyf M, Weaver IC, Champagne FA, Diorio J, Meaney MJ (2005) Maternal programming of steroid receptor expression and phenotype through DNA methylation in the rat. Front Neuroendocrinol 26:139–162

Tapia PC (2006) Sublethal mitochondrial stress with an attendant stoichiometric augmentation of reactive oxygen species may precipitate many of the beneficial alterations in cellular physiology produced by caloric restriction, intermittent fasting, exercise and dietary phytonutrients: "mitohormesis" for health and vitality. Med Hypotheses 66:832–843

Tobler M, Sandell MI (2009) Sex-specific effects of prenatal testosterone on nestling plasma antioxidant capacity in the zebra finch. J Exp Biol 212:89–94

Ugochukwu NH, Figgers CL (2007) Dietary caloric restriction improves the redox status at the onset of diabetes in hepatocytes of streptozotocin induced diabetic rats. Chem Biol Interact 165:45–53

Vaiserman AM (2010) Hormesis, adaptive epigenetic reorganization, and implications for human health and longevity. Dose Response 8:16–21

Ward IL (1972) Prenatal stress feminizes and demasculinizes the behavior of males. Science 175:82–84

Webster Marketon JI, Glaser R (2008) Stress hormones and immune function. Cell Immunol 252:16–26

West-Eberhard MJ (2003) Developmental plasticity and evolution. Oxford University Press, Oxford

Wingfield JC, Maney DL, Breuner CW, Jacobs JD, Lynn S, Ramenofsky M, Richardson RD (1998) Ecological bases of hormone–behavior interactions: the "emergency life history stage". Am Zool 38:191–206

Wingfield JC, Hunt KE (2002) Arctic spring: hormone-behavior interactions in a severe environment. Comp Biochem Physiol B 132:275–286

Yahav S, McMurtry JP (2001) Thermotolerance acquisition in broiler chickens by temperature conditioning early in life–the effect of timing and ambient temperature. Poult Sci 80:1662–1666

Yates FE (1994) Order and complexity in dynamical systems: homeodynamics as a generalized mechanics for biology. Math Comput Model 19:49–74

Index

G. Laviola and S. Macrì (eds.), *Adaptive and Maladaptive Aspects of Developmental Stress*, Current Topics in Neurotoxicity 3, DOI 10.1007/978-1-4614-5605-6,

R

T

U

V

W

Y

Printed by Publishers' Graphics LLC
LSI20130405.15.01.18